Developing Adaptation Policy and Practice in Europe: Multi-level Governance of Climate Change

Developing Adaptation Policy and Practice in Europe: Multi-level Governance of Climate Change

Edited by

E. Carina H. Keskitalo

Political Science/Department of Social and Economic Geography,
Umeå University,
Sweden

Springer

Editor
Dr. E. Carina H. Keskitalo
Political Science/Department of Social
and Economic Geography
Umeå University
901 87 Umeå
Sweden
carina.keskitalo@geography.umu.se

ISBN 978-94-017-8168-8 ISBN 978-90-481-9325-7 (eBook)
DOI 10.1007/978-90-481-9325-7
Springer Dordrecht Heidelberg London New York

Softcover re-print of the Hardcover 1st edition 2010

Cover illustration: Cover image © 2010 JupiterImages Corporation
Photo text: Trees submerged by a flood

Printed on acid-free paper

Springer is part of Springer Science+Business Media (www.springer.com)

Acknowledgements

This work largely began in 2006, at which time I was a visiting scholar at the Environmental Change Institute at Oxford University, and had the opportunity to join in some of the UK Climate Impacts Programme (UKCIP) consultations with regional climate change partnerships. What surprised me at that time was the advanced stage of adaptation at regional and in some cases even local levels – and the relatively broad awareness of adaptation needs in many of these cases. The experience led me to develop a research project application to study the differences in adaptation to climate change in European countries. The project was funded by the Swedish Research Council in 2007 under the title 'Organising Adaptation to Climate Change in Europe' (EUR-ADAPT). For David Ellison's contribution (Chapter 2), we are also grateful for the support of the Future Forests research programme, funded under a call by the Mistra Foundation for Strategic Environmental Research. In addition to this volume, the projects outputs have also included several articles.

Both the editor and authors would like to extend sincere thanks to each of the over 100 policy-makers, civil servants and NGO representatives at different levels in each case study country who agreed to support the study with their knowledge on adaptation policy and processes – without you this study would have been impossible to undertake. We are thankful to all the interviewees who commented on our interpretation of the data and on the way adaptation policy had developed in the time after the interviews had been conducted. We also want to thank the many reviewers and patient readers who helped us to develop the volume into its final format, who commented on full chapters as well as individual case study sections in the comparative chapter (Chapter 7), and who supported us with their knowledge of European and non-European contexts. Thanks especially to Chris West and Diana Liverman (for my initial experience at UKCIP), Johanna Wandel, Derek Armitage, Svante Ersson, Olle Stjernström, Anders Lidström, Dominic Standish, Sabine Kuhlmann, Emily Boyd, Simo Haanpää, Nathalie Behnke, Stefan Bouzarovski, Kate Lonsdale, Gábor Bálint, Timothy Carter, András Horányi, Maria Pettersson and Rik Leemans. Thanks also to commentators at workshops and seminars, including the project special session at the Bonn Human Dimensions of Global Environmental Change conference in April 2009, the Governance and Adaptation panel at the ECPR General Conference in Potsdam September 2009, and participants at climate change and climate policy PhD schools held at Norrköping, Sweden, and Waterloo, Canada.

Finally, warm thanks to the team of authors: you went above and beyond what was expected! Lisa Westerhoff and Sirkku Juhola both contributed hours of discussions of the material, underlying both this book and related articles, and patiently reviewed chapters on the framework and conclusion and their relationship to their case studies. Thanks also to Johanna Liljenfeldt, who worked as a field assistant and with map development, to Åsa Eriksson who undertook a pre-study on mitigation in Sweden and a comparative study of the regional climate change partnerships in England and to the many transcribers who made the transformation of hours of interviews into analysable data possible.

Umeå, Sweden Carina Keskitalo
November 2009

Contents

1 Introduction – Adaptation to Climate Change in Europe: Theoretical Framework and Study Design 1
E. Carina H. Keskitalo
1.1 Introduction 1
1.2 Theoretical Background 4
1.2.1 Multi-Level Governance and the Capacity to Act on Adaptation 4
1.2.2 Adaptive Capacity 7
1.2.3 Governance and Adaptive Capacity on Different Levels . 11
1.2.4 Attempts to Bridge Sectors and Levels: Connecting Governance and Adaptive Capacity 22
1.3 Method 28
1.3.1 National Case Study Selection 28
1.3.2 Nested Regional and Local Studies in the Four Main Cases 30
1.3.3 Research Material 31
1.4 Outline of the Volume 32
References 34

2 Addressing Adaptation in the EU Policy Framework 39
David Ellison
2.1 Introduction 40
2.2 From the ECCP to the Green and White Papers on Adaptation . . 42
2.2.1 EU-Funded Research and Resources 45
2.2.2 The ECCP and the Development of National Adaptation Strategies 47
2.2.3 Potential Consequences of the National Level Approach . 49
2.2.4 Sectoral Policies and Early Warning Systems in the EU 51
2.3 From Policy Linkage to Ecosystem Preservation 56
2.3.1 Water and Policy Linkage 56
2.3.2 The Water Management Policy Framework 64
2.3.3 Biodiversity and Ecosystem Preservation 73

2.3.4 Permanence vs. Flexibility 76
2.3.5 The Biodiversity Policy Framework 80
2.4 From the Water Towers of Europe to Ecosystem Preservation . . 82
2.5 On the Governance of Adaptation: Goal Conflict And Institutional Confusion 85
2.6 Conclusions 90
References 91

3 Climate Change Adaptation in the United Kingdom: England and South-East England 97
E. Carina H. Keskitalo
3.1 Introduction 97
3.2 National Level 99
3.2.1 Environmental Policy in the UK and the Establishment of Adaptation as a Priority 99
3.2.2 National Legislative, Policy and Planning Frameworks on Adaptation 103
3.2.3 Creating Multi-Level Linkages: The UK Climate Impacts Programme (UKCIP) 110
3.3 Regional Level 113
3.3.1 Regional Climate Change Partnerships: Climate South East 114
3.3.2 Regional Planning and Administration 116
3.4 Local Level 119
3.4.1 Adaptation and the Structure of Local Government 119
3.4.2 Actions at Selected Local Authorities in the South East . 121
3.5 The Role of the EU 130
3.6 The Participation of Voluntary and NGO Networks 132
3.7 Adaptation as an Issue: Shifts in Thinking, Policy Transfer and Integration Across Levels 133
3.7.1 Policy Transfer Between Contexts? 133
3.7.2 Adaptation as a Shift in Thinking 138
3.8 Conclusion 140
References 143

4 Mainstreaming Climate Change Adaptation: The Case of Multi-Level Governance in Finland 149
Sirkku Juhola
4.1 Introduction 149
4.2 Background 150
4.2.1 Environmental Policy in Finland 152
4.3 National Climate Policy 153
4.3.1 National Climate and Energy Strategy 2001 154
4.3.2 Revised National Climate and Energy Strategy 2005 . . . 155
4.3.3 Revised National Climate and Energy Strategy 2008 . . . 156
4.4 National Adaptation Policy 157

4.4.1 Preparation of the National Strategy 157
4.4.2 Key Projected Impacts of Climate Change 158
4.4.3 The National Adaptation Strategy 159
4.4.4 Implementation of the NAS 160
4.4.5 Evaluating the Implementation of the NAS 162
4.4.6 Role of Research in Climate Change Adaptation 166
4.5 The Role of the European Union in Climate Policy 168
4.6 Other Actors in Climate Change Adaptation 169
4.7 Sub-national Level Adaptation Work 170
4.7.1 Uusimaa Regional Council Climate Strategy 172
4.7.2 Climate Strategy for Helsinki Metropolitan Area Council 2030 174
4.7.3 Climate Strategy for the KUUMA-partnership 175
4.7.4 Espoo Preparedness Strategy 176
4.7.5 Integration of Adaptation into Policy-Making 178
4.8 Challenges and Opportunities of Multi-Scale Adaptation 179
4.9 Conclusions 183
References 183

5 Adapting to Climate Change in Sweden: National Policy Development and Adaptation Measures in Västra Götaland 189
E. Carina H. Keskitalo
5.1 Introduction 189
5.2 National Level 192
5.2.1 Swedish Work on Climate Change of Particular Relevance for Adaptation 192
5.2.2 The Commission on Climate and Vulnerability 195
5.2.3 National Legislation, Regulation and Measures Taken on Adaptation 198
5.3 Regional Level 205
5.3.1 Impacts and Authority at the Regional Level 205
5.4 Local Level 209
5.4.1 Municipal Climate Change Networks and Responsibilities 209
5.4.2 Municipal Case Studies 212
5.5 Actors Beyond Government and Administration 218
5.5.1 The Role of the Private Sector 218
5.5.2 Impacts of the EU Level 220
5.6 The Distribution of Responsibility for Adaptation to Climate Change 221
5.6.1 Responsibility for Adaptation Across Governmental Levels 221
5.6.2 Policy Transfer and 'Lesson-Drawing' 225
5.7 Conclusion 226
References 229

6 'Planning for Today': The Nature and Emergence of Adaptation Measures in Italy . . . 233
Lisa Westerhoff
6.1 Introduction . . . 233
6.2 Italian Environmental Policy: Evolution and Current Issues . . . 234
6.3 Entry into the Climate Change Arena: Mitigation Policies and Programmes . . . 239
6.4 Adaptation and the Italian State . . . 240
6.4.1 From Impacts to Adaptation: 1995–2007 . . . 241
6.4.2 The National Climate Change Conference: 2007 . . . 245
6.4.3 National Adaptation to Climate Change: The Present . . . 248
6.5 Adaptation at Sub-national Scales . . . 250
6.5.1 Adaptation in the Region of Emilia-Romagna . . . 252
6.5.2 Adaptation Actions in the Province of Ferrara . . . 258
6.5.3 Adaptation at the Municipal Level: the Municipality of Ferrara . . . 262
6.6 Conclusion: Multi-Level Governance in Italian Adaptation . . . 265
References . . . 268

7 Planned Adaptation Measures in Industrialised Countries: A Comparison of Select Countries Within and Outside the EU . . . 271
Lisa Westerhoff, E. Carina H. Keskitalo, Heather McKay, Johanna Wolf, David Ellison, Iosif Botetzagias, and Bertrand Reysset
7.1 Introduction . . . 271
7.2 Approach and Methods . . . 272
7.3 Background: Adaptation Actions Across Europe . . . 273
7.4 Adaptation Across Scales: Examples from the European Context 275
7.4.1 Austria . . . 277
7.4.2 France . . . 280
7.4.3 Germany . . . 285
7.4.4 Greece . . . 291
7.4.5 Hungary . . . 293
7.4.6 Netherlands . . . 297
7.4.7 Norway . . . 302
7.4.8 Spain . . . 308
7.5 Beyond Europe: Adaptation Actions in Canada and Australia . . . 312
7.5.1 Australia . . . 312
7.5.2 Canada . . . 318
7.6 Discussion: The Emergence of Adaptation . . . 322
7.6.1 Diverging Steering Abilities in Federal and Unitary States . . . 323
7.6.2 Multi-Level Governance: Decentralisation and the Participation of Sub-national Authorities . . . 325

7.6.3 The Relevance of National Environmental Policy Implementation 327
7.6.4 Stakeholder Participation and the Role of NGOs and the EU 329
7.7 Conclusions 331
References 332

8 Conclusion: The Development of Adaptive Capacity and Adaptation Measures in European Countries 339
E. Carina H. Keskitalo
8.1 Introduction 339
8.2 National Level and General Political System Features 341
8.2.1 Structure of the Political and Planning System 341
8.2.2 The Role of Environmental Policy Institutionalisation . . 344
8.2.3 The Role of Focusing Events and Other Contextual Features 347
8.3 Regional Level 350
8.4 Local Level 352
8.5 The Role of the EU 354
8.6 Actors Beyond Government and Administration 356
8.7 Types of Adaptation 358
8.8 Lessons for Current and Future Adaptation 361
8.9 Conclusion: Does Adaptation Require a Change in Thinking? . . 363
References 365

Index 367

Contributors

Iosif Botetzagias Department of Environment, University of the Aegean, Centre for Environmental Policy & Strategic Environmental Management, University Hill, 81100 Mytilini, Greece, iosif@aegean.gr

David Ellison Department of Social and Economic Geography, Umeå University, Umeå 90187, Sweden, Ellisondl@gmail.com

Sirkku Juhola Department of Social and Economic Geography, Umeå University, Umeå 90187, Sweden; Centre for Urban and Regional Studies, Helsinki University of Technology, Helsinki 02015, Finland, sirkku.juhola@tkk.fi

E. Carina H. Keskitalo Political Science/Department of Social and Economic Geography, Umeå University, Umeå 90187, Sweden, Carina.Keskitalo@geography.umu.se

Heather McKay Department of Social and Economic Geography, Umeå University, Umeå 90187, Sweden, heather@protectforests.com

Bertrand Reysset Observatoire national sur les effets du réchauffement climatique (ONERC), Grande Arche, Tour Pascal A et B, 92055 La Défense CEDEX, France, bertrand.reysset@developpement-durable.gouv.fr

Lisa Westerhoff Department of Social and Economic Geography, Umeå University, Umeå 90187, Sweden, lisa.westerhoff@gmail.com

Johanna Wolf Tyndall Centre for Climate Change Research, School of Environmental Sciences, University of East Anglia, Norwich NR4 7TJ, UK, j.wolf@uea.ac.uk

Abbreviations

ACACIA	Consortium for the Application of Climate Impact Assessments
ACC	Adapting to Climate Change (UK)
ACCC	Austrian Council on Climate Change
ADAGIO	Adaptation of Agriculture in European Regions at Environmental Risk under Climate Change
ADAM	Adaptation and Mitigation Strategies – Supporting European Climate Policy
ALFRA	Association of Finnish Local and Regional Authorities
ANPA	National Agency for Environmental Protection (Italy)
AMICA	Adaptation and Mitigation – an Integrated Climate Policy Approach
ARK	National Programme on Climate Adaptation and Spatial Planning (Netherlands)
ARPA	Regional Environmental Protection Agency (Italy)
BaltCICA	Climate Change: Impacts, Costs and Adaptation in the Baltic Sea Region
BC	British Columbia
BMLFUW	Federal Ministry of Agriculture, Forestry, Environment and Water Management (Germany)
BRANCH	Biodiversity Requires Adaptation in Northwest Europe under a Changing Climate
CAP	Common Agricultural Policy
CC	City Council
C-CIARN	Climate Change Impacts and Adaptation Research Network (Canada)
CCIRG	Climate Change Impacts Review Group (UK)
CCP	Cities for Climate Protection
CCPCC	Coordination Commission of Climate Change Policies (Spain)
CcSP	Climate changes Spatial Planning (Netherlands)
CFMP	Catchment Flood Management Plans (UK)

CICERO	Centre for International Climate and Environmental Research (Norway)
CIPE	Interministerial Committee for Economic Planning (Italy)
CIRCE	Climate Change and Impact Research: the Mediterranean
ClimChAlp	Climate Change, Impacts and Adaptation Strategies in the Alpine Space
CMCC	Euro-Mediterranean Centre for Climate Change
CO	Carbon dioxide
CSIRO	Commonwealth Scientific and Research Organisation (Australia)
DECC	Department for Energy and Climate Change (UK)
DEFRA	Department of Environment, Food and Rural Affairs (UK)
DETR	Department for Environment, Transport and the Regions (UK)
DG	Directorate General
DMP	Drought Management Plans (EU)
DPC	Italian Department for Civil Protection
DSB	Directorate for Civil Protection and Emergency Planning (Norway)
EAFRD	European Agricultural Fund for Rural Development
ECCE	Effects of Climate Change in Spain
ECCP WG II	Working Group II on Impacts and Adaptation (EU)
ECCP II	Second European Climate Change Programme
ECSC	Energy Centre for Sustainable Communities (Sweden)
EEDA	East of England Development Agency
EEA	European Environment Agency
ENEA	Italian National Agency for New Technologies, Energy and the Environment
EPA	Environmental Protection Agency
ESPACE	European Spatial Planning: Adapting to Climate Events
ESRC	Economic and Social Research Council
EU	European Union
EURISLES	European Islands System of Links and Exchanges
FIGARE	Finnish Global Change Research Programme
FINADAPT	Assessing the Adaptive Capacity of the Finnish Environment and Society under a Changing Climate
FINESSI	Integrated assessment modelling of global change impacts and adaptation
GDP	Gross Domestic Product
GHG	Greenhouse Gas
GMES	Global Monitoring for Environment and Security
GO	Government Office
GOL	Government Offices London
GOSE	Government Office South East

GR	Gothenburg Region Association of Local Authorities
GRaBS	Green and Blue Space: adaptation for urban areas and eco towns
HMAC	Helsinki Metropolitan Area Council
ICID	International Commission on Irrigation and Drainage (Hungary)
IDeA	Improvement and Development Agency for Local Government (UK)
IESE	Improvement and Efficiency South East (UK)
IMC	Hydro-Meteorological and Climate Service (Italy)
IPCC	Intergovernmental Panel on Climate Change
ISPRA	Italian Institute for Environmental Protection and Research
ISTO	Climate Change Adaptation Research Programme (Finland)
JRC	Joint Research Centre
KLIMZUG-NORD	Strategic approaches to climate change adaptation in the Hamburg Metropolitan Region
LA21	Local Agenda 21
LAA	Local Area Agreement (UK)
LCLIP	Local Climate Impacts Profile (UK)
LGA	Local Government Association (Sweden)
LIP	Local Investment Programme (Sweden)
LRAP	Local and Regional Adaptation Partnership (UK)
LSP	Local Strategic Partnerships (UK)
MATTM	Italian Ministry of Environment and Territory
MCPFE	Ministerial Conference on the Protection of Forests in Europe
MNC	Multi-National Corporations
NARP	National Adaptation Research Plan (Australia)
NAS	National Adaptation Strategy
NCCS	National Climate Change Strategy (Hungary)
NCCARF	National Climate Change Adaptation Research Facility (Australia)
NGO	Non-Governmental Organisation
NI	National Indicator
NORKLIMA	Climate change and its impacts in Norway
NPACC	National Plan for Climate Change Adaptation (Spain)
NPM	New Public Management
OECC	Spanish Climate Change Office
OECD	Organisation for Economic Cooperation and Development
ONERC	French National Observatory of the Effects of Climate Warming
PCAN	Portsmouth Climate Change Action Network
PEER	Partnership for European Environmental Research
PESETA	Projection of Economic impacts of climate change in Sectors of the European Union based on boTtom-up Analysis
PPG	Planning Policy Guidance (UK)

PPS	Planning Policy Statement (UK)
PSA	Public Service Agreements (UK)
PSAG	Portsmouth Sustainability Action Group
PUSH	Partnership for Urban South Hampshire
RA	Regional Assembly (UK)
RAC	Regional Adaptation Collaboratives (Canada)
RBMP	River Basin Management Plans
RCCP	Regional Climate Change Partnership (UK)
RDA	Regional Development Agency (UK)
REC	Regional Environment Centre (Finland)
REGKLAM	Integrated Regional Climate Adaptation Programme (IRCAP) for the Model Region of Dresden
RIEP	Regional Improvement and Efficiency Partnerships (UK)
SALAR	Swedish Association of Local Authorities and Regions
SEEDA	South East England Development Agency
SEERA	South East England Regional Assembly
SEK	Swedish kronor
SGU	Geological Survey of Sweden
SHI	Swedish Geotechnical Institute
SILMU	Finnish Research Programme on Climate Change
SMHI	Swedish Meteorological and Hydrological Institute
SSNC	Swedish Society for Nature Conservation
TÖOSZ	Hungarian National Association of Local Governments
UK	United Kingdom
UKCIP	United Kingdom Climate Impacts Programme
UKIRCCG	United Kingdom Interregional Climate Change Group
UN	United Nations
UNEP	United Nations Environment Programme
UNFCCC	United Nations Framework Convention on Climate Change
VAHAVA	Weather and Climate: Changes-Impacts-Answers (Hungary)
VROM	Ministry of Housing, Spatial Planning and Environment (Netherlands)
VTT	New Vásárhelyi Plan (Hungary)
WFD	Water Framework Directive
WWF	World Wildlife Fund

List of Figures

1.1 A simplified figure of the impacts on development of adaptation policy and practice in a multi-level context 23
2.1 Policy overlap in water resource management, reproduced from EEA (2009a) 57
2.2 Change in summer temperatures, precipitation and river flow compared to 1961–1990 reference period 62
3.1 Map of the South-East England case study region and the selected case study local authorities in the region 99
4.1 The Finnish case study areas 172
5.1 Map of the Västra Götaland case study region and the selected case study municipalities in the region 191
6.1 The Italian case study area 253

List of Tables

1.1 National level case study selection and criteria 29
1.2 Nested case study areas 31
2.1 Differences between the EU's Green and White Papers on adaptation 44
2.2 Potential and actual targets of individual sectoral policies 84
3.1 National adaptation policy initiatives in the UK 103
3.2 Climate change focus in regional bodies 114
3.3 Particular adaptive capacity strengths and limitations at local authority level 122
3.4 Summary of main types of adaptation policy development on different levels 134
4.1 Details of selected sub-national cases 173
5.1 Main adaptation policy development and initiatives on different levels 222
6.1 Expected climate impacts in Italy as outlined in the Fourth National Communication to the UNFCCC 243
7.1 Country characteristics and state of adaptation policy 276
7.2 Preliminary adaptation options identified in the national climate change adaptation plan 310
7.3 Progress on sectoral adaptation research and measures 315
8.1 Examples of adaptation development in the case studies 359

List of Boxes

3.1 The progress levels in the NI 188 indicator with technical definitions of each level 106
4.1 Measures for implementation at the Municipal level 155
4.2 Sectors considered in the National Adaptation Strategy 159
4.3 Adaptation measures in the environment administration 161
4.4 Indicator of progress on adaptation 164
6.1 Priority areas for adaptation intervention identified by the NCCC (2007) 246
6.2 Examples of planned climate change adaptation activities in Italy . . . 251
7.1 The nine axes of the French National Adaptation strategy 282
7.2 Sub-national responsibilities under the Dutch National Adaptation Strategy 300

Chapter 1
Introduction – Adaptation to Climate Change in Europe: Theoretical Framework and Study Design

E. Carina H. Keskitalo

Abstract As mitigation will not likely be sufficient to hinder climate change, adaptation to the consequences of climate change will be needed. The impacts of climate change will include such phenomena as increased flooding and sea level rise, which will in turn have significant effects on densely populated and infrastructurally-developed areas in advanced industrial states. Despite the potential for serious consequences, very little of the existing climate change adaptation literature has focused on adaptation in the EU or the industrialised world in general. This chapter and the volume at large address this gap. This chapter describes the governance system of public and private actors and bodies that set the context for adaptive capacity at local, regional, national and EU levels, and argues that adaptive capacity can largely be seen as related to the resource distribution and prioritisation processes within such systems. The chapter further outlines the comparative approach taken by the volume, including a common methodology for the presented multi-level studies.

Keywords Adaptation · Adaptive capacity · Climate change · Multi-level governance

1.1 Introduction

Climate change will pose major challenges for adaptation in Europe. Even if greenhouse gas emission outputs were to cease completely (an unlikely accomplishment), existing levels of greenhouse gases in the atmosphere indicate that mitigation alone would be insufficient in preventing the effects of climate change. As a result, adjustments and adaptations to cope with the effects of climate change will be required. Given this need, the assessment of vulnerability to climate change and the possibility for adaptation has been identified as a priority area for research (IPCC, 2007).

E.C.H. Keskitalo (✉)
Political Science/Department of Social and Economic Geography, Umeå University, Umeå 90187, Sweden
e-mail: Carina.Keskitalo@geography.umu.se

E.C.H. Keskitalo (ed.), *Developing Adaptation Policy and Practice in Europe: Multi-level Governance of Climate Change*, DOI 10.1007/978-90-481-9325-7_1,

Projected impacts of climate change for Europe include shorter winters, the earlier onset of spring with a corresponding increase in precipitation and flood risk, warmer and drier summers with an increase in risk of heat waves, and a later autumn, with large variations across countries and regions (IPCC, 2007; EEA, 2008). Adaptations to such changes may take the form of a number of responses at local, national and even international levels, including early warning systems, changes in planning systems and the development of adaptation strategies at different levels. In some cases, the need for adaptation may even prompt a greater awareness of the limitations of spatial planning based on the assumption of a steady state (i.e. one based on the existing situation), as climate change may come to change the distribution of species and patterns of flood and drought.

While knowledge of likely changes in climate is fairly well developed, less research has been conducted on how such changes can and will impact societies, or on the actual or perceived possibilities for and obstacles to further adaptation. So far, mainly localised, single-sector case studies have been conducted (cf. IPCC, 2001; Keskitalo, 2008), often either on a community scale (e.g., Ford & Smit, 2004) or as an overview of general national vulnerability (e.g., O'Brien et al. 2004). Very little has been done to evaluate the different ways in which institutions in different national contexts or at different organisational scales may adapt to climate change (Adger, Brown, & Tompkins, 2004a). So far, adaptation to climate change has also been viewed principally as a question for developing countries, while the treatment of adaptation in advanced industrial states has been limited (Gagnon-Lebrun & Agrawala, 2007). As a rule, adaptation in developed countries has been framed separately and developed into policy later than mitigation, indicating that a state can be both a leader on adaptation and a follower on mitigation (cf. Lorenzoni, O'Riordan, & Pidgeon, 2008).

In 2007, Gagnon-Lebrun and Agrawala (2008) noted that 'no developed country ha[d] yet formulated a comprehensive approach to implementing adaptation and the "mainstreaming" of such measures within sectoral policies and projects, although the UK might be coming close' (p. 401). Over the last couple of years, however, this has begun to change. A frontrunner to date, the United Kingdom (UK) began its climate impacts programme in 1997 and implemented a Climate Change Act in 2008, including requirements and economic incentives for adaptation at the municipal level as well as by national government. This leadership is particularly notable as the UK so far has not been a strong leader in environmental policy in the European Union (EU), but is now taking a leading role on adaptation (cf. Börzel, 2002).

Another relatively early mover among industrialised countries, both in adaptation and to some extent in environmental policy, is Finland. In 2007, Gagnon-Lebrun and Agrawala (2008) noted that Finland was 'moving towards implementing adaptation in many sectors' (p. 402), and in 2008 Finland developed a national adaptation strategy that mainstreams adaptation across governmental sectors. By contrast, Sweden – traditionally seen as a leader in environmental policy – has been more of a slow mover on adaptation and has instead focused principally on mitigation. While adaptation is included in Sweden's 2009 National Bill on climate and energy, adaptation has so far been relatively fragmented, with adaptation measures emerging mainly in areas where particularly significant risks have been identified. The differences between these two countries are particularly interesting given the considerable similarities between Sweden and Finland with regard to their national political, administrative and planning systems. Finally, in countries that are traditional followers in environmental policy, such as Italy, adaptation has been developed in select local cases as a response to existing risks, but with limited future-oriented planning as a result of the absence of any structured national adaptation policy.

This volume describes the development of adaptation to climate change in the above-mentioned European Union (EU) countries at the national scale, as well as in select nested regional and local cases where development of adaptation policy has been relatively early. Case studies include countries with varying characteristics across a number of spectra: political and planning systems, environmental policy tradition, and extent of policy and practice on adaptation. The volume also includes a chapter on the role of the EU, including the impact of EU policy that is not explicitly linked to adaptation but that may impact the potential for adaptation in different countries. Examples of these include the EU common agricultural policy, EU projects that support climate research, and EU environmental policy (e.g., the Habitats and Water Framework Directives) that direct policy in sectors with the potential to be strongly impacted by climate change. The book further includes a comparison with industrialised countries beyond the four main cases, selected to exemplify broad groups of varying environmental policy and planning systems. This comparison is undertaken with the aim to discuss the potential impact of different political system characteristics on adaptation policy development. Countries included in the comparative chapter include both those with centralised and decentralised planning systems and federal and unitary states within Europe (Germany, Austria, France, Spain, Greece, Hungary, the Netherlands and Norway) as well as industrialised country cases outside the EU (Canada and Australia).

The aim of this volume is to further the understanding of the concept of adaptive capacity in a governance context through narrative description and analysis of the factors that have allowed for adaptation in each of these national to local cases. The study problematises the issue of governance and adaptation across levels and sectors with a basis in the following questions:

1. What is the governance context for climate change adaptation? In other words, what policies and action programmes exist on different levels and how well are these coordinated across levels and sectors? This question includes a focus on multi-level governance and the extent to which actions on local and regional levels are dependent on the national context, as well as the extent to which local actors are able to respond independently or even to 'jump scales' (Jones, 1998) by drawing on EU frameworks or funding.
2. To what extent have differences in political and planning systems and determinants of adaptive capacity, such as access to information, economic resources and institutional capacity, influenced the form of adaptation policy and adaptation measures that have developed in response to identified vulnerabilities? This question includes a focus on the extent to which different capacities can

compensate each other: for instance, whether there are examples where local political leadership has been able to develop local adaptation policy and measures, even in the face of limited economic resources or in cases where national level responses are less prominent.

3. To what extent can examples be found that are considered by actors in different countries to be transferable between contexts, and that may support policy transfer or 'lesson drawing' on adaptation? Given the ongoing development of adaptation policy, this question aims to respond to the need identified by many interviewees to access examples of processes and models elsewhere that may speak to local needs – an aim also expressed in the EU Green Paper on adaptation (2007).

Adaptation is thus inherently viewed as taking place within a political context on multiple levels, within which responses are formed by multiple interests, including those in the existing political and administrative systems. This chapter first outlines the theoretical background for the book in terms of multi-level governance, describing the factors that may impact adaptive capacity within a political context and on national, regional, local and EU levels, respectively, as well as in relation to policy transfer and lesson-drawing. The study thus explicitly aims to define adaptive capacity within a governance context and connect adaptive capacity literature to a broader political science literature. The chapter then describes the methodology for the studies and concludes with an outline of the volume's contents.

1.2 Theoretical Background

1.2.1 Multi-Level Governance and the Capacity to Act on Adaptation

Multi-level governance is defined as decision-making that is steered not only by public but by private and other interests, and as a process that takes place across multiple geographic scale levels and sectors (Boland, 1999; Hooghe & Marks, 2003). While national governments have generally been seen as the principal actors in decision-making, discussions have emerged in recent years of the role of complementary or competing systems such as the EU, and of an increased devolution to the local level: 'what has emerged ... in recent years, is a complex set of overlapping and nested systems of governance involving European, national, regional, and local actors, groups and networks' (Loughlin, 2001, p. 20).

Climate change is a problem that poses high requirements for governance by requiring the coordination of demands and needs across international, national, regional and local scales, as well as coordination between sectors (e.g., across departments that deal with environmental, energy and financial issues, or between private and public sector actors). In addition, the way in which climate change impacts and adaptation are treated by actors on different levels is to a large extent dependent on differential adaptive capacities, including financial resources, access to information, decision-making structures and other institutional features (Smit & Wandel, 2006). The fact that there may be sufficient access to resources for adaptation at the national scale does not necessarily translate to a high adaptive capacity at local scales. Similarly, high resource access at the local scale (for example, in municipalities with significant economic resources) does not necessarily lead to the development of adaptation unless adaptations have been defined as urgent in the local context or the implementation of adaptation measures is required by national legislation (Næss, Thorsen Norland, Lafferty, & Aall, 2006).

An assessment of adaptive capacity in a governance context requires an assessment of the attribution of responsibility for adaptation to climate change. Vulnerability literature has often argued that adaptation to climate change needs to take place at the local level (cf. Næss, Bang, Eriksen, & Vevatne, 2005), where vulnerability to specific stresses (e.g., the flooding of specific areas) and adaptive capacity (the resources to deal with these threats) can be defined. For instance, an area with extensive economic resources may respond to climate change in very different ways than an area with fewer. Such differences could lead to responses as different as abandoning low-lying areas or defending them against floods, even where rise in sea level may be the same or the areas in question may be situated in close proximity along the coastline. However, this fact problematises in particular the responsibility attributed to different scales within the political system, both in terms of responsibility and the corresponding allocation of resources, as well as in terms of the potential for policy development and implementation specifically on adaptation. As McConnell (2003) notes on crisis and emergency management, adaptation may be considered '[a]t heart. . .a political activity' (p. 409).

Relevant roles and the distribution of responsibilities in the context of existing institutional structures thus include, among others, the role of the regional level and local authorities in relation to the national level. The policy style concept (e.g., Richardson, 1982) has been used to demonstrate that 'each nation's regulatory style is a function of its unique political heritage' (Andersen, 1999, p. 25), indicating that 'policy actors in different governance systems do not necessarily propose the same course of action when faced with similar policy problems' (Wurzel, 2002, p. 17). The policy style concept also indicates that an emerging issue such as adaptation will most likely come to be regulated in ways similar to those applied to existing issues.[1] Different countries and institutions can thus be seen as exhibiting a certain degree of path dependence, which is often seen as an indication that 'initial social outcomes concerning institutional, organisational, or policy design – even suboptimal ones – can become self-reinforcing over time' (Pierson, 2002, p. 372). Historical

[1]Similar mechanisms are also present in theories of governmentality, which note that specific mentalities may govern the selection of programmes and instruments to regulate particular fields (cf. Rose, 1996; Keskitalo, Juhola, & Westerhoff, in prep). In relation to policy style literature, Wurzel (2002) adds that sectoral and sub-sectoral differences will also influence the regulation of environmental issues (e.g., within different branches of government).

choices thus contribute to form the options available today and the institutions that are available to handle these. However, some leeway in terms of paths chosen does exist, particularly as new issues come onto the political agenda. For example, Andersen (1999) notes that while the distribution of competences and resources may be seen as a constant struggle between actors and administrative levels, the potential for changes in standard operating procedures may be greatest when a new issue emerges.

Liefferink and Andersen (1997) note that 'the most obvious opportunity to initiate innovations in a certain policy field is probably the process of agenda-setting ... innovations may entail either the introduction of wholly new issues onto the agenda, or the definition and re-definition of problems and potential solutions' (p. 11). The agenda-setting literature has emphasised that the rise of an issue on the agenda (e.g., to become institutionalised in policy and legislation) depends on the simultaneous existence of several factors, including policy entrepreneurs who push a given issue; indicators and events (e.g., storms or floods) that serve to problematise an issue to policy-makers, the media and the public; and existing politics and policy development that provide an entryway to the issue (cf. Kingdon, 1995; Baumgartner & Jones, 1993). Together, these factors support the development of a 'policy window' for the issue where adaptation or other forms of policy have the chance to develop until the window is closed (e.g., by new crises or events in unrelated areas) and other issues are moved onto the agenda (cf. Keskitalo, Westerhoff, & Juhola, in prep.).

Other authors have noted a number of additional factors relevant to the agenda-setting literature, particularly in relation to environmental policy development and crisis response. Wurzel (2002) describes a number of explanatory variables for domestic environmental policy action, including ecological vulnerability or the state of the environment; economic capacity to deal with problems; the political salience of issues (including public environmental awareness and media attention to environmental issues); and environmental regulatory style, including modes of policy-making. Drawing upon examples from the UK, McConnell (2003) notes that the nature of crisis response depends on whether crises are sudden, creeping or chronic (prompting responses that range from improvised to business-as-usual); the perceived seriousness of the threat (where serious threats result in more centralised responses); the need for immediate action vs. longer-term consultation; and the political structure of the government and its individual departments. Other factors affecting agenda-setting include dominant political ideas, personalities, the media and public opinion, interest representation and the international and EU context (McConnell, 2003).

In much of the agenda-setting and crisis response literature, 'focusing events' have been particularly emphasised as powerful catalysts of policy development. However, focusing events also emphasise the degree of randomness in policy-making as potential long-term important actions may to some extent need to rely on events to push them and there is no guarantee that actions following an event will sufficiently take long-term planning into account. Johnston, Tunstall, and Penning-Rowsell (2005) note:

> It is arguably the case that it takes a severe and damaging flood to place flooding on the political agenda, at a time when the public and media response is such that a failure to act is politically unacceptable. There is, however, no guarantee that the nature of the policy issues raised by a major flood disaster will offer anything more than post-event response and recovery. (p. 561)

Other literature additionally notes that responses to crises may include 'passing the buck' or non-action, often as a result of infighting between interests (McConnell, 2003). Policy implementation may fail as a result of diverse actors and perspectives, which makes for difficulties in reaching an agreement or result in the separation of policy design from implementation (Schuct, 2001).

These factors all describe the importance of attention being drawn to an issue and the need to be able to draw upon different types of resources (similar to those constituting adaptive capacity) in institutionalising or acting upon an issue such as climate change. In addition, agenda-setting, policy development or implementation of adaptation measures will often require action on different levels. For example, McConnell (2003) notes that elements of decentralisation are necessary in crisis response: '[m]ost crises or emergencies require those individuals close to the impact of the crises to take "local" decisions' (p. 401). Given the requirement of different responses in different localities, adaptation may be considered an even more compelling example of the need for such decentralisation as well as coordination. Adaptation may thus be seen as an issue characterised by the need for multi-level responses, which are realised to different extents in different cases depending on whether the issue is able to rise on the agenda and the form in which this takes place in different countries, regions and localities.

1.2.2 Adaptive Capacity

Any assessment of vulnerability to change must be grounded in the sensitivity of a system to a certain exposure to change, as well as the capacity of the system to adapt to change without limiting important functions (Smit & Wandel, 2006). Exposure to climate change indicates the extent and type of climate change effects relevant for in the area in question, while sensitivity indicates the sensitivity of, for instance, ecosystems to such exposure. The term adaptive capacity is used to capture the ability of any geographical or organisational entity (e.g., county, region, community or individual) to cope with, adjust to, or recover from external stresses. Within adaptive capacity, particular adaptations are undertaken, often as uncoordinated responses among numerous actors spanning the individual, community, enterprise, state and international levels (IPCC, 2007; Smit & Wandel, 2006).

In many cases, impacts will be felt at levels other than those at which decisions on adaptation are taken, which raises the possibility of discrepancies between perceived needs for adaptation and the resultant decisions (Keskitalo, 2008). For instance, local level administrators or entrepreneurs may perceive problems but may not have the decision-making or policy-making capacity to act on them (e.g., if taxes and therefore funding are controlled by the state, or if the national policy framework does not allow for independent definition of local actions). Adaptive capacity is therefore defined here to include the present ability to cope (Adger, Brooks, Bentham, Agnew, & Eriksen, 2004b), as well as the capacity for extending such actions into the future using novel adaptation approaches that may be specifically tailored to a given stress (Smit & Wandel, 2006). Adaptive capacity is considered to be dependent on the underlying resources that make up the capacity to adapt to any change (a perspective in accordance with social vulnerability literature, e.g. Adger, 2000; Adger et al., 2004b), and to include coping capacities as well as the ways in which existing resources can be marshalled to deal with new stresses (for a typology, see e.g., Smit, Burton, Klein, & Wandel, 2000). Such a definition is used here as it is often difficult to draw precise distinctions between coping and more novel adaptation measures, particularly as coping may develop into adaptation as a result of innovation in existing practices (Brooks, 2003).

Thus, both measures taken within existing frameworks and emerging practices that may be re-framed as adaptation in light of an emerging adaptation problematique could be defined as adaptations. The crucial delineation would be whether these measures respond to events and occurrences likely to increase with climate change. Novel strategies and reactive or planned measures in response to potentially climate change-related events would similarly also be defined as adaptations. Adaptive capacity is thereby reflected in a unit's management of current and past stresses, its ability to anticipate and plan for future change, and its resilience to perturbations (Smit & Wandel, 2006). Improved adaptive capacity – increased by, for instance, adaptive planning, the allocation of resources in response to projected threats, and/or the identification of possible ameliorative actions – can improve an area's resilience or robustness to external stress.[2]

Adaptive capacity therefore requires a comprehensive understanding of the system, including its capacities in terms of decision-making and regulative (governance) frameworks on multiple levels. The adaptive capacity of different actors is generally the result of their established priorities, resources (financial, knowledge and other) and readiness for learning in response to change (IPCC, 2001). This capacity may also be increased or decreased by responses to simultaneous stresses, indicating that adaptive capacity should be viewed in relation not only to climate change, but to other stresses that impact stakeholders' adaptation decisions (such as globalising market pressures or economic transition) (Keskitalo, 2008; Smit & Wandel, 2006).

Adaptive capacity is thus a very broad concept, highlighting factors that are of importance also to determine mitigative capacity, i.e. the ability to limit emissions at the source (e.g., Kane & Shogren, 2000). A number of broad but similar frameworks try to distinguish the different determinants of adaptive capacity (e.g., Eakin

---

[2]Resilience can broadly be defined as the possibility for a system to be able to absorb disturbances while still retaining its basic functions (cf. Walker & Salt, 2006). Resilience is thereby related to adaptive capacity.

& Lemos, 2006; Smit & Pilifosova, 2001; Tol & Yohe, 2007; Leary et al., 2007). Determinants are here broadly understood as the underlying factors or components that contribute to adaptive capacity. For instance, Smit and Pilifosova (2001) define the determinants of adaptive capacity under the headings of economic resources, technology, information and skills, infrastructure, institutions, and equity. Drawing upon Smit and Pilifosova (2001) and Yohe and Tol (2001), Eakin and Lemos (2006) instead link determinants of adaptive capacity to different types of capital, such as human, organisational and social, political, and wealth and financial capital, together with factors such as information and technology, material resources and infrastructure, and institutions and entitlements.

In all of these frameworks, an important component of adaptive capacity is economic or financial resources. In a governance context, such resources may be interpreted also in relation to their impact on staffing, and thus on the ability of an administration to develop and maintain knowledge and skills on a particular issue. Financial resources may also increase or decrease in response to phenomena such as the increasing 'hollowing out' of the financial ability of the state in response to globalisation (cf. Rhodes, 2000). Other important resources in a governance context are leadership and political resources (e.g., Smit & Pilifosova, 2001; Eakin & Lemos, 2006) and political mobilisation, which are necessarily developed in an institutional context (i.e. within institutionalised decision-making systems that determine the distribution of resources).[3] As such, the resources that exist within a decision-making body or at a particular scale may be determined through political priorities at various levels, and potentially include elements of public and media influence on these priorities (such as those described in agenda-setting literature, cf. Baumgartner & Jones, 1999). Leadership at the international level may also support actions at lower levels by influencing agendas and political priorities. The importance of such resources is highlighted more generally in political science conceptions – for example, through the truism that 'organisation is the mobilisation of bias' (Schattschneider, 1960, p. 71).

Closely related to such political mobilisation and political resources are issues of information and technology brokerage, or the cross-sectoral/actor capacities that serve to make processes, technologies or knowledge accessible. This highlights the fact that the existence of information and technologies cannot be assumed to imply utilisation (although they are often treated with such an implication, cf. Smit & Pilifosova, 2001); rather, information and technologies need to be made available or accessible to actors for utilisation, a requirement briefly discussed by Eakin and Lemos (2006) in terms of technology transfer and innovation capacity. With regard to the multi-level governance context, information and technology-related capacities may be associated with the successful development of 'epistemic communities', or policy-science communities (Haas, 1990), or with the concept of 'knowledge

[3] Institutions are also seen as an independent factor in some descriptions of adaptive capacity (e.g., Smit & Pilifosova, 2001), but are discussed here within a multi-level governance and political context. Resources related to knowledge and skills or human resources (Smit & Pilifosova, 2001, Eakin & Lemos, 2006) can be seen as integrated both in this category and in other categories.

brokers' as intermediaries between science and policy (Litfin, 1994). In an institutionalised context, the broker or community may be constituted by a formal organisation, reflecting the need for 'boundary organisations' (cf. Schneider, 2009) that serve to translate scientific findings for specific target groups and associated organisations.

In addition, physical infrastructure, including access to building infrastructure, transport, water, and the extent to which these may be used to support local development and industry, is also a parameter that is used to define vulnerability and adaptive capacity. This concept relates to the idea that the protection of material resources may be a crucial issue for adaptation in developed countries (cf. Gagnon-Lebrun & Agrawala, 2008), and that access to infrastructure may serve to support and make specific types of adaptation accessible for specific groups or areas. For instance, a well-funded area with dense infrastructure and high population pressure may choose to construct tunnels or pumping systems as part of flood protection, while areas with more limited infrastructure and possibilities for development may select less costly measures or even abandon certain areas unless significant value is ascribed to them.

On the whole, decisions taken within a system will impact what has often been seen as a dimension of equity of adaptation on several levels (Smit & Pilifosova, 2001; Eakin & Lemos, 2006). For instance, in the context of measures taken to protect areas from flooding, issues of equity and fairness that potentially affect the transfer of vulnerability between actors may come into play, particularly with regard to the selection of areas to be abandoned and where measures should be focused. It should also be noted that the different types of resources interact with each other; for instance, an increase in demand and funding for an area or sector could result in an increase in the political prioritisation of that area, and vice versa. Thus, political resources may be drawn upon to marshal an extension of economic resources in the face of identified risks, while the use of economic resources will largely depend on existing priorities and policy; an array of responses, from business-as-usual to more long-term and demanding strategic adaptation responses, are then possible.

Determinants of adaptive capacity thereby touch upon a broad scope of resources, several of which may only be identified contextually and will play out differently depending on case-specific parameters.

From a political science viewpoint, the concept of capacity (and specifically policy capacity) has sometimes been defined in a way that makes it possible to draw parallels to adaptive capacity in the context of policy development. Painter and Pierre (2005) note that capacity draws 'attention to the structural characteristics and resource stocks of a governing system' (p. 3), within which policy capacity has been defined as 'the ability to marshal the necessary resources to make intelligent collective choices about and set strategic directions for the allocation of scarce resources to public ends' (p. 2). Policy capacity is thus built on features such as 'funding, managerial skills, human resource development and professionalisation in government' (p. 10), a definition that may be related to the broader definition of the political factors that affect adaptive capacity.

The concept of policy capacity is further related to the effectiveness of political and administrative structures, where more generic institutional arrangements may influence the chances of policy success in a particular field. The concept of policy capacity is therefore related to administrative capacity – the capacity to effectively manage resources for delivering governmental output – as well as to state capacity, defined as the state's ability to mobilise resources for the achievement of public aims (Painter & Pierre, 2005). In addition, policy capacity is related not only to governmental or administrative functioning, but also to the nature of state-society exchange and the extent to which existing institutions are able to implement policy among differing constituencies and interest groups (Painter & Pierre, 2005). Jahyasuriya (2005) notes that capacity is largely the result of the 'development of the strategic capacity of agents' (p. 32), where new domains for governance need to be facilitated by the development of capable agents or agencies.

Capacity – both adaptive capacity and policy capacity – is thus largely related to the interaction of and prioritisation within different processes. It has been noted that policy capacity of the state could be strengthened by participation and other more inclusive forms of governance, including multiple levels of the state, to 'enhance the capacity of a system to mobilise resources and to leverage action' (Peters & Pierre, 2005, p. 49). However, the complexity and number of actors in multi-level governance also result in greater demands than those required of traditional hierarchical steering, which may decrease the overall capacity to govern across the diversity of levels.

1.2.3 Governance and Adaptive Capacity on Different Levels

1.2.3.1 The Role of the National Level

The mechanisms that influence governance and adaptive capacity at different levels are influenced by relevant system, level and actor characteristics. Adaptive capacity and the development of adaptation policy and measures in effect depend upon the abilities accorded to different levels within existing decision-making and market structures, which may also more broadly impact access to economic or other resources. The ability of different levels and actors to act is impacted by their positions relative to each other and by the mechanisms that govern these relations within the multi-level framework.

The role of the national level has been problematised in governance literature in particular. The state is often viewed as having become more 'hollowed-out' (Rhodes, 2000), often as a result of the increasingly transnational or globalised economy and the mobility of capital as well as the transfer of policy competence to other levels such as the EU. This indicates that the state must increasingly rely on steering specific actions together with other actors, for instance in partnerships with the private sector, as governments may no longer be able to ensure compliance with regulations or sufficient funding for specific measures on their own. While this may be seen as increasing participation and legitimacy by increasing the number of participating actors, thereby potentially contributing to 'good governance', the other side of the coin presents a decrease in accountability, where important political, economic and social decisions may be made by others than those elected in representative democracy.

Rhodes (2000) discusses several ways of defining governance that relate to aspects of this phenomenon. One is the linkage of governance with New Public Management (NPM), defined as the increasing importance of private sector management methods such as performance measures in conjunction with the increased marketisation or contracting out of services (Rhodes, 2000).[4] However, NPM doctrines are being implemented differently in different countries, resulting in a complex mix of their modification, supplement and strengthening according to context (Pollitt, 2006). Rhodes also defines governance through public and private networks as an emerging form of governance. This form of governance has emerged from the state's rendering of 'interorganisational linkages [into] a functional set of service delivery' (Rhodes, 2000, p. 60). Governance through networks highlights that 'government cannot impose its policy but must rather negotiate both policy and implementation with partners in public, private and voluntary sectors' (Stoker, 2000, p. 98).

Neither NPM nor governance through networks exist in their ideal forms in reality, but rather are parts of an increasingly complex context that influences the way governance is conducted across actors and levels. The formation of governance and the operationalisation of performance measurement and networks, or the increased marketisation and devolution of power from the national level depend on differences in the national context, including the existing organisational setting, culture, and power distribution, as well as on the ways the state manages processes of change. As such, state government is not supplanted, but rather modified by the addition and influence of other mechanisms: '[t]he purpose for comparative analysis appears, therefore, to investigate the capacity of the centre to govern, rather than to define it away' (Peters, 2000, p. 42). Peters (2000) further notes that variance in governance may be,

> by country, with the state in some countries (Singapore, Iraq, but also the United Kingdom) having a great deal of capacity to achieve compliance from society. The variance may also be by policy arena, with governments generally being better placed to achieve compliance in areas such as defence and immigration than in policy areas with stronger domestic interest organisations. (p. 42)

Societal traditions may also differ; for instance, Scandinavia has a strong tradition of corporatism, dense networks of interest groups established especially in the labour area, and a history of working towards consensus (Peters, 2000). Similarly,

---

[4]New public management is generally defined by the integration of a number of neo-liberal measures, including, according to Torres and Pina (2004): 'downsizing, privatisation, accountability for performance, replacement of input control by output control, accrual accounting, performance measurement, decentralisation, corporatisation, contracting-out, competition, management devolution' as well as empowerment of citizens and employees and the separation of politics and administration (p. 450).

the extent to which public management reform has been institutionalised differs: while the UK is often seen as a 'regulative state' where NPM has been strongly institutionalised, reform in Finland and Sweden has been driven by senior public servants and without the strong demands for reform that have been present in the UK (Pollitt & Bouckaert, 2000).

Such differences between countries have sometime been treated by dividing countries into rather broad 'state traditions' or 'families of states'. Such divisions distinguish between, for instance, Southern European systems, where informal networks are important, and the more administratively steered northern European countries (Loughlin, 2001; Newman & Thornley, 2002). These definitions are, however, rather sweeping and relatable only to the factors they define: using other parameters, distinctions between countries may result in other 'families'. Many of the attempts to classify states into groups include a focus on the state system (e.g., federal or unitary) and the level of decentralisation. One well-cited example is Ljiphart's (1999) division of states along a federal-unitary axis. One of his divisions also includes the level of decentralisation, along which countries can be grouped as federal and decentralised (e.g., Australia, Canada and Germany), semi-federal (e.g., Netherlands and Spain), unitary and decentralised (e.g., Norway, Sweden, Finland, Denmark) and unitary and centralised (e.g., the UK, Greece and Italy). This is to some extent in agreement with Andersen's (1999) description of policy styles as dependent on legal-constitutional systems (e.g., federalism) and administrative structures of competence between national, regional and local level. Newman and Thornley (1996) further emphasise the degree of decentralisation as 'an important distinguishing feature between different planning systems' (p. 27), defined as the delegation of powers, formal legal responsibilities and political and financial resources.

In line with such a distinction, Newman and Thornley (1996) divide the types of planning systems in Europe into different groups or planning families, based principally on legal (including constitutional) and administrative systems. They first distinguish between Northern and Southern Europe,

> ...with countries in the North more likely to conform closely to legal and formal arrangements. In the South there is a greater tradition of alternative informal mechanisms and greater flexibility in conforming to the law. Thus in certain parts of Europe, such as Italy or Greece, it is quite common to find a disparity between the formal laws and regulation and implementation (Newman & Thornley, 1996, p. 39).[5]

Beyond this basic distinction, they define five groups: the British, Napoleonic, Germanic, Scandinavian and broad East European families. The British system is seen as a centralistic system with limited independence of local government, while in the Napoleonic family, comprised of France, the Netherlands, Belgium,

[5]For instance, Bull and Rhodes (2007) describe Italy as a country that has retained a number of 'features common to the "southern type" of politics – described by Sapelli (1995, p. 18) as "collusion, a lack of sense of state and the ubiquity of clannish parties" with a weak embrace of the Weberian concept of "belief in law"' (quoted in Bull & Rhodes, 2000, p. 658).

Luxembourg, Italy, Portugal and Spain, and Greece,[6] local authorities often have strong (though potentially fragmented) representation alongside strong central government controls. The Germanic family builds upon the Napoleonic approach and includes Germany, Austria and Switzerland, all distinguished by a strong federal state approach adopted as a result of the State's inability to impose a unitary system, often leading to tension between national and regional responsibilities. The Scandinavian family encompasses Denmark, Sweden, Norway and Finland, decentralised unitary states where the regional level often represents an implementing body of central government while traditions exhibiting strong local autonomy are pronounced. Finally, the East European family is a diverse group that draws upon different historical legacies and central planning systems derived from a post-communist legacy (Newman & Thornley, 1996). However, later work by Altrock, Günther, Huning, and Peters (2006) describes differences in national context, administrative and fiscal decision-making structures, selected reform paths and planning situations. These have prompted the authors to claim that 'it makes little sense to speak of "one Eastern European (planning) family"' (p. 3).

A number of authors further note that the characteristics of political systems (unitary vs. federal states) and administrative systems (level of decentralisation) have different effects across scales. For instance, Glachant (2001) suggests that 'decentralised political systems are more likely to adjust easily when unexpected changes occur' (pp. 7–8). Peters and Pierre (2005) suggest that federal states experience fewer problems than unitary states in shifting from hierarchical to multi-level governance, such as in the context of the EU, as they are already familiar with multi-level arrangements. However, they also note that unitary states 'have tended to be more decentralised than one might expect from conventional wisdom' (Peters & Pierre, 2005, p. 48).

Other attempts to identify systematic differences between groups of states along other parameters have distinguished between laggards, followers and leaders in EU environmental policy (Börzel, 2002; cf. Liefferink & Andersen, 1997). While such delineations have generally been made in relation to the EU level and whether states are uploading their policies and are able to gain a leading role on that level, distinctions are also often used to describe regulatory structures more broadly. Here, Sweden, Finland, Denmark, Germany and the Netherlands (and sometimes Austria) have been seen as leaders or 'pace setters', while followers and laggards are often defined as 'industrial latecomers whose regulatory structures are less developed than those of the firstcomers' (Börzel, 2002, p. 203). These countries often have less-developed structures of ecological activism, resulting in higher requirements for new structures to deal with environmental issues than in countries that can draw upon developed administration and systems. The UK, Italy, Spain, Greece, and Portugal – to a large extent, southern European countries – have variously been described as

[6] However, Spain has retained specific local and regional features and lacks the legally unitary base of the other countries, while Italy exhibits strong regional differences. Greece draws upon Germanic legal traditions but has an administrative approach similar to the Napoleonic group (Newman & Thornley, 1996).

followers (a more intermediate position) or laggards (more reluctant member states) in this literature (Börzel, 2002, cf. Liefferink & Andersen, 1997; Koutalakis, 2004).

While these categorisations draw on relatively supra-level and generalised country descriptions, such policy traditions and their related established decision paths, bodies and resulting resources for environmental policy could presumably impact the way in which resources or a broader adaptive capacity are developed and distributed to specific issues related to the environment at the state level. Similar processes could also be expected to influence adaptive capacity at local and regional levels, where institutionalised environmental policy may support the development of adaptation in the cases where adaptation is seen as an environmental issue. However, categorisations of environmental policy in relation to the EU level describe neither the national processes behind this development nor any regional or local dynamics. Environmental policy orientation may also vary over time and with individual environmental policy issues more than is apparent in general categorisations. Characterizing environmental policy in relation to the EU can thus only serve as a very general basis on which countries may be contrasted with each other and for which a more process oriented assessment is necessary in order to understand the different underlying dynamics.

We can, however, distinguish a number of broad institutional features that may impact decision-making on adaptation and the extent to which a state is able to steer adaptation. These include the state political, administrative and planning systems, within which states may range from unitary to federal states with varying degrees of (de-)centralisation of power to national and sub-national levels (i.e. variation in the degree to which they bestow powers to sub-national levels or retain decision-making power at national level). In addition, decision-making on adaptation may be influenced by environmental policy traditions and the degree of environmental policy establishment, where one might assume that states with an established strong focus on environmental policy may act earlier on adaptation. That this does not so far necessarily seem to be the case seems to indicate, however, that mechanisms of agenda-setting and issues of perceived vulnerability may also influence the extent to which national and other levels act on adaptation. The manner and extent to which such broad systemic differences may influence the treatment and development of adaptation (policy and practice) in different states is an open question for this book.

1.2.3.2 Regional Level

The role and significance of the regional level varies across countries, ranging from being the level responsible for the implementation of national legislation and regulation to a scale at which relatively independent decisions are taken by democratically-elected bodies. The nature of a region also differs largely according to context. One distinction centres on the region as a territorially defined administrative division of a national state, defined in relation to the national state. This level may in some countries have fully-fledged regional governments, in which case they may be defined as political regions of the state. In this understanding based on the definition of the region within the state system, the region may range from being a state within a federal system to being for instance the regional administrative level in a unitary state (Keating & Loughlin, 1997). This definition of region is the one generally used in the case studies in this book, where administrative or political regions of the state have been selected as targets of study. However, this definition of a region does not necessarily overlap with the different types of functional definitions of, for instance, economic regions within which networks and cooperation may exist, or with historical or ethnic regions to which a number of actors may feel allegiance. A number of definitions focus on for instance the employment market region as a framework for market interaction and identity regions as foci for political mobilisation (Keating & Loughlin, 1997; Paasi, 2002). In cases where administrative and functional regions do not coincide, it is possible that sub-regions or information cooperation networks may over time develop to the point that they limit the power of the state to implement policy through its regional administrative divisions. Such developments may in extension even result in changes in regional delineations within the national context.[7]

Changes in governance (such as marketisation) that are prominent at the national level can also be seen at the regional level, where they have often been described in terms of a 'new regionalism'. These processes may impact existing administrative or political regions, but can also serve to support economic regions where these do not overlap with administrative or political delineations (Keating & Loughlin, 1997). New regionalism is generally described as a form of regionalism that is steered by three different scales – national governments, European level, and the globalised market – and can be seen as an arena for the implementation of neo-liberal ideas in a multi-level governance context (Veggeland, 2000). Privatisation and liberalisation may here simultaneously restrict administrative or state management at the regional level while increasing the region's growth potential (Veggeland, 2000), and may thereby have implications for the power of the region relative to the national level.

Where the increasing importance of the EU level has resulted in limited decision-making power (or a 'hollowing out') at the state level, new network relationships may be created between Brussels and the regions. This may hold true especially for larger, more competitive, and economically powerful regions that have access to lobbying capacity, European offices and representation in Brussels, and that may in themselves be important links to national policy. Regions in such a position may be well placed to influence the national level, and may also see advantages in implementing EU policy early – even prior to national enforcement – in order to increase their own competitiveness. As has previously been suggested for the national level in relation to the EU (Börzel, 2002), such regions may also act independently to 'upload' environmental policy ideas to the EU and in this case also national level. As a result, the 'hollowing-out' of the state level may increase decentralisation to the regional level, both formally or in the case of direct interaction between regions in a more centralised state and Brussels, functionally.

---

[7]Transnational regions may also exist, such as those often described in the term 'a Europe of Regions' (cf. Paasi, 2002); however, the transnational regional context will not be treated here.

Processes such as these may also create greater standardisation and institutional convergence through the integration of economic, welfare and environmental policy as regions increasingly orient towards the EU level, with both positive and negative results. While institutional convergence may support interaction through the removal of some of the 'interaction costs' for integration across widely differentiated systems, it could also result in a more limited organisational diversity of models for policy innovation. While stronger regions could integrate more directly and beyond the state, less powerful and peripheral regions may be increasingly left behind and effectively suffer a decrease in resources given increasing limitations in the 'hollowed-out' state's capacity for resource re-distribution.

With regard to decentralisation, the subnational regional level has been strongly impacted by the decentralisation processes that have been taking place over the last decades, seen by Jeffrey (2009) as '[o]ne of the most striking political trends of the last thirty-plus years' (p. 290). Such changes include reforms to the regional structure of Belgium, Spain and Italy, as well as challenges (although less actual reform) to federalism in Canada, Germany, Austria and Switzerland (Jeffrey, 2009). Reforms at the level of the region have perhaps been most pronounced in federal states, for example, in Canada, where some observers have noted a movement toward an increasingly co-operative arrangement between federal and state levels (Cameron & Simeon, 2002). In Germany, some authors have noted that the federal system is defined by inter-linkages across different actors that limit the ability to institutionalise change (Bandelow, 2007). In other countries, decentralisation has largely taken the form of devolution to the regions. This is for instance the case in Italy, where the relatively large decision-making rights at regional level have recently been discussed in terms of 'regional federalism' (Cotta & Verzichelli, 2007), but where financial devolution has not always followed the devolution of responsibilities.

The role of regional actors in relation to adaptation may thus depend largely upon the formal responsibilities and resources at the regional administrative level, which may in turn determine the possibilities for the development of regional approaches to adaptation. The role of regional actors in relation to adaptation will also depend on the economic and network capacities of the region (for instance, where, in accordance with new regionalism theories, highly-populated and economically strong regions may hold particular potential for development). Such developments will also have implications on the equity dimension of adaptive capacity, where economically stronger areas may be able to act earlier on adaptation even as other areas may exhibit greater sensitivity or exposure to climate change.

1.2.3.3 Local Level

In adaptation literature, the local level is often seen as the level at which effects will fundamentally manifest and which will ultimately need to respond to climate change (cf. Næss et al., 2005). This is particularly true as adaptation needs may differ between regions and localities, where a low-lying locality may have fundamentally different adaptation needs than a neighbouring mountainous municipality.

In a governance context, many actions at the local scale are steered from, decided or constrained by actions at higher levels. The local authority has a crucial role as the level at which national policy is fundamentally implemented (more directly in centralised states) or at which service decisions are made directly by elected local politicians (especially in countries with large local self-government). In countries such as the Netherlands, Italy and the Scandinavian countries, local governments have a power of general competence to undertake any actions perceived as being in the interest of their citizens, as long as these are permitted by law. In contrast, local government in the UK only has the right to fulfil its statutory aims; going beyond these is to operate *ultra vires*, or 'beyond the powers' and therefore beyond the law (Wilson & Game, 2006). Thus, the intersections between scales and the ways in which national and local government and governance are connected, as well as the possible combinations of top-down, bottom-up, mandatory or voluntary actions, are vital.

For an issue such as adaptation to climate change that has largely been regulated through planning systems, the level at which planning power is situated is paramount (Newman & Thornley, 2002). For instance, Sweden has what is often called a local 'planning monopoly', where the local level exercises decision-making power over local development and associated local planning and levies tax locally. As a result, the local level both manages an extensive range of activities and is given relatively high autonomy from the central state (Lidström, 2001), characteristics that typify both the Scandinavian planning system and the unitary decentralised state (cf. Ljiphart, 1999; Newman & Thornley, 1996). The UK, on the other hand, typifies the British planning system and the centralised unitary state, where the national government may cap local taxes. Some of the funding to the local level may also be distributed through mechanisms such as performance assessments by which the national level may ultimately set local targets and control implementation (Wilson & Game, 2006; Newman & Thornley, 1996).

As with regional actors, local actors may also be able to 'jump scale' (e.g., through direct connection with the EU level), potentially extending local adaptive capacity. This may take place through institutionalisation in a specific issue area, for instance, through the development of steering together with international non-governmental organisations (NGOs) or state actors with the aim to bypass the specific government (cf. Bulkeley, 2005; Gupta, 2008), or by tapping into EU funding mechanisms. Local actors may also take part in international or national networks between cities and often have dedicated local government interest organisations that support local government aims at both EU and national levels. Local authorities may impact national policy through coordinated action in local government networks.

Even national level policy, for instance on climate change, may thus be developed in ways akin to network governance, i.e. through complex networks of local, regional and national actors both within and outside the state. Factors that may prompt a state to develop leadership on an issue may also be developed to some extent from below, which may in turn facilitate the acceptance of binding state regulations and make them easier to implement at the local scale.

The extent to which national and local levels act on adaptation, either jointly or separately, will thus likely differ according to systemic characteristics of the political and administrative system. In states with greater local self-government, local actors may undertake certain actions or develop policy independently. This may in turn foster the development of national structures (and relevant funding) to motivate further local implementation. On the other hand, if actions within a decentralised system are taken only by local actors in communities that are both sensitive to climate change and relatively well funded, this may result in considerable inequity across areas and do little to raise municipalities to a lowest common denominator with regard to adaptation. In contrast, a centralised state committed to adaptation may be able to bring municipalities up to a lowest common denominator, limiting local self-determination but retaining potentially greater equity with regard to risk protection. However, were a centralised state to express low commitment to adaptation, it would likely limit the agency and ability of the local government level to respond to threats to a greater extent than in a decentralised system. Agency at the local level in general would also be impacted by processes similar to those described for governance in general (e.g., agenda-setting mechanisms and focusing events) that determine the attention and priority given to adaptation.

1.2.3.4 The Role of the EU

Finally, the EU level is particularly relevant to multi-level governance, both with regard to countries in the EU and Associated States (such as Norway). While the EU is not a government in the same sense as national governments in the nation-state system, it is 'a system of decision-making and governing, that is, a system of "governance" with distinct patterns of institutions, actors and processes' (Loughlin, 2001, p. 20) that influences the national, regional and local levels.

The ways in which the EU may influence these scales have been discussed in the preceding sections on national, regional and local governments. As some national decision-making is transferred to the EU, a larger emphasis is placed both on the implementation of EU directives and on the provision of structural funds that may support actors on different levels. Decisions made at the EU-level to move ahead with broad policy approaches have immediate impact and effect on Member and occasionally also Associated states. In addition, different sub-national governments may forge their own relationships with Brussels, creating diversified patterns of lobbying and differentiated access to funding and influence comparable to regional-national patterns within the nation state. With regard to its impact on state structure, the EU's structural funding mechanisms are of particular note. The EU has been referred to as a 'potent agent of rationalisation and reform' in Greece (Diamanduros, 1994, p. 43, quoted in Featherstone, 2005, p. 231) that to some extent influences the decentralisation of power from Athens. Similar examples have been noted for Italy (Quaglia & Radaelli, 2007), indicating that state structure and actors' use of European policy may be what Featherstone (2005) calls 'key intervening variables' in the extent to which different states Europeanise or incorporate features drawn from European-level policy-making.[8]

The EU also has a strong influence over legal and regulative requirements at the national level. With respect to environmental governance, these include the Water Framework Directive and the new Floods Directive, the Habitats and Birds Directives, and the development of the EU Natura 2000 network of protected sites. In communication with national and regional actors, the EU has also developed policy on adaptation through the Green Paper (2007) and White Paper (2009) on adaptation. However, with regard to adaptation, the EU has so far not been a driving actor and Europeanisation on adaptation has consequently been limited.[9]

Despite limited direct development of adaptation policy, however, both the provision of state input (at the formulation stage) and the implementation of EU directives may influence (and potentially delimit) national policy and adaptation. Directives set a context within which actions can be taken with regard to such issues as flooding and water use and may sometimes support action relevant to adaptation (in states where national policy and support may be limited) or hinder the development of measures (in fields where concerns relevant to adaptation have not been included in the Directives). Wilby, Orr, Hedger, Forrow, and Blackmore (2006) note that the ability to implement the Water Framework Directive will be impacted by climate change. While the directive does not explicitly mention risks posed by climate change, climate change may result in that water bodies may change their form and flow at the same time as temperature patterns and land management and use may also shift. Existing measures such as the Water Framework Directive are thus predicated on more of a steady state than is likely in the future. However, the EU structure to some extent also contributes to capacity-building through research support and research projects that may serve to highlight environmental changes in different countries, build the scientific basis on climate change and influence policy processes.

The EU organisational and bureaucratic context may also either limit or foster accessibility for different actors, depending on their organisational form, preferences and resources. Agenda-setting at the EU scale is more complex than at the domestic level given the large number of potential players (including private actors), the involvement of a number of national cultures and policy styles, and the potential involvement of several complex Directorates General (Liefferink & Andersen, 1997). In some literature, the degree to which different EU countries act

---

[8]Quaglia and Radaelli note that Europeanisation can be defined as 'processes of (a) construction, (b) diffusion and (c) institutionalisation of norms, beliefs, formal and informal rules, procedures, policy paradigms, styles, "ways of doing things" that are first defined and then consolidated in the EU policy processes and then incorporated in the logic of domestic (national and sub-national) discourse, political structures, and public policies' (Radaelli, 2003, p. 30, quoted in Quaglia & Radaelli, 2007, p. 925).

[9]For instance, Quaglia and Radaelli (2007) note that in order to produce effects in terms of Europeanisation, interactions at the EU level must 'become a reference point in domestic political action, either via socialisation or the production of resources and policies that modify the logic of political interaction at home' (p. 925).

as leaders, followers or laggards with regard to environmental performance and their compliance with EU environmental directives is assumed to depend on their perception of the costs and benefits of adjustment (Koutalakis, 2004). This distinction is founded on the idea that '[e]conomically advanced countries are more likely to act as pace-setters and policy-makers at the European level since they have strict regulations and a strong incentive as well as the necessary resources to upload them' (Börzel, 2002, p. 208; cf. Liefferink & Andersen, 2005). Börzel further argues that

> [t]he better the fit between European and domestic policies, the lower the implementation costs at the national level. Since Member States have distinct institutions, they compete at the European level for policies that conform to their own interest and approach. (p.194)

It could be assumed that some of these features of benefits or costs of uploading may also apply to adaptation in the future. On the other hand, given the specific nature of adaptation and the variation in terms of risks at different localities, member states may also support less binding, framework legislation on adaptation, in which case EU level impacts on adaptation may largely occur through the existing body of policies and directives and their potential modification in the future.

In general, the ability of the EU to deal with member state non-compliance is largely constrained by the limitations in monitoring mechanisms and depends to a large extent on the alerting mechanisms from investigations, complaints by stakeholders (e.g., industry, NGOs and citizens), and petitions from the European Parliament (Koutalakis, 2004). However, non-compliant states may be taken to the European Court of Justice on the basis of such complaints or petitions (Börzel, 2002; this has already occurred for instance in the UK as a result of a complaint by a domestic environmental NGO alleging non-compliance with the Habitats Directive).

1.2.3.5 The Role of Non-governmental Actors

Taking action on adaptation to climate change will require input from actors beyond government and administration alone, as is apparent from the above discussions on networks and the potential roles of the private and other sectors within a 'hollowed-out' state context. Relevant actors in adaptation include both industry and private actors, as well as NGOs. In the UK, a focus has been placed on the need to achieve local targets by local authorities in partnership with industry and other stakeholders (Wilson & Game, 2006). While this is to some extent a result of limited local authority resources and the high degree of privatisation in the UK system, it also indicates the explicit inclusion of a larger group in the implementation of targets (functionally, developing local governance). Thus, industry, national or EU fora for lobbying or stakeholder engagement, as well as local public-private partnerships may play a large role (cf. Mörth and Sahlin-Andersson, 2006). In traditionally corporatist states such as the Netherlands or the Nordic countries, a focus may instead be placed on state authorities in cooperation with actors traditionally included through a corporatist system (labour unions in particular). With regard to flooding issues, for instance, the owners of industrial water rights and water companies may play a large role, as defined within the regulative and legislative framework of the state. Given differences between institutional contexts, the way in which the inclusion of actors takes place and which actors are included are important questions.

Within a globalising context, multi-national corporations (MNC) and to a lesser extent local small- and medium-sized enterprises may also have a significant impact on the adaptive capacity of states (in the case of MNC) or localities. In an industrialised context, adaptive capacity at any level is largely related to the market in that the market forms the basis for the formal employment through which differential access to economic resources for groups and individuals is established. The hollowing out of a state is strongly linked to the power of transnational capital, which among other things limits the extent to which a state can control employment and thus tax income and its redistribution, as well as the provision of services. In areas that depend on single companies that are increasingly connected to a world market, the impact of the relocation or bankruptcy of such companies can be considerable (Keskitalo, 2008).

On all levels, the presence of NGOs may also play a significant role and may in turn affect the role and resource extraction practices of industry. Actions by NGOs have traditionally been a way of raising social and environmental concerns in the eyes of national policy-makers. Even despite limited resources or inclusion into national policy frameworks, NGOs have been able to 'name and shame' formal actors at national and international scales into addressing their demands (cf. Keskitalo, Sandström, Tysiachniouk, & Johansson, 2009). With regard to adaptation policy, NGOs may be able to highlight certain issues at the local and national scales, among other things through cooperation networks with other stakeholders. NGOs may also exert influence internationally through their acceptance into specific fora (for instance, within the United Nations framework).

1.2.4 Attempts to Bridge Sectors and Levels: Connecting Governance and Adaptive Capacity

In light of the above analysis, it becomes clear that an exploration of adaptation to climate change within a governance context necessarily entails defining a complex network of actors and options. Figure 1.1 below illustrates the ways in which the preceding framework describes potential influences on the development of adaptation policy and measures at different levels.

Smit et al. (2000) have attempted to summarise the requirements with regard to typifying adaptation in terms of identifying who is to adapt (i.e. which actors at which scales), what changes or impacts actors adapt to, and by what possible means adaptation is undertaken. Climate change adaptation can in these respects be related to the above described dimensions of the governance context that influence adaptive capacity at different levels. An important additional characteristic may be that of the transferability of adaptation options: to what extent can different areas and actors learn from each other – and to what extent may adaptations be too locality- or context-specific to transfer?

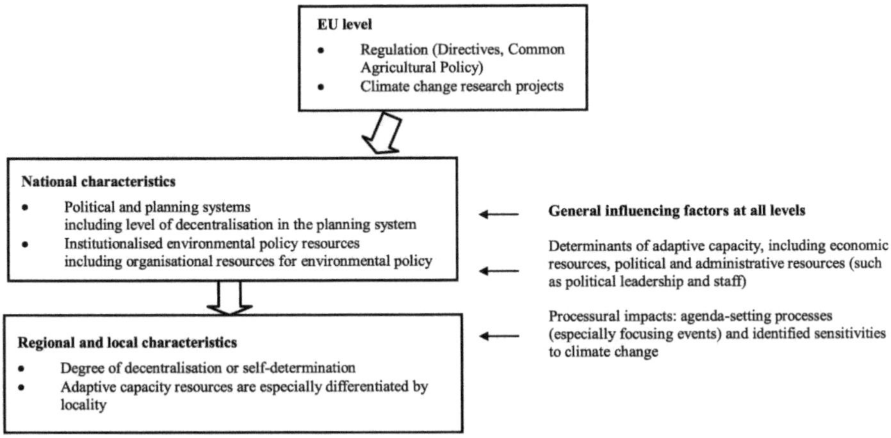

Fig. 1.1 A simplified figure of the impacts on development of adaptation policy and practice in a multi-level context

1.2.4.1 Who Is to Adapt?

The question of who is to adapt directly relates to the issue of agency at the different scales and actors described above, and the extent to which they interlink. As stated by Berkhout (2005), '[m]uch adaptation will draw on resources (including capital, knowledge, technology, and consent) that are not held by the adapting agents themselves' (p. 388). This indicates that actors cannot be seen as independent units that act in isolation, but must instead be seen as impacted by and exerting an impact on other actors. One actor's successful adaptation to change may result in poorer chances for adaptation for others, often defined as the transfer of vulnerability. Given the unequal and diversified capacities of actors – ranging from individual rural actors with limited monetary resources to states or transnational companies that command resources to steer policy or market actions – this often means that those most vulnerable and least able to affect their surroundings become even more vulnerable. This also renders adaptation a political and distributional equity issue within multi-level governance, where adaptation actions or changes in frameworks that steer or allocate resources may provoke conflict and requirements for trade-offs (as well as win-win situations, cf. Berkhout, 2005).

Despite the acknowledgement of adaptation as a multi-sectoral issue, the actors involved in adaptation may also be drawn from certain sectors more than others, framing adaptation as an issue only for certain actors. For instance, as shown in this volume, the Ministry of the Environment is often the lead actor for adaptation at the national scale, acting horizontally drawing upon financial resources available at the government level. Departments at regional and local levels with environmental risk management responsibilities may also take on adaptation issues within accessible frameworks at these levels, often related to planning. The assumption of responsibility by such authorities, however, may result in the positioning of adaptation as an 'environmental' issue which as a result may be addressed through leadership and political resources specifically associated with environmental issues without necessarily gaining broader representation or resources. Adaptation may also remain an issue for municipalities that have either specifically defined themselves as environmentally-oriented, or that have identified and are able to respond to particularly climate-related sensitivities, such as high susceptibility to flooding or storms.

1.2.4.2 What Are Actors Adapting to, and by What Means?

The question of what adaptation occurs in response to is similarly complex. One of the major outcomes of work on adaptation to climate change has been the recognition that adaptation to climate change cannot be seen in isolation. Actors seldom attribute their adaptive actions to discrete stimuli, but instead see them as responses to an entire array of issues (Hertin, Berkhout, Gann, & Barlow, 2003), including both climate change and processes of globalisation. While climate change may be one factor in response to which adaptation – for instance, flood risk management – is developed, adaptation decisions will necessarily be taken both on the basis of underlying adaptive capacity and with regard to existing priorities and processes such as economic change (cf. O'Brien & Leichenko, 2000). Consequently, adaptation by different actors 'will be taken up with the challenge of learning how much adaptation space is available to them and which adaptation strategy is most appropriate to their internal capabilities, corporate goals, and market and regulatory context' (Hertin et al., 2003, p. 289). Thus, adaptation may involve 'poorly-defined choices between complex sets of measures, often made up of chains of adjustment that may involve several actors' (Hertin et al., 2003, p. 289).

As a result of the complexity of actors and adaptation options, the definition of the possible means for adaptation is a moving target. In order to remain adaptive or able to attain resources for adapting to change, actors need to continuously adjust, to larger-scale processes as well as the competitive and organisational demands that these pose. A number of typologies of adaptation exist in the literature, some of which are detailed by Smit et al. (2000). Planned (strategic or active) adaptations are those most relevant to policy, and can be understood as those that have resulted from deliberate policy-making. Such adaptations can, among other things, reduce vulnerability to change by enhancing the capacity for autonomous adaptation, i.e. adaptation that may be undertaken spontaneously upon the occurrence of an unforeseen event (Smit et al., 2000; Nicholls & Klein, 2000). Adaptations can also be classified across a range of factors somewhat similar to determinants of adaptive capacity, including structural or infrastructural, legislative and regulatory, institutional and administrative or organisational, financial incentives or subsidies, market mechanisms or research-oriented measures (Smit et al., 2000). Potential measures may also range from an increase in existing measures such as risk response, to new responses to changing threats that move beyond existing practices.

The range of measures that may be undertaken in response to adaptation needs may thus be very wide. Gagnon-Lebrun and Agrawala (2008) note three stages of adaptation: the establishment of institutional mechanisms for steering and implementing adaptation, the formulation or modification of policies for adapting to climate change or taking adaptation into account, and the explicit integration of adaptation measures at project level. Adaptation measures may therefore be developed either through the development of issue-specific organisation or by 'mainstreaming' adaptation into ongoing policies or processes (see e.g., Smit & Wandel, 2006). In political science literature, a focus has been placed on the ways in which policy may be steered: by the state or other actors, or through different policy instruments using institutional, regulatory, economic or informational mechanisms (Appelstrand, 2007; Nilsson, 2005; Holmgren, Keskitalo, & Lidestav, 2010). Institutional instruments include those that regulate property or resource rights, while regulatory instruments may include legislation and enforcement procedures. Economic instruments include taxes and subsidies, while informational methods (which could be related to building adaptive capacity) encompass the broad range of communication and awareness-raising mechanisms. In relation to the development of adaptation, a state, regional or local government may utilise any of these at their disposal, given the distribution of authority within the national framework.

A commonality between these frameworks is the distinction between more and less formalised measures in terms of whether these enhance prioritisation through the formulation of policy, or go further to set mandatory requirements. Beyond such measures, actions that center on research-oriented measures or informational methods could be defined as part of adaptive capacity-building to understand and respond better to future events. Significant differences between policy systems could be expected. In an adaptation-committed polity where climate change impacts are made known within policy systems and dedicated adaptation needs are formulated, policy for planned adaptation measures will likely be developed in an explicit form. Policy may also over time be translated into binding legislation and regulation. However, polities where adaptation is less emphasised may also exist: for instance, where only non-binding priorities are formulated, or where actions are undertaken on an ad hoc basis within existing and not explicitly adaptation-related policy frameworks. In such cases, the focus with regard to risk may instead be placed on present risks or emergency management rather than on prevention and may be based solely on the existing risk response system. Adaptive capacity is in such cases delimited by the existing system, within which more autonomous adaptations or planned adaptations in sub-systems such as local authorities may emerge.

Broad distinctions may thus be drawn between cases where *policy defines priorities* and where it also sets *binding targets*. In addition may differing *adaptive capacity-building measures* to support the development or implementation of adaptations be delineated, some of which may be independently developed for instance within projects. As noted above, the organisation of adaptation may also differ between developing *issue-specific organisation* that institutionalises the issue, or utilising *mainstreaming* approaches where the implementation of adaptation is undertaken through and integrated within existing administrative structures.

Finally, these different policy developments may also result in different options for adaptation over shorter or longer terms. In the case of flooding, for instance, Nicholls and Klein (2000) note that future (and possibly present) adaptations will entail a number of choices. These range from protection (where engineering measures are taken to control environmental changes) to accommodation (where environmental changes are allowed to occur but human use of the coastal zone is minimised) and in the worst cases, planned retreat (where high socioeconomic costs are avoided by abandoning areas completely) (Nicholls & Klein, 2000; cf. Johnston et al., 2005). The types of adaptation policy that are developed may result in such options over the longer term, and may be integrated in policy making in the present to varying extents.

1.2.4.3 The Potential for Policy Transfer of Adaptation Options

Beyond the actors, processes and means directly involved in adaptation at any location or scale, 'lesson-drawing' may provide a basis for comparison and learning between contexts and a possibility for policy transfer, or importing examples from elsewhere. The terms 'lesson-drawing' or sharing 'best practice' denote similar processes, between or within national contexts, although the latter has been particularly promoted in NPM approaches (Cameron & Simeon, 2002; Holzinger & Knill, 2005). Elements of 'lesson- drawing' and 'best practice' are relatively general and can be found in for instance the EU level context and the UK. To some extent have possibilities for lesson-drawing also provided a basis for local participation in international climate networks such as ICLEI and its Cities for Climate Protection campaign (Bulkeley & Betsill, 2005). In the political science literature, such processes are often treated as policy transfer, defined as 'the process by which actors borrow policies developed in one setting to develop programmes and policies within another' (Dolowitz & Marsh, 1996, p. 357, quoted in Turner & Green, 2007, p. 5). Transfer of policy can be imposed or entirely voluntary, and in the latter case is often discussed as lesson-drawing. For instance, state level lesson-drawing is considered a 'voluntaristic process whereby government A learns from government B's solution to a common problem what to do ... or not to do' (Holzinger & Knill, 2005, p. 783).

The term of policy transfer (or diffusion) can be seen as falling under the broader category of policy convergence, defined as any increase in similarity between certain policy characteristics across selected jurisdictions during a specific time period, without regard to causality (Bandelow, 2007). However, Bandelow (2007) notes that as different causes may result in policy convergence without policy transfer, the term 'policy transfer' should be used only to refer to explicit references to foreign or external lessons. A reference to another country is thus not necessarily seen as a major cause for policy change, but rather as an indication of the prestige of the relevant policy within the constituency in question (Bandelow, 2007). Such policy emulation may be the result of wide-spread adoption that signals the importance of a policy; the possible increase in social legitimacy of an organisation through adopting a specific policy; an attempt to avoid being left behind as other countries adopt policy in an emerging arena (mimetic isomorphism); or, if coercive elements are present, the exertion of informal or formal pressures on dependent organisations (coercive isomorphism). The adoption of 'off the shelf solutions' will also allow for quicker implementation and a nation or body may thus borrow ready-made solutions for any of the above reasons (Cameron & Simeon, 2002; Holzinger & Knill, 2005).

However, given the difficulties in identifying general design variables and the importance of context in the efficiency or competitiveness of institutional design, the transferability of 'best practice' is certainly not straightforward. Bulkeley (2006) sees best practice development as based on the assumptions that knowledge or information is a crucial lacking component, that lessons are transferable between contexts and 'good practices will lead to the promotion of sustainable development either through direct transfer or through an instrumental form of lesson-drawing' (Bulkeley, 2006, p. 1035). However, the extent of convergence that might manifest between states as a result of policy transfer depends on the degree of similarity between the states in question, including cultural, institutional and economic structures (Cameron & Simeon, 2002). This coheres both with an understanding of systems in terms of different policy styles, and with an attempt to distinguish different system features that may impact the adoption of certain policy measures (such as national political and administrative systems). Lesson transfer may additionally require both considerable investments in terms of time and resources as well as changes in the way problems are framed by different institutional and political interests or rationalities (Bulkeley, 2006). With regard to adaptive capacity, this variability 'means that it is very difficult to argue for a "best" adaptation strategy for any given adapting agent' (Berkhout, 2005, p. 386). Elements of adaptation drawn upon from outside may thus potentially be practical tools or organisational forms (such as coordinating bodies for specific levels) which are more easily adjusted to other national contexts than more far-reaching strategies may be.

However, Gagnon-Lebrun and Agrawala (2008) note that even without necessarily being *directly* transferable, examples may constitute inspiration far beyond the context with which they are associated. The potential for policy transfer should therefore not be underestimated. For instance, fields such as regulation studies highlight the degree to which transnational and cross-national approaches may develop that transcend a particular context (cf. Djelic and Sahlin-Andersson, 2006). The 'best practice' concept may here be compared with processes of 'pushing by example' that actors may undertake in order to 'develop innovations in environmental policy which, if successful, are uploaded' (Börzel, 2002, p. 203). By extension, 'best practice' or 'lesson-drawing' could also result in an increase in isomorphism as a part of a more general trend of increasing similarity (e.g., Europeanisation). However, while such processes could limit the scope of institutional diversity over time, it may be unlikely that practices that conflict with national systems will be implemented (cf. Hollingsworth, 1997).

In full, the extent to which convergence or policy transfer may play a role is largely related both to the specific policies in question and to the linkages between different actors. To determine the extent of any convergence, as well as any development of adaptation, comparative studies may be seen as constituting

> an extraordinary laboratory for beginning to consider the policy architecture of the next generation of climate policy. Systematic study of actual experience in policy development and implementation might help move the debate from a feckless quest for the optimal toward a

more realistic exploration of what policy tools do – and do not – hold considerable promise (Rabe, 2007, p. 442–443).

1.3 Method

This volume aims to illustrate and analyse the broad range of factors that may have supported or limited the development of adaptation policy and practices in different multi-level cases, and that thereby contribute to or limit adaptive capacity. On this basis, national case studies have been selected for a comparative study using two types of criteria. Cases were first selected to represent a range of development of national adaptation policy and support structures for adaptation. Secondly, national cases were targeted to include an array of structural characteristics with respect to political, administrative and planning systems, and to represent a scope ranging from traditional leaders to followers (late-adopters) and laggards in EU environmental policy.

Within each case study country, a focus was placed on the interaction between levels, thus targeting nested cases with a focus on interaction between levels, rather than each level in isolation.

The selection of regional and local case studies within the different national systems represented in the study implies a bias towards the development of adaptation. Varying levels of development of national adaptation approaches are present, within which positive local and regional cases of apparent adaptation policy or measures were selected based on policy literature review. As a result the study may illucidate the potential for development of adaptation at local and regional levels in relation to the national level, including the extent to which a national framework is a prerequisite for local action (cf. Næss et al., 2005). The local cases may thus highlight aspects with regard to adaptation policy development in a multi-level system, as well as the potential for regions and localities in countries where a national adaptation framework is absent to 'jump scale' to attain support (cf. Bulkeley, 2005; Gupta, 2008).

Given the extremely broad character of adaptive capacity and adaptation as concepts, the study is necessarily explorative and indicative in nature. It centres on qualitative and narrative description of the development in the different cases by drawing upon a substantial body of interviews and a study of policy documents on adaptation at the focal levels. In order to elucidate the impact of the multi-level governance context of the EU, the primary focus in the volume is placed on countries inside the EU; some brief examples of non-EU countries (with a focus on federal multi-level systems) are included in a comparative chapter.

1.3.1 National Case Study Selection

A large section of the volume (four chapters) draws on case studies undertaken with common methodology at national to local levels over the same time period in four countries: the UK, Sweden, Finland and Italy. A centralised unitary state typifying the British planning system, the UK is one of the recognised world leaders in adaptation but has traditionally been a follower in European environmental policy development. Sweden and Finland, both unitary decentralised states within the Scandinavian planning system, have traditionally played leading roles in European environmental policy development, although Sweden has been comparatively slower in developing climate change adaptation than Finland. Finally, Italy exemplifies a southern planning system in which the region plays a large role despite being part of a unitary state. Italy has also been relatively slow with regard both to environmental policy development in an EU context and in national adaptation policy development, providing a contrast to the Nordic countries regarding the implementation of environmental policy, and a case where financial assistance granted by the EU Cohesion Fund and environmental programmes could play a particularly significant role.

A comparative chapter further presents the results of a policy document review and interviews with national coordinating bodies in adaptation for selected additional countries. This chapter broadens the comparison with a focus on political and planning systems and environmental policy traditions (see Table 1.1). Within this chapter, Norway is selected to provide comparison with the Swedish and Finnish cases and further differentiation within the Scandinavian planning family. The Italian example of a southern or Napoleonic planning family is further added to with examples from France, Spain and Greece, which have generally been laggards or followers in environmental policy with regard to the EU context (cf. Koutalakis, 2004; Börzel, 2002). Hungary provides an example of the differentiated planning context in Eastern Europe, while the Germanic planning family is included through a treatment of the federal states of Germany and Austria, as well as the Netherlands,

Table 1.1 National level case study selection and criteria (italicised text indicate a main case study)

Administrative and planning characteristics	National case studies
British planning family, centralised unitary state, follower in environmental policy	*The UK*
Scandinavian planning family, unitary decentralised states, often leaders in environmental policy	*Sweden* *Finland* Norway
Southern planning family (Napoleonic), unitary decentralised states, often laggards or followers in environmental policy	*Italy* France Spain Greece
Germanic planning family, federal decentralised states, often leaders in environmental policy	The Netherlands Germany Austria
East European planning legacies	Hungary
Non-European comparisons	Australia Canada

a group which has to different extents also acted early on in environmental policy in the EU. Examples are also taken from Australia and Canada to provide some comparison with development of adaptation outside Europe, and some indications for multi-level (federal) systems outside the EU context.

1.3.2 Nested Regional and Local Studies in the Four Main Cases

Within each of the four main case studies, nested multi-level (national to local) cases were studied in order to elucidate the mechanisms discussed above. In this respect, the book draws on policy literature and field work undertaken during the spring of 2009 using a common interview guide and interview design. In the UK, Sweden, Finland and Italy, the study targeted nested case studies on national, regional, and municipal levels in order to assess the ways in which climate change policy at one level may influence others. Case studies at regional and local levels have been chosen on the basis of their development of planned adaptation, including policies based on existing risk response and planning systems. Case studies were thus selected where actors demonstrated some interest in and development of adaptation to climate change, with a preference for cases where this was apparent on several levels through the development of plans, publications or policy documents, as well as through the participation in relevant international or national networks or partnerships. Especially in some of the local cases, the development of adaptation policy was not clearly forthcoming in literature review. In these instances, cases were selected for their demonstrated susceptibility to impacts that may increase with climate change and the development of policy and practice in relation to such risks (operationally, most developed cases regarded water-related hazards such as flooding and drought) or the development of mitigation policies in order to research the potential issue linkage of adaptation to mitigation as a part of common climate change policy. This latter criterion is based in earlier studies at the regional and local levels that demonstrate that a focus on adaptation can develop out of mitigation with the aim of developing coherent climate change policies (Riberio et al., 2009). In each chapter, cases are discussed in detail. Case studies were also selected to be comparable in terms of relative size and situation within the national context. Thus, regions were selected to have similar natural prerequisites (in particular, coastline) (Table 1.2). Where possible, a range of local authorities of different size were included, in order to highlight potential differences in sensitivity and adaptive capacity.[10]

As a result of the above case study selection principles, the adaptation issues that are treated in this volume focus on issues of water stress: flooding and sea level rise, as well as in Italy, drought. These were the issue areas where adaptation in

[10] As the selection of local cases was based on policy literature as well as limited in the numbers of local authorities that could be included in the study, this has in some cases (notably the UK) meant that not all potentially relevant local authorities could be included.

Table 1.2 Nested case study areas

Country	Region	Sub-region	Local level
United Kingdom	South East England	Hampshire County, Surrey County	Winchester, Portsmouth, Woking Borough
Finland	Uusimaa	Helsinki Metropolitan Area, KUUMA municipality cooperation	Espoo, Tuusula, Kevava, Mäntsälä, Pornainen
Sweden	Västra Götaland	n/a	Göteborg, Mölndal, Trollhättan, Munkedal
Italy	Emilia-Romagna	Province of Ferrara	Ferrara

these cases had been developed the furthest, potentially mirroring a general concern regarding adaptation to flooding as an immanent issue with high impact and risk perception (cf. Nicholls & Klein, 2000). In addition, the case study areas all represent relatively wealthy areas of their respective countries, which may reflect the traditional preference for living by the coast or large rivers, as well as the possibility that such regions have sufficient resources to respond to sensitivities once they have been identified.

1.3.3 Research Material

The material presented in this book is based on a policy literature study (for all chapters) and a total of 94 semi-structured interviews for the four main cases. In addition, eight interviews (as well as an additional interview in Hungary and a larger interview selection in Canada, with somewhat differing methodology) were undertaken for the supporting cases described in the comparative chapter. All interview references are listed in the end of each respective chapter.

The literature study was undertaken during 2008 and 2009 and targeted climate change adaptation networks and policy development in the selected countries and case studies. The survey included state policy developments (policy and legislation) and existing networks at local, regional and national levels in each country, as well as international networks of relevance (cf. Bulkeley & Betsill, 2005).

Interviewees in the main cases were selected to represent each scale (national, regional and local) and focused on policy and administrative actors and bodies with a role in climate change adaptation according to policy literature. Organisations targeted for interview were selected on the basis of their role in policy development with regard to adaptation to climate change. The choice of specific interviewees in cases where this could not be identified from the literature review was based on referrals received out of inquiries on climate change adaptation in the targeted organisations. These selection criteria resulted in a somewhat diverse selection between the different countries, reflecting national organisational diversity. In addition, snowball sampling was used to verify organisational and interviewee selection. While snowball sampling most often identified actors already identified through the policy literature review, on some occasions it resulted in the inclusion of prominent NGO or private sector actors (i.e. units outside administration or not described in policy literature). Interviews were generally 1–1.5 hours long, conducted on location and in the national language, recorded and transcribed in full. All interviews targeted the description of policy development, policy instruments and actions taken with regard to adaptation, cooperation with other levels, motivations for action and steering from other levels, perceived vulnerability and the role of extreme events, existence of scientific knowledge, political leadership, and funding and staff resources for adaptation, and the perceived role of internal or external examples of adaptation measures for lesson drawing.[11]

Interviewees for the studies in the comparative case study chapter (Chapter 7) were selected to represent coordinating bodies with regard to adaptation to climate change on the national level, and in some cases, across levels. These cases can thus mainly be seen as providing a broader context for the four main case studies, and illustrate some of the factors that may support or limit the development of adaptation policy and measures on different scales. Interviews lasted an average of 35 min and were carried out over the telephone and in English (with the exception of the Norwegian interview, conducted in Swedish and Norwegian). Interviews were recorded and transcribed in full. The interview guide used here was an abbreviated version of that used for the main case studies.

Coding of the interviews focused on existing adaptation policy and measures on national, regional, local and EU levels, the existence and impact of policy and networks on climate change adaptation (both at their own and other levels), the development and perceived supporting factors for development of adaptation (including available resources), and issues of policy transfer and lesson drawing. All quotes used that were not originally in English have been translated by the authors.

1.4 Outline of the Volume

Chapter 2 focuses on the EU context for adaptation. It illustrates, as is later underlined in the national case study chapters, the EU's considerable influence on adaptation despite its currently limited adaptation-specific policy. The chapter outlines the context of EU research projects, agricultural policy and regional

[11] In some cases, several people were interviewed together at the main interviewee's request; in these cases, multiple participants are still treated as a single interview.

development support, and directives that may influence adaptation, with a focus on water management and habitat distribution issues. The chapter illustrates that directives such as the Habitats Directive influence the designation of protected areas without sufficiently including potential future impacts from climate change that may change habitat distribution. The chapter as a whole suggests a closer inter-linkage of ecosystem- and water-related regulation and a more integrated treatment of adaptation at the EU level. The UK case study (Chapter 3) describes the relatively advanced multi-level development of adaptation in the UK, illustrating both impacts from the EU context as above and the strong development of a legislative and implementation framework for adaptation at the national level. Here, the centralised national state has been able to develop mechanisms to steer adaptation both at regional and local levels, but largely in partnership with and in relation to bottom-up approaches on adaptation.

In comparison, Finland (Chapter 4) illustrates a more nationally-focused policy development, where more limited resources for adaptation have been provided at local levels. However, largely unfunded local voluntary cooperation between municipalities illustrates the development of adaptation approaches even in areas where focusing risk events have been more limited. In Sweden (Chapter 5), national level development has been taking place mainly within a state commission and part of a national Bill; however, this development has at least partly been motivated by perceived sensitivity in flood risk influenced municipalities and counties in the heavily decentralised Swedish system, indicating that regional and local levels have impacted national policy development. In Italy (Chapter 6), the case study illustrates a largely 'absent state' on adaptation, creating a policy vacuum in which regional and local actors rely on existing policy; however, these have largely focused on present rather than future risks. In this absence of state policy, the EU provides a comparatively large impetus for environmental policy development and implementation through funding for projects and the Common Agricultural Policy.

The comparative chapter (Chapter 7) illustrates a multiplicity of approaches to adaptation, ranging from more to less centralised, including varying roles for the national level. In general, differences in accordance with a broad differentiation into planning and political systems including the level of decentralisation can be noted, though direct ties to environmental policy traditions are less evident. For instance, federal states relate more closely to a multi-level governance problematique that may allow for large differentiation among provinces. However, distinctions are not clear-cut, as significant differentiation also exists among unitary states, though these could exhibit a larger dependency on national level funding for decision-making on adaptation at multiple levels in the future.

Drawing upon the broad theoretical framework presented in this introduction, the conclusion (Chapter 8) summarises the development of adaptation policy and measures at the different levels and draws parallels with the theoretical models raised in this chapter. The volume concludes that despite the necessarily explorative nature of the study, the broad multi-level governance characteristics outlined for different levels constitute a relevant path for further development of a governance perspective that has the potential to better define and contextualise adaptive capacity.

References

Adger, W. N. (2000). Social and ecological resilience: Are they related? *Progress in Human Geography*, *24*(3), 347–364.

Adger, W. N., Brown, K., & Tompkins, E. L. (2004a). *Why do resource managers make links to stakeholders at other scales?* (Working Paper No. 65). Manchester: Tyndall Centre for Climate Change Research.

Adger, W. N., Brooks, N., Bentham, G., Agnew, M., & Eriksen, S. (2004b). *New indicators of vulnerability and adaptive capacity*. Technical Report 7. Manchester: Tyndall Centre for Climate Change Research.

Altrock, U., Günther, S., Huning, S., & Peters, D. (2006). Spatial planning and urban development in the New EU Member States – Between adjustment and reinvention. In U. Altrock, S. Günther, S. Huning & D. Peters (Eds.), *Spatial planning and urban development in the New EU Member States. From adjustment to reinvention* (pp. 1–18). Aldershot: Ashgate.

Andersen, M. S. (1999). Ecological modernisation capacity: Finding patterns in the mosaic of case studies. In M. Joas & A.-S. Hermanson (Eds.), *The Nordic environments. Comparing political, administrative, and policy aspects* (pp. 15–46). Aldershot: Ashgate.

Appelstrand, M. (2007). *Miljömålet i skogsbruket – styrning och frivillighet*. Lund Studies in Sociology of Law 26. Media-Tryck Sociologen, University of Lund.

Bandelow, N. (2007). Health policy: Obstacles to policy convergence in Britain and Germany. *German Politics, 16*(1), 150–163.

Baumgartner, F. R., & Jones, B. (1993). *Agendas and instability in American politics* Chicago: University of Chicago Press.

Berkhout, F. (2005). Rationales for adaptation in EU climate change policies. *Climate Policy*, *5*(3), 377–391.

Boland, P. (1999). Contested multi-level governance: Merseyside and the European Structural Funds. *European Planning Studies*, *7*(5), 647–664.

Börzel, T. A. (2002). Pace-setting, foot-dragging, and fence-sitting: Member State responses to Europeanisation. *JCMS, 40*(2), 193–214.

Brooks, N. (2003). *Vulnerability, risk and adaptation: A conceptual framework* (Working Paper No 28). Manchester: Tyndall Centre for Climate Change Research.

Bulkeley, H. (2005). Reconfiguring environmental governance: Towards a politics of scales and networks. *Political Geography, 24*, 875–902.

Bulkeley, H. (2006). Urban sustainability: Learning from best practice? *Environment and Planning A*, *38*, 1029–1044.

Bulkeley, H., & Betsill, M. (2005). Rethinking sustainable cities: Multilevel governance and the "urban" politics of climate change. *Environmental Politics, 14*(1): 42–63.

Bull, M., & Rhodes, M. (2007). Introduction – Italy: A contested polity. *West European Politics* 30(4), 657–669.

Cameron, D., & Simeon, R. (2002). Intergovernmental relations in Canada: The emergence of collaborative federalism. *Publius: The Journal of Federalism, 32*(2), 49–71.

Cotta, M., & Verzichelli, L. (2007). *Political institutions in Italy.* New York: Oxford University Press.

Djelic, M.-L., & Sahlin-Andersson, K. (2006). Introduction. A world of governance: The rise of transnational regulation. In M.-L. Djelic & K. Sahlin-Andersson (Eds.), *Transnational governance. Institutional dynamics of regulation* (pp. 1–28). Cambridge: Cambridge University Press.

Eakin, H., & Lemos, M. C. (2006). Adaptation and the state: Latin America and the challenge of capacity-building under globalisation. *Global Environmental Change*, *16*, 7–18.

EEA European Environment Agency. (2008). *Impacts of Europe's changing climate – 2008 indicator-based assessment. Joint report of European Environment Agency*, Joint Research Commission of European Commission and World Health Organisation. EEA Report No 4/2008. Copenhagen: European Environment Agency.

Featherstone, K. (2005). Introduction: "Modernisation" and the structural constraints of Greek politics. *West European Politics, 28*(2), 223–241.
Ford, J. D., & Smit, B. (2004). A framework for assessing the vulnerability of communities in the Canadian Arctic to risks associated with climate change. *Arctic, 57*(4), 389–400.
Gagnon-Lebrun, F., & Agrawala, S. (2008). Implementing adaptation in developed countries: An analysis of progress and trends. *Climate Policy*, *7*, 392–408.
Glachant, M. (2001). Introduction: A policy perspective on the implementation of the Community environmental legislation. In M. Glachant (Ed.), *Implementing European environmental policy. The impacts of directives in the Member States* (pp. 1–11). Cheltenham: Edward Elgar.
Gupta, J. (2008). Analysing scale and scaling in environmental governance. In O. R. Young, L. A. King & H. Schroeder (Eds.), *Institutions and environmental change: Principal findings, applications, and research frontiers*. MIT Press.
Haas, P. M. (1990). *Saving the Mediterranean: The politics of international environmental cooperation.* New York: Columbia University Press.
Hertin, J., Berkhout, F., Gann, D., & Barlow, J. (2003). Climate change and the UK house building sector: Perceptions, impacts and adaptive capacity. *Building Research & Information, 31*(3), 278–290.
Hollingsworth, J. R. (1997). Continuities and changes in social systems of production: The cases of Japan, Germany and the United States. In J. R. Hollingsworth & R. Boyer (Eds.), *Contemporary capitalism. The embeddedness of institutions* (pp. 265–310). Cambridge: Cambridge University Press.
Holmgren, E., Keskitalo, E. C. H., & Lidestav, G. (2010). Swedish forest commons – A matter of governance? *Forest Policy and Economics, 12*, 423–431.
Holzinger, K., & Knill, C. (2005). Causes and conditions of cross-national policy convergence. *Journal of European Public Policy, 12*(5), 775–796.
Hooghe, L., & Marks, G. (2003). Unravelling the central state, but how? Types of multilevel governance. *American Political Science Review, 97*(2), 233–243.
IPCC. (2001). *Climate change 2001: Impacts, adaptation, and vulnerability.* Contribution of Working Group II to the Third Assessment Report of the Intergovernmental Panel on Climate Change (IPCC). Cambridge: Cambridge University Press.
IPCC. (2007). *Climate change 2007: Impacts, adaptation and vulnerability.* Contribution of Working Group II to the Third Assessment Report of the Intergovernmental Panel on Climate Change (IPCC). Cambridge: Cambridge University Press.
Jahyasuriya, K. (2005). Capacity beyond the boundary: New regulatory state, fragmentation and relational capacity. In M. Painter & J. Pierre (Eds.), *Challenges to state policy capacity* (pp. 19–37). London: Palgrave.
Jeffrey, C. (2009). Devolution in the United Kingdom: Problems of a piecemeal approach to constitutional change. *Publius: The Journal of Federalism, 39*(2), 289–313.
Johnston, C. L., Tunstall, S. M., & Penning-Rowsell, E. C. (2005). Floods as catalysts for policy change: Historical lessons from England and Wales. *International Journal of Water Resources Development, 21*(4), 561–575.
Jones, K. T. (1998). Scale as epistemology. *Political Geography*, *17*(1), 25–28.
Kane, S., & Shogren, J. F. (2000). Linking adaptation and mitigation in climate change policy. *Climatic Change, 45*, 75–102.
Keating, M., & Loughlin, J. (1997). Introduction. In M. Keating & J. Loughlin (Eds.), *The political economy of regionalism* (pp. 1–15). London: Routledge.
Keskitalo, E. C. H. (2008). *Climate change and globalisation in the Arctic. An integrated approach to vulnerability assessment.* London: Earthscan.
Keskitalo, E. C. H., Juhola, S., & Westerhoff, L. (in prep). Climate change adaptation as governmentality: Technologies of government in the development of adaptation policy in four countries. 23p.
Keskitalo, E. C. H., Sandström, C., Tysiachniouk, M., & Johansson, J. (2009). Local consequences of applying international norms: Differences in the application of forest certification in northern Sweden, northern Finland, and northwest Russia. *Ecology and Society, 14*(2), online.

Keskitalo, E. C. H., Westerhoff, L., & Juhola, S. (in prep.). Agenda-setting on the environment: The development of climate change adaptation as an issue in European States. 25p.

Kingdon, J. W. (1995). *Agendas, alternatives & public policies* (2nd ed.). New York: HarperCollins.

Koutalakis, C. (2004). Environmental compliance in Italy and Greece: The role of non-state actors. *Environmental Politics, 13*(4), 754–774.

Leary, N., Adejuwon, J., Barros, V., Batimaa, P., Biagini, B., Burton, I., et al. (2007). A stitch in time: General lessons from specific cases. In N. Leary, J. Adejuwen, V. Barros, I. Burton, J. Kulkarni & R. Laseo (Eds.), *Climate change and adaptation* (pp. 1–26). London: Earthscan.

Lidström, A. (2001). Sweden: The end of the 'Swedish model'? In J. Loughlin (Ed.), *Subnational democracy in the European Union. Challenges and opportunities* (pp. 319–342). Oxford: Oxford University Press.

Liefferink, D., & Andersen, M. S. (1997). The innovation of EU environmental policy. In M. S. Andersen & D. Liefferink (Eds.), *The innovation of EU environmental policy*. (pp. 9–35). Oslo: Scandinavian University Press.

Liefferink, D., & Andersen, M. S. (2005). Strategies of the 'green' Member States in EU environmental policy making. In A. Jordan (Ed.), *Environmental policy in the European Union. Actors, institutions and processes* (2nd ed., pp.49–66). London: Earthscan.

Lijphart, A. (1999). *Patterns of Democracy*. New Haven and London: Yale University Press.

Litfin, K. T. (1994). *Ozone discourses: Science and politics in global environmental cooperation*. New York: Columbia University Press.

Lorenzoni, I., O'Riordan, T., & Pidgeon, N. (2008). Hot air and cold feet. The UK response to climate change. In H. Compston & I. Bailey (Eds.), *Turning down the heat. The politics of climate policy in affluent democracies* (pp. 104–124). Houndmills, Basinstoke: Palgrave Macmillan.

Loughlin, J. (2001). Introduction: The transformation of the democratic state in Western Europe. In J. Loughlin (Ed.), *Subnational democracy in the European Union. Challenges and opportunities* (pp. 1–36). Oxford: Oxford University Press.

McConnell, A. (2003). Overview: Crisis management, influences, responses and evaluation. *Parliamentary Affairs, 56*, 393–409.

Mörth, U., & Sahlin-Andersson, K. (Eds.) (2006). *Privatoffentliga partnerskap. Styrning utan hierarkier och tvång?* Stockholm: SNS Förlag.

Næss, L. O., Bang, G., Eriksen, S., & Vevatne, J. (2005). Institutional adaptation to climate change: Flood responses at the municipal level in Norway. *Global Environmental Change, 15*, 125–138.

Næss, L. O., Thorsen Norland, I., Lafferty, W. M., & Aall, C. (2006). Data and processes linking vulnerability assessment to adaptation decision-making on climate change in Norway. *Global Environmental Change, 16*, 221–233.

Newman, P., & Thornley, A. (1996). *Urban planning in Europe. International competition, national systems and planning projects*. London and New York: Routledge.

Newman, P., & Thornley, A. (2002). *Urban planning in Europe. International competition, national systems and planning projects*. London and New York: Routledge.

Nicholls, R. J., & Klein, R. J. T. (2000). *Adaptation frameworks for sea-level rise impacts*. Retrieved June 21, 2009, from http://www.survas.mdx.ac.uk/pdfs/romepap.pdf

Nilsson, S. (2005). Experiences of policy reforms of the forest sector in transition and other countries. *Forest Policy and Economics, 7*, 831–847.

O'Brien, K., Leichenko, R., Kelkar, U., Venema, H., Aandahl, G., Tompkins, H., Javed, A., Bhadwal, S., Barg, S., Nygaard, L. & West, J. (2004). Mapping Vulnerability to Multiple Stressors: Climate Change and Economic Globalization in India. *Global Environmental Change, 14*(4), 303–313.

Paasi, A. (2002). Place and region: Regional worlds and words. *Progress in Human Geography, 26*(6), 802–811.

Painter, M., & Pierre, J. (2005). Unpacking policy capacity: Issues and themes. In M. Painter, & J. Pierre (Eds.), *Challenges to state policy capacity. Global trends and comparative perspectives* (pp. 1–18). Houndmills Basinstoke: Palgrave Macmillan.

Peters, B. G. (2000). Governance and comparative politics. In J. Pierre (Ed.), *Debating governance. Authority, steering and democracy* (pp. 36–53). Oxford: Oxford University Press.

Peters, B. G., & Pierre, J. (2005). Swings and roundabouts? Multilevel governance as a source of and constraint on policy capacity. In M. Painter & J. Pierre (Eds.), *Challenges to state policy capacity. Global trends and comparative perspectives* (pp. 38–51). Houndmills Basinstoke: Palgrave Macmillan.

Pierson, P. (2002). Coping with permanent austerity: Welfare state restructuring in affluent democracies. *Revue Française de Sociologie, 43*(2), 369–406.

Pollitt, C., & Bouckaert, G. (2000). *Public management reform. A comparative analysis.* Oxford: Oxford University Press.

Pollitt, C. (2006). Discipline and punish – or trust? Contrasting bases for performance management in executive agencies. In T. Christensen & P. Laegreid (Eds.), *Autonomy and regulation. Coping with agencies in the modern state* (pp. 301–327). Cheltenham: Edward Elgar.

Quaglia, L., & Radaelli, C. M. (2007). Italian politics and the European Union: A tale of two research designs. *West European Politics, 30*(4), 924–943.

Rabe, B. G. (2007). Beyond Kyoto: Climate change policy in multilevel governance systems. *Governance: An International Journal of Policy, Administration, and Institutions, 20*(3), 423–444.

Rhodes, R. A. W. (2000). Governance and public administration. In J. Pierre (Ed.), *Debating governance. Authority, steering and democracy* (pp. 54–90). Oxford: Oxford University Press.

Ribeiro, M. , Losenno, C., Dworak, T., Massey, E., Swart, R., Benzie, M., Laaser, C. (2009). Design of guidelines for the elaboration of Regional Climate Change Adaptations Strategies. Study for European Commission - DG Environment - Tender DG ENV. G.1/ETU/2008/0093r. Ecologic Institute, Vienna.

Richardson, J. (1982). *Policy styles in Western Europe.* London: Allen & Unwin.

Rose, N. (1996). Governing 'advanced' liberal democracies. In A. Barry, T. Osbourne & N. Rose (Eds.), *Foucault and political reason*, (pp. 37–64). UCL Press, London.

Schattschneider, E. E. (1960). *The semisovereign people: A realist's view of democracy in America.* New York: Holt, Rinehart & Winston.

Schneider, A. L. (2009). Why do some boundary organisations result in new ideas and practices and others only meet resistance? Examples from juvenile justice. *American Review of Public Administration, 39*(1), 60–79.

Schuct, S. (2001). What can we learn from economics and political science analysis on the efficiency and effectiveness of policy implementation? In M.Glachant (Ed.), *Implementing European environmental policy. The impacts of directives in the Member* States (pp. 30–58). Cheltenham: Edward Elgar.

Smit, B., Burton, I., Klein, R., & Wandel, J. (2000). An anatomy of adaptation to climate change and variability. *Climatic Change, 45*, 223–251.

Smit, B., & Pilifosova, O. (2001). Adaptation to climate change in the context of sustainable development and equity. In J. J. McCarthy, O. F. Canziani, N. A. Leary, D. J. Dokken & K. S. White (Eds.), *Climate change 2001: Impacts, adaptation, and vulnerability.* Contribution of Working Group II to the Third Assessment Report of the Intergovernmental Panel on Climate Change (IPCC) (pp. 877–912). Cambridge: Cambridge University Press.

Smit, B., & Wandel, J. (2006). Adaptation, adaptive capacity and vulnerability. *Global Environmental Change, 16*, 282–292.

Stoker, G. (2000). Urban political science and the challenge of urban governance. In J. Pierre (Ed.), *Debating governance. Authority, steering and democracy* (pp. 91–109). Oxford: Oxford University Press.

Tol, R. S. J., & Yohe, G. W. (2007). The weakest link hypothesis for adaptive capacity: An empirical test. *Global Environmental Change, 17*, 218–227.

Torres, L., & Pina, V. (2004). Reshaping public administration: The Spanish experience compared to the UK. *Public Administration, 82*(2), 445–464.

Turner, E., & Green, S. (2007). Understanding policy convergence in Britain and Germany. *German Politics, 16*(1), 1–21.

Veggeland, N. (2000). *Den nye regionalismen. Flernivåstyring og europeisk integrasjon.* Bergen: Fagbokforlaget.

Walker, B., & Salt, D. (2006). *Resilience thinking.* Washington: Island Press.

Wilby, R. L., Orr, H. G., Hedger, M., Forrow, D., & Blackmore, M. (2006). Risks posed by climate change to the delivery of Water Framework Directive objectives in the UK. *Environment International, 32*, 1043–1055.

Wilson, D., & Game, C. (2006). *Local government in the United Kingdom* (4th ed.). Houndmills Basingstoke: Palgrave Macmillan.

Wurzel, R. K. W. (2002). *Environmental policy-making in Britain, Germany and the European Union. The Europeanisation of air and water pollution control.* Manchester: Manchester University Press.

Yohe, G., & Tol, R. S. J. (2001). Indicators for social and economic coping capacity – moving toward a working definition of adaptive capacity. *Global Environmental Change, 12*, 25–40.

Chapter 2
Addressing Adaptation in the EU Policy Framework

David Ellison

Abstract Though the EU's climate change mitigation strategy has taken precedence over adaptation, there are signs this is beginning to change. With the publication of both a Green (2007) and a White Paper (2009) on an EU Adaptation strategy, the European Commission has taken the important step of initiating broad discussion and encouraging the *mainstreaming* or integration of adaptation strategies into the existing EU and Member state policy framework. Still, without extensive revision – in particular in the direction of ecosystem preservation, improvement and creation – and the parallel introduction of a Climate Change Commission mandated to pursue mitigation *AND* adaptation strategies in the Community interest, policy outcomes are likely to remain fragmented and suboptimal. Institutional divisions at the EU and national levels reinforce sectorally-driven climate strategies that only partially address the goals of either mitigation or adaptation. Among other policy suggestions, this chapter makes two broad recommendations. First, the EU and the Member states should seriously re-evaluate the approach to such policies as the water framework directive, Natura 2000 sites and biodiversity, afforestation, ecosystem services and ecosystem preservation. Second, significant institutional reforms could heighten EU commitment to the climate change agenda, reinforce its already significant international bargaining authority and broaden the focus and impact of the EU's growing mitigation and adaptation agenda. Rapidly changing climate dynamics leave little room to dally.

Keywords Adaptation · Climate change · EU · Water framework directive · Natura 2000 · Afforestation

---

D. Ellison (✉)
Department of Social and Economic Geography, Umeå University, Umeå 90187, Sweden
e-mail: Ellisondl@gmail.com

E.C.H. Keskitalo (ed.), *Developing Adaptation Policy and Practice in Europe: Multi-level Governance of Climate Change*, DOI 10.1007/978-90-481-9325-7_2,

2.1 Introduction

Rising temperatures, changing precipitation, increased flooding, droughts and other extreme weather events, the rise of invasive species, all conspire to encourage and ultimately require renewed attention to the challenges of climate change. Yet European Union (EU) strategies to address climate change – apart from its rapidly growing mitigation agenda – remain in their infancy. Thus far, the EU has focused almost exclusively on the mitigation of greenhouse gas (GHG) emissions with the goal of limiting the increase of the average global temperature to +2°C. To do so, the EU has set a target for emissions reductions of 20% based on 1990 levels by the year 2020 (30% in the event of an international agreement), to be met through reductions in carbon intensity, reductions in energy use, increases in energy efficiency and a rise in the use of renewable energy resources. Though a considerable amount of research on climate impacts and adaptation needs is gradually becoming available – much of it EU-funded – the ambitiousness of EU mitigation and research efforts has not been equally matched. To date, the EU has neither defined nor set a clear strategy for adaptation to climate change. Whether these initial observations foretell the future weighting of EU climate strategies remains an open question. While mitigation may continue to receive the greater amount of attention, the current rapidity of climate change suggests this adaptation must be addressed with equal vigour.

Currently at the White Paper stage in the development of an EU Adaptation strategy, the European Commission – along with the help of stakeholders and other experts from EU Member States – has already gone through several rounds of consultation and discussion. Launched under the framework of the Second European Climate Change Programme (ECCP II) in 2005, the adaptation agenda has gradually gained momentum, leading first to the Commission's publication of a Green Paper *Adapting to Climate Change in Europe* in 2007 and then a White Paper *Adapting to Climate Change: towards a European Framework for Action* in April 2009. With the publication of the White Paper, the EU has formally announced its intention to develop and formulate a formal Adaptation Strategy over the period 2009–2012 and to implement this EU Adaptation Strategy by 2013.

This chapter explores how adaptation to climate change is being approached at the EU-level and possible implications for Member States. Generally speaking, EU Member States have been encouraged to develop adaptation strategies on their own but have no legal obligations to do so. However, the completion of (and the *failure* to complete) national adaptation strategies will likely influence the relative success of an EU-level effort. Due in part to tremendous variation in climate impacts expected across the European continent, many more EU Member States must first complete the task of developing independent strategies in order to contribute meaningfully to the development of an appropriate EU-level strategy. Thus far, only eight or nine of the 27 EU Member States have done so.[1]

---

[1]For up-to date information on which countries have completed Adaptation strategies and which have not, see the EEA's National Adaptation Strategy registry: http://www.eea.europa.eu/themes/

In addition to a number of smaller claims, this chapter has two principal recommendations. The first is that the EU should go much further in developing an adaptation strategy. Though the intention of mainstreaming adaptation strategies into the existing policy framework as rapidly as possible represents a courageous effort, in many ways the EU needs to think far more broadly and carefully about the interconnected relationships across different sectors and to develop more comprehensive and deliberate strategies for responding to the climate challenge. Most importantly in this context, the EU should seriously consider extending its focus on an ecosystem services approach to an ecosystems approach which, in addition to defining the value of its services, is also designed to protect, improve and create ecosystems. Though there is already an extensive literature on the importance of ecosystems for human survival, the implementation of relevant policy strategies currently falls short. Moreover, piecemeal sectoral efforts, such as raising water use efficiency, afforestation, or shifting to renewable energy use – important goals in their own right – should be considered in the general context of the far grander goal of protecting, maintaining and creating ecosystems.

The second recommendation is that the EU should move rapidly to create a Climate Change Commission. There are many reasons for this recommendation, not the least of which is to strengthen and heighten the symbolic impact of the EU's already significant international role in the promotion of the climate agenda. In addition to this, the EU needs to devise a strategy to overcome the current fragmentation of policy efforts. This is necessary in order for the EU (and ultimately other countries) to pursue successful mitigation and adaptation strategies. Though increased coordination and communication across the different institutions of the European Commission and with national, regional and local level interests in the Member States are helpful goals, without the centralisation of a mandate to address mitigation and adaptation in a single institution, relevant but competing interests will be neglected or ignored. The outcome is a less coordinated, comprehensive or targeted adaptation strategy.

This chapter is organised as follows: the first section addresses the current state of play regarding the EU adaptation strategy, discussing first the Green and the White Papers on the EU Adaptation strategy, the role played by the Member States, current efforts at building a knowledge-base on climate impacts in the EU, and initial EU sectoral level efforts. The second section addresses the interconnectedness of adaptation needs across a wide range of policy areas and addresses policy choices with regard to water and biodiversity management. The third section provides a more detailed discussion of the importance of ecosystem protection, improvement and creation. The fourth section elaborates the reasoning behind the recommendation for establishing a Climate Change Commission and provides detail on the consequences of not pursuing this strategy. The fifth section concludes.

---

climate/national-adaptation-strategies. See also Swart et al. (2009), Massey (2009) and Massey and Bergsma (2008).

2.2 From the ECCP to the Green and White Papers on Adaptation

With adaptation to climate change, the EU faces a moving target fraught with considerable uncertainty. While IPCC (Intergovernmental Panel on Climate Change) scenarios up through 2007 projected atmospheric concentration levels and potential temperature change through 2100, *near-term* global warming and climate change was frequently not considered as severe or threatening. This view is changing. In September 2009, UNEP (the United Nations Environment Programme) came out with the Climate Change 2009 Science Compendium (McMullen & Jabbour, 2009), a publication intended to raise the level of awareness of recent literature suggesting climate change is happening far more rapidly than originally predicted. The actual extent and magnitude of warming and its related impacts are potentially much greater than originally reported. Under a *business-as-usual* scenario, Sokolov et al. (2009) project the world could reach temperatures of +5.2°C by 2100. Though this recent estimate is twice as high as one from 2003, Hansen et al. (2008) suggests similar prospects and the Global Carbon Project recently suggested the world is currently on course for the 6°C mark by 2100.[2] In particular, both the failure to address climate feedback mechanisms in the IPCC's 2007 Synthesis Report and an assessment process based on broad consensus presumably constrained many of the baseline IPCC conclusions, rendering them more conservative than some of today's findings.[3]

Many of the assumptions upon which the European Commission's analysis is based may thus be open to criticism – in particular regarding the global GHG reduction target required in order to achieve the EU's proposed +2ºC ceiling on global warming. Though the Commission's Green and White Papers on adaptation are based on IPCC guidelines, these are now considered conservative and out-dated. Authors focus, in particular, on discussion of the proposed atmospheric concentration target. Hansen et al. (2008) and Hansen et al. (2009), for example, argue humanity should aim far lower than the IPCC's proposed maximum atmospheric concentration target of 450 ppm, suggesting instead that 350 ppm (2008) or 300–325 ppm (2009) is advisable. A much lower atmospheric concentration target also means countries will have to undertake far more significant emission reductions than originally proposed.

Despite the increasing urgency of climate change and global warming, to-date no single policy in the EU has been specifically designed to address adaptation to climate change. Though a relatively broad range of EU policies have potential relevance for adaptation or could be conceived as indirectly addressing adaptation, for the most part this has never been the direct intent of current EU legislation.

---

[2]See 'World on course for catastrophic 6° rise, reveal scientists' (*The Independent*, Nov. 18th 2009).

[3]On climate change feedbacks and potential tipping phenomena, see in particular Lenton et al. (2008).

Though the findings of the broad range of climate impact studies cannot be reproduced here, thinking from the Green to the White Paper on the potential climate impacts and thus the related EU policy responses has evolved dramatically. Based primarily on the findings of the PESETA study,[4] the climate impact assessments in the Green Paper are relatively rudimentary compared to the analysis that emerges in the White Paper and the flurry of impact studies discussed below.

It is expressly difficult to point to individual triggers of the push toward an EU-level policy on climate adaptation. The emphasis on an EU-level strategy wells up from various directions in the EU and international policy arena (see also Swart et al., 2009: Ch.3). For one, the research community has contributed significantly to a rapidly expanding understanding of climate impacts and the potential need for policy-related responses. For another, the United Nations Framework Convention on Climate Change (UNFCCC) has likewise provided strong impetus and motivation for both individual signatory states and also the European Union to make progress on the development of adaptation strategies through the requirement of National Communication reporting to the UNFCCC on adaptation efforts. Finally, both the rapidity of climate change and the increasingly frequent occurrence of potentially climate change-related events such as floods, droughts and other extreme weather events provide powerful motivation for the pursuit of adaptation strategies at the sectoral, regional and national levels, as well as the EU level. In this regard, sectoral level actors and stakeholders, related research communities and NGO's all provide potential pressure points for EU action on issues such as water management, flood control, the development of forestry policies and in other areas.

The current White Paper strategy is composed of four basic pillars that define a future course for the preparation and formulation of an EU Adaptation strategy. Compared to the original four pillars in the 2007 Green Paper, the White Paper reveals several shifts in emphasis (see Table 2.1).

The transition from the Green to the White Paper has resulted in an upgrading of the agenda to incorporate or 'mainstream' adaptation strategies into the framework of existing EU policies and thus a downgrading of the participatory agenda and the inclusion of stakeholders in the adaptation debate. Additional changes in emphasis between the Green and the White Papers involve the upgrading of the need for research and a solid information database on the impacts of climate change and the slight downgrading (at least in terms of the order of priorities) in the external dimension. In the White Paper, uncertainty is pushed into the background, perhaps paving the way for more resolute EU level action on adaptation research and the development of adaptation strategies. Thus, an emphasis on the development of a solid knowledge base has been pushed to the foreground of EU activities.

The shift away from the 'early action' feature in the first pillar of the Green Paper may be partly explained by the changing degree of urgency. It is awkward to classify policy efforts as 'early action' when the rate of climate change is rapidly turning

---

[4]The PESETA study was in fact incomplete at the time. The final conclusions of the PESETA study have only just recently been made available. See Ciscar (2009).

Table 2.1 Differences between the EU's Green and White Papers on adaptation

Green Paper 2007	White Paper 2009
(1) Early action to integrate adaptation into existing and new policy structures and funding programs and develop new policies	(1) Building a solid knowledge base on the impacts and consequences of climate change in the EU
(2) Integrate adaptation in *external* dimension	(2) Integrating adaptation into key EU policy areas
(3) Fill knowledge gaps: reduce uncertainty by improving the knowledge base and by integrating climate research	(3) Employing a combination of policy instruments to ensure effective delivery of adaptation
(4) Participatory, inclusive framework involving major stakeholders (European society, business, public sector) in preparation of comprehensive and coordinated strategies	(4) Increasing international cooperation on adaptation

these into 'reactionary' rather than 'anticipatory' measures. In a very real sense, the EU (along with the rest of the world) is significantly behind the game of climate mitigation and adaptation. Climate change and global warming are well upon us.

On the other hand, the Annex to the EU White Paper on Adaptation (European Commission, 2009b) outlines the next steps in the EU strategy and, in particular, details where and in what sectors 'early efforts' – perhaps we should now be referring to the EU's 'first' efforts – at adaptation should be made. The points raised in the Annex provide a strong foundation from which to begin the work of integrating ('mainstreaming') adaptation into the existing EU policy framework. Annex 2 develops, in particular, three general 'cross-cutting issues', water, land and biodiversity/ecosystems while Annex 3 goes on to discuss the importance of eight different sectors and the EU's external dimension (or foreign policy concerns); Agriculture, forests and forestry, fisheries and aquaculture, energy, infrastructures and buildings, industry and services, health, coastal areas and finally the external dimension.

In the framework of Pillar II efforts, Annex 5 of the White Paper outlines a series of potential action points where the Commission and the EU more generally could immediately dig in and begin to elaborate an adaptation strategy. Without providing an exhaustive list, the White Paper notes that strategies for addressing the effects of climate change should be mainstreamed into the Natura 2000 framework, the consideration of River Basin Management Plans (RBMP) and the Floods Directive, the EU Maritime Policy and the Marine Strategy Framework Directive, as well as the Strategic Energy Review and the Common Transport Policy (European Commission, 2009b, pp. 127–8). In this regard, the Commission has also committed to proposing guidelines to assist Member States in considering adaptation-related goals in their implementation of a number of EU policies. In particular, the Commission has committed to proposing guidelines for mainstreaming or integrating adaptation in the Water Framework Directive (WFD) (by 2009), on health impacts (by 2011), on RBMP (by 2009), Natura 2000 sites (by 2010), for coastal marine areas (no date specified), as well as other areas (European Commission, 2009a).

2.2.1 EU-Funded Research and Resources

As suggested by the rapidly expanding amount of research on climate impacts and adaptation needs, just beneath the surface quite a bit has been going on at the EU and other levels. Both the assessment of vulnerability to climate impacts in Europe and the development of strategies for adapting to climate change are currently under significant scrutiny. The research community has been fully engaged in the research and development of a solid knowledge base for an adaptation agenda, both through the European Commission, as it clearly draws upon the work of the research community, as well as through individual Member state research projects on climate impacts and adaptation.

In important ways the EU, like the Member States, has been spurred forward by the UNFCCC requirement of reporting on adaptation efforts. The first large-scale EU level study on climate impacts and adaptation needs was coordinated by Martin Parry (2000) and conducted through the Jackson Environment Institute of the University of East Anglia.[5] The ACACIA project ran from 1998 to 2000 and provided the foundation for the impacts and adaptation section of the EU's 2001 Third Communication to the UNFCCC. The ACACIA work was further extended and broadened by Kunzewicz, Parry et al. (2001) as part of the Adaptation in Europe contribution to the IPCC Third Assessment report. Though the Third Communication to the UNFCCC and the IPCC reports provide the first significant emphasis of EU reporting on impacts and adaptation, it precedes both ECCP WG II work on this topic as well as the Commission's later work on the Green and White Papers.

The study of climate impacts resulting from global warming and climate change has evolved significantly in recent years at the broad EU level, though in some cases research investments at the national level have been scaled down.[6] Though there is ultimately a relatively large gap between the ACACIA project (1998–2000) and later large-scale EU research projects, studies of regional and local level climate change impacts have been more numerous in recent years. This gap is acknowledged by the Commission in its Fourth Communication to the UNFCCC. Since the ACACIA project,

> there have been no large-scale studies on adaptation at the EU level, although the European Commission recognised the need to undertake further research and to develop adaptation strategies in their working paper 'Winning the battle against global climate change'. Most of the policy action currently in this area is undertaken by individual Member States and will be reported in their Fourth National Communications. (European Commission, 2006a, p. 106)

More recently, the EU has made significant progress with the preparation and publication of extensive background assessments. In particular, very substantial EU

[5]The only other somewhat larger scale project prior to the ACACIA project was the ESCAPE project completed in the early 1990s. See Rotmans, Hulme, and Downing (1994).

[6]One commentator argued that this has been the case in particular in Finland in recent years.

level reports on vulnerability and adaptation requirements have been prepared by the Commission's Joint Research Centre (JRC),[7] and the European Environment Agency (EEA, 2008). In addition, studies have been sponsored by the Commission for Agriculture and Rural Development on adaptation to climate change in the agricultural sector (AEA, 2007) and on impact and adaptation requirements in forestry (EFI-BOKU-INRA-IAFS, 2008). A second report on Adaptation issues in Europe was also completed by Alcamo et al. (2007) in the framework of the IPCC's Fourth Assessment. All of these studies have informed Commission work on the Green and White Papers. The earlier PESETA study by the JRC formed the principal foundation for the Green Paper and the EEA, JRC and DG Agriculture reports formed the basis of White Paper.

Apart from national level projects however, there were a number of smaller scale research projects not mentioned in the EU's national communication that nonetheless deserve note. The PRUDENCE project, for example, aimed at developing better temperature and precipitation prediction maps of Europe, at significantly higher resolution (from 300 km grids down to 50 km grids, see Christensen, Carter, Rummukainen, & Amanatidis, 2007).[8]

In the longer run, a significant amount of EU funded research is engaged in the analysis of climate change, its impacts and adaptation requirements. In preparation for the 15th UN Conference of the Parties meetings in December 2009 and an earlier conference in Geneva at the end of August, the European Research Commission compiled a document on EU-funded 6th and 7th Research Framework Programs on climate change. At 357 pages, the document lists a substantial number of EU projects. Though not all of these projects deal specifically with adaptation, the opening foreword explicitly points out that more knowledge is required in particular on, 'understanding of the climate system, on the evaluation of the impacts and on the identification and assessment of options for mitigation and adaptation' (DG Research, 2009, p. 1). In total, the report covers 134 research projects to which €543 million have been dedicated. Moreover, this tabulation is not exhaustive, since research in other areas such as energy and transport may have spill-over or overlapping implications. Approximately half of this report on research covers projects dealing with impacts (Chapter 5), natural hazards and extreme events (Chapter 6) and adaptation (Chapter 7). Though the balance of research investment may still favour climate change mitigation related studies, adaptation has become a significant focus of EU research.

---

[7]In particular, the JRC's then unfinished version of the PESETA project provides much of the foundation for the Green Paper and the Annex to the Green Paper. See for example the website of the PESETA project: http://peseta.jrc.ec.europa.eu/ and also the final report published in 2009 (Ciscar, 2009).

[8]The ENSEMBLES project, which followed up on the PRUDENCE project, has further refined the level of resolution to 25 km and some regional projects manage even higher resolution projection maps. See Van der Linden and Mitchell (2009).

2.2.2 The ECCP and the Development of National Adaptation Strategies

Rather than the EU, the Member states have taken on primary responsibility for the development of adaptation strategies. The EU however has played an important part in this process and the decision to move ahead first with national-level adaptation strategies (NAS) was at least in part the result of decisions made within the framework of discussions organized by the European Commission. In fact, one of the express goals of the ECCP II initiated in 2005 was to oversee the process of developing National Adaptation Strategies in the Member States. The ECCP II was intended to help 'define the role of the community in adaptation', 'encourage Member States to draft national adaptation strategies' and to help Member states identify optimal patterns of resource allocation and efficient resource use (ECCP WG II, 2007; European Commission, 2006a, pp. 107–8).

The second European Climate Change Programme (ECCP II) addressing climate change was launched in October 2005. Organized by the European Commission for the Environment, this discussion forum included six sub-groups and five stakeholder working groups. Though the first ECCP round from 2000 to 2003 did not address adaptation, the fifth sub-group of ECCP II was responsible for addressing adaptation and was named Working Group II on Impacts and Adaptation (ECCP WG II).[9] Building in part on EEA impact and adaptation research,[10] ECCP WG II proceeded to organise stakeholder consultations based on a broad sectoral breakdown. These consultations covered nine separate economic sectors (from water resource management, agricultural and forestry, biodiversity, to the role of the insurance industry). In addition, a tenth sub-group addressed national level adaptation strategies.[11] Results for these sectoral reports were published in March 2007.

Among the more interesting outcomes of these meetings was the general expression of a stakeholder interest in limited EU level action and a preference for national level action on adaptation. ECCP WG II's sectoral stakeholder report on Building National Adaptation Strategies points out, 'in keeping with the subsidiarity principle, the development of National Adaptation Strategies falls within the remit of Member States, not the EU' (ECCP WG II, 2007, p. 7). The report goes on to note that the, 'stakeholder meetings on climate change impacts and adaptation *recommended that the EU should not introduce any compulsory strategies on adaptation at this stage*'; it was instead suggested the EU should play a '*key supporting role* in providing a long-term view and in encouraging information sharing, and providing tools' (ECCP WG II, 2007, p. 7, author's emphasis). Though multiple options were

---

[9]See the Commission's webpage: http://ec.europa.eu/environment/climat/eccpii.htm, and http://ec.europa.eu/environment/climat/eccp_impacts.htm.

[10]The most important reports in this regard were EEA Report No. 2/2004 on *Impacts of Europe's Changing Climate: An Indicator Based Assessment*, and EEA Technical Report No. 7/2005 on *Vulnerability and Adaptation to Climate Change in Europe*.

[11]For more detail and access to the sectoral reports, see: http://ec.europa.eu/environment/climat/eccp_impacts.htm.

considered – including mandating that Member States develop National adaptation strategies – the final stakeholder recommendation was for the EU to 'lend support to Member States developing National Adaptation strategies'.

A second round of public and stakeholder consultations was undertaken after the June 2007 publication of the Commission's Green Paper. A series of three international workshops were organised in 2007, a web-based consultation ran for a period of six months from July to December 2007 and written submissions were encouraged as part of a public consultation.[12] The public and stakeholder consultation process identified that a fairly substantial share of respondents thought more attention should be paid to adaptation needs and more recognition should be granted to variation in impacts across sectors and geographic regions. Though many felt it was unlikely that one response could be crafted to fit the needs of all Member States and regions, there was support for the idea that the EU could provide a 'framework for action'. In one category, only 44% of respondents thought the pace of action was sufficient.

In what ways stakeholders will be involved in future discussions and the elaboration of an EU-specific adaptation strategy is unclear. The EU's past record of involving stakeholders has however been favourable. Moreover, the logic of encouraging Member States to first develop their own NAS in advance of the EU suggests, these strategies will then somehow play into the development of an EU strategy. Some of the work of reviewing and assessing national-level strategies has already begun (cf. Swart et al., 2009; Massey, 2009; Massey & Bergsma, 2008).[13] Annex Four to the Commission's White Paper on Adaptation likewise provides a preliminary and very brief assessment and overview of the various National Adaptation strategies introduced to date in the Member States (European Commission, 2009b).

Though the adoption of National Adaptation Strategies has not been mandated either by the European Commission or the European Council, this process has been encouraged and promoted by the Commission. Since signatory members to the UNFCCC are required – based on Article 4 – to develop national adaptation strategies, the UNFCCC has presumably also played a role – both with respect to individual Member states as well as with respect to the EU. In addition, the UNFCCC requires states to provide information on their adaptation efforts as part of their National Communications to the UNFCCC. To date, UNFCCC signatories – including both the EU and the Member states – have completed four National Communications, many of which have dedicated individual chapters to adaptation.

---

[12]These points outlined in Commission documentation of the consultation process at: http://ec.europa.eu/environment/climat/adaptation/stakeholder_consultation.htm.

[13]Other projects have also attempted to catalogue NAS strategies across countries. One of the more interesting in the context of this chapter is the attempt to catalogue the NAS approach to biodiversity strategies (see the MACIS study paper; University of Oxford 2008). The CIRCLE group, though still at a very preliminary stage, is also involved in assessing approaches to adaptation strategies: http://www.circle-era.net/. The ADAM project (http://www.adamproject.eu/) has likewise attempted to assess and report on adaptation efforts in individual countries.

Finally, some EU Member States – though a considerably smaller number – have also begun integrating adaptation strategies into their National Forest Plans (See e.g., Swart et al., 2009; Roberts, Parrotta, & Wreford, 2009).

The EU has thus embarked on a deliberate strategy of encouraging Member States to develop national-level adaptation strategies. The Annex to the 2009 White Paper likewise commits to pushing Member States – in particular those that have not done so – to develop adaptation strategies (European Commission, 2009b, p. 132).

2.2.3 Potential Consequences of the National Level Approach

On the one hand, the specifically local nature of adaptation requirements in individual EU Member States argues strongly in favour of a strategy based first and foremost on the subsidiarity principle and the perceived requirements of individual states. On the other hand, at least three distinct factors point to the advantage of organising intervention and shared burdens at higher levels of governance. First, at least two different logics argue strongly in favour of a burden-sharing arrangement across the Members states of the European Union and for wealth and resource transfers to states that bear a higher share of the adaptation burden. For one, the specifically transboundary nature of CO_2 and GHG pollution means that the countries responsible for creating emissions are not always those that must pay the highest adaptation costs – in particular because these are at least to some extent geographically determined. For another, the uneven nature of climate impacts and their occasional geographic unpredictability – in particular in the case of extreme weather events – further argues in favour of burden-sharing across states and reliance on higher levels of governance (i.e. EU and international). Second, increasing returns to information sharing and the centralisation of some features of adaptation management also have likely benefits.

As arguments in favour of the centralization of adaptation policy at the EU level suggest, an emphasis on the early development of national-level adaptation strategies has some potentially negative side-effects. First, countries with adequate resources will tend to do a better job of developing strategies and will thus find it easier to gain strong footing in the development of any future EU-level policy strategy. Moreover, EU-level strategies will most likely be strongly influenced and perhaps determined by the negotiating positions of individual Member States. Where the interests of a specific set of Member states are better developed and further along, these will likely carry the day over the potential policy interests and adaptation needs of other Member states. Second, though the current EU strategy is focused to some degree on 'effectiveness and efficiency', this approach may not have that effect. Some countries – in particular late-movers – that are not successful in introducing their agendas may have to revise, rewrite or even reverse existing strategies in view of the adoption of an EU-level strategy. Both of these points can likewise incur significant implementation costs.

Though many authors have identified the less advanced, southern EU Member States as more vulnerable to climate change (Massey & Bergsma, 2008; Alcamo et al., 2007; Parry, 2000), many of the more developed and northern Member States are much further along in the process of thinking about and developing adaptation strategies. This fact harbours at least two potential but very different threats. On the one hand, since many of the Southern Member states are not as far along in the development of adaptation strategies, EU-level policy outcomes could potentially ignore many or some of the issues that are of particular importance to the adaptation needs of the southern EU Member States. On the other hand, the view that northern states face fewer threats from climate change could unwittingly lull some Member States into inaction. Many in the northern EU Member states tend to downplay the threat of climate change and frequently think of the climate impacts as bringing potential benefits. Higher levels of precipitation in conjunction with warmer temperatures, shorter winters, longer summers and higher CO_2 levels are all seen as potential advantages. Thus, for example, growing seasons will likely be longer and both forestry and agriculture may benefit in the long run. This could, for example, lead northern Member states to be less enthusiastic about the need for Community action and community funded programs.

Other considerations are enough however to suggest however Northern states could be severely impacted as well and thus should have a strong interest in Community action. For one, greater precipitation in the north is also likely to mean a higher incidence of floods in northern regions and potentially also mudslides. For another, invasive or newly competitive species, for example, may represent serious threats in regions that undergo significant climate change but are highly dependent on a relatively small number of plant and tree species – such as the northern timber industry. The strongest example of this type of risk is represented by the mountain pine beetle in the Northwest American continent. Previously not competitively favoured, the mountain pine beetle now thrives on the slightly warmer temperatures and has destroyed many million hectares of timber across the US and Canada. Though no such widely destructive pest currently affects the forest industry in Europe, shifting temperatures and biomes mean the likelihood of such outbreaks will rise. Where these lead to widespread forest devastation, they have potential spill-over effects on the likelihood of forest fires and the reduced carbon sink potential of forests.

For a third, the likelihood of severe weather events does not appear to favour either northern or southern regions. Though events like the Gudrun storm of 2005 in Sweden (which damaged 85 million cubic meters of forest, almost one annual forest cutting) and similar but far less extensive events in Finland (two separate storms in 2001 damaged approximately 7.3 million cubic meters of forest) are relatively uncommon occurrences, the predictions are that their likelihood and frequency will increase (BFH-EFI, 2007, p. 33). To-date we know relatively little about what this might mean for the future of forest-based industries in these countries, for the carbon sequestration potential of Europe's forests more generally, or even for the increasing pressure being placed on the bioenergy potential of Europe's forests.

To some extent, individual Member States must accept responsibility for helping the EU develop an adequate EU-level adaptation strategy. The decision to first pursue the development of national-level adaptation strategies is a core feature of the current policy development process and was apparently endorsed by the wishes of individual Member states. In this sense, the failure of some Member states to carry through with the development of national-level adaptation strategies cannot easily be blamed upon the EU. On the other hand, the relatively slow development of adaptation strategies in some Member states is likely to weigh significantly on the breadth and quality of the final EU-level outcome.

On the other hand, as suggested above and as argued at much greater length below, the strong coordination of this process at the EU level is potentially beneficial to the development of a successful adaptation strategy. Moreover, though there was previously resistance to an EU-driven approach, it is presumably time for the EU to significantly upscale its adaptation efforts and to centralize the policy-making process and lend a more significant mandate to national and regional efforts. After several significant rounds of information gathering and stakeholder consultation, the EU appears ready to pursue more significant efforts at implementation.

2.2.4 Sectoral Policies and Early Warning Systems in the EU

The EU is most advanced when it comes to building a broad range of knowledge and awareness of the potential climate impacts and the development of early warning and emergency response systems/mechanisms for handling climate-related events (such as severe weather events, forest fires and other emergencies). The EU is least advanced, on the other hand, when it comes to the detailed integration of Adaptation-related policy options and strategies both at the sectoral and of course also at the broader EU level.

In addition to the Green and White Papers, much initial work is also currently being done within individual Commissions in the EU. As the Green Paper points out, some individual Commissions have already begun analysing how and when adaptation to climate change can be integrated into the EU policy framework. Mid-term reviews, for example, were conducted in various Commissions with the goal of addressing the potential integration of climate impacts into sectoral level strategies. Though the word 'adaptation' is never actually mentioned in many of these reviews – an exception is the EU Action Plan on Climate change and Development – there is considerable attention to climate change and its potential impacts.

Of the mid-term reviews noted in the Green Paper, apart from the Action Plan on Climate Change and Development, the Commission's 2007 'Mid Term Review of the European Environment and Health Action Plan 2004–2010' goes furthest in discussing and attempting to address the impacts of climate change. However, at this somewhat early stage, the mid-term review only notes that work is being done to investigate how adaptation strategies can be incorporated into EU health sector policies and points to the fact that several EU-funded research projects investigating the health impacts of climate change were underway and would be emphasised in future funding rounds. A quick glance at the Commission Health and Consumer Protection's webpage on the dissemination of health information and data yields information on a considerable number of climate and adaptation-related projects, as well as links to a number of disaster and potentially weather-related early warning systems (see also above). The announced 2007–2010 research focus on the human and animal health effects of climate change suggests this will be the next area where the Health Commission will focus future policy efforts.[14]

According to the Green Paper, the mid-term review of EU industrial policy was supposed to address 'how industrial policy can contribute to adaptation efforts' (European Commission, 2007a, p. 15). Yet not a single mention of adaptation or of addressing climate impacts in any other than the mitigation context appears in the final report. Though many possible climate impacts requiring potential adaptation efforts are likely in industry – related for example to water resource management, flood management or sensitivity to temperature changes – these receive no discussion or analysis. The closest the report comes to noting adaptation concerns is in its expression of support for the Global Monitoring for Environment and Security (GMES) project and its potential to provide 'monitoring and control of climate change impacts' (European Commission, 2007c, p. 13). However, as witnessed by a recent report from the World Business Council for Sustainable Development (WBCSD, 2009), the business community is clearly interested in and concerned about potential future climate impacts.

The EU Action Plan on Climate Change and Development, on the other hand, devotes considerably more time and attention to adaptation. The Action Plan's second pillar aims to provide support for adaptation in developing countries and the plan highlights specific areas for adaptation-related projects (in particular related to forests, agriculture, water resources and coastal areas).[15] According to the Green Paper, the EU will spend a total of €50 million between 2007 and 2010 to aid developing countries in promoting dialogue and developing mitigation and adaptation measures (European Commission, 2007a, p. 23). According to recent reports, the EU is considering dedicating considerably more financial resources to adaptation efforts in the developing world in the context of the Copenhagen international climate negotiations scheduled for December 2009. The EU's Copenhagen negotiation strategy does include measures intended to address adaptation in the developing world and the EU is currently offering to spend between €2 and 15 billion per year on developing countries for both mitigation and adaptation efforts. According to the UNFCCC Secretariat, the developing world faces potential adaptation costs of between €23 and 54 billion per year in 2030. Developing countries are not satisfied with this amount and are requesting far more.[16]

---

[14]The Health Commission's webpage clearly outlines its future emphasis on the health effects of climate change (see: http://ec.europa.eu/health/ph_threats/climate/climate_en.htm).

[15]In highlighting these areas for focused attention, the Commission's Action Plan leans on the UNEP/IVM *Handbook on Methods for Climate Change Impact Assessment and Adaptation Strategies* (1998).

[16]On the EU bargaining position, see (European Commission, 2009d). On bargaining developments at the negotiating table, see e.g.; 'Commission Unveils First Climate Aid Blueprint' (*Euractiv.com*, Sept. 11th 2009).

In general, the EU's major spending programs – the Common Agricultural Policy and Rural Development Funds, the Structural and Cohesion Funds and INTERREG (the cross-border cooperation fund) – all allow for spending on adaptation-related measures. What appears to matter more concerning whether or not individual Member States make use of these resources is the degree to which Member States and/or the EU have successfully highlighted and/or prioritised potential adaptation strategies at the national, local and in particular the sectoral level.

The timing of the EU's 2013 target date for the introduction of an official policy deserves some discussion. Though it is difficult to know how pre-meditated this is, the 2013 target date corresponds well both with the introduction of the EU's second major climate strategy (covering the years 2013–2020) and also more or less with the next EU Framework perspective period from 2014 to 2020. Thus in potentially interesting ways, the EU's timing both pairs efforts at mitigation with the introduction of an official adaptation agenda and simultaneously prepares the way for a potential shift of the EU's regional development and cohesion agenda over to one addressing both mitigation and adaptation. Debate over the future use of the EU's structural and cohesion funding is already firmly underway (see e.g., Begg, 2009).

The EU's 2020 Climate Package concluded in December 2008 and addressing EU climate policy for the period from 2013 to 2020 has likewise dedicated resources to funding for Adaptation and the development of adaptation strategies. Countries will be permitted to use 50% of revenues from the sale (auction) of unused carbon allowances for goals related to adaptation. In addition, current EU rural development policy also has some funding available for the development of national adaptation strategies. The EU's current 2007–2013 Framework Perspective allows rural development spending for a broad range of different adaptation-related measures in forestry. These include: improving human potential (in particular with respect to active forest management), developing physical potential (in particular with regard to improvements in forest stand management), harvesting (such as efforts to improve machinery or modify harvesting to improve resilience, or reduce damage), forest management planning (in particular concerning change in management planning related to adaptation) and the sustainable use of forestry land (such as establishing and sustaining forest ecosystems with diverse tree composition, age and structure) (see European Commission, 2009c, pp. 72–3).

As detailed in part by the EU's Fourth National Communication to the United Nations Framework Convention on Climate Change (European Commission, 2006a), the EU has developed a number of early warning systems intended to aid Member States in keeping abreast of and responding to natural, weather and of course potentially *climate-related* phenomena. Whether the strategy of developing early warning systems is only *good practice* or is specifically related to an increasing sense of urgency, attention to the development of early warning systems has spread almost like wildfire across the EU.

Thus, the EU's Fourth National Communication notes the development of the European Flood Alert System (EFAS) and the European Forest Fire Information

Systems (EFFIS). Since the Fourth National Communication, however, the EU has introduced the Network of European Meteorological Services (the EU METNET), which provides up-to-date information on weather alerts in all European countries; the European Environment Agency's Ozone Map, which provides up-to-date information on ozone pollution for all European countries; the Water Information System for Europe (WISE) database on water and water quality in Europe; the DAISIE and NOBANIS projects, which catalogue the extent of invasive species in Europe; and the project on the Future Development and Implementation of an EU-level Forest Monitoring Systems (FutMon), which, according to its website, collects both 'quantitative and qualitative forest data related to climate change, air pollution, biodiversity, and forest condition'.[17] With the so-called SEBI indicators program (Streamlining European 2010 Biodiversity Indicators) the European Environment Agency (EEA) is attempting to improve the cross-country comparability of biodiversity indicators for a broad range of European countries, including the current 27 EU Member States.[18]

Finally, some of these monitoring projects, as for example the WISE database and the FutMon project, have been specifically extended to include more *climate-related* information (WISE), or have evolved out of similar projects introduced with a different intent (FutMon). The FutMon project, for example, is essentially the outgrowth and extension of the ICP Forest Monitoring project, initially introduced in order to track the effects of pollution on the natural habitat.[19] Among other things, this project has closely tracked the incidence of defoliation, initially the result of acid rain and excess nitrogen use. Likewise of interest with regard to climate change and the incidence of droughts, the monitoring of forest condition (of which defoliation is one component) is being transferred to the FutMon project.

The EU's Fourth National Communication likewise lists a number of other EU projects as adaptation-related. In particular, the communication points to a Community scheme to develop preventative activities against fires (previously through the Forest Focus Regulation and now through Life+, DG Environment's environmental funding mechanism). Though strategies to improve forest fire protection have a relatively long history in the EU and clearly precede attempts to address adaptation, they have received renewed attention as a result of climate change and increasingly warmer temperatures, in particular in Southern Europe. The EU Common Agricultural and Rural Development strategy likewise now provides funding for the restoration of forest resources that have been damaged by natural disasters.

Increased attention has likewise been focused on civil protection in the EU. Council Decision 2002/792/EC (amended in 2007) ensures the coordination of assistance intervention in cases where this is necessary (including forest fires, floods). Moreover, the EU has provided assistance both in Europe and beyond.

---

[17] See the FutMon project website: http://www.futmon.org/Project.htm.

[18] See e.g.: 'Europe Must Grasp the True Value of Biodiversity' (EEA Highlight, Apr. 27th 2009).

[19] See the website of the Forest Monitoring Project: http://www.icp-forests.org/index.htm.

According to the Community Mechanism for Civil Protection's website, within Europe, assistance was provided to Portugal to help fight forest fires in 2003, 2004 and 2005, and help manage flooding in Romania and Bulgaria in 2005. Assistance has been granted to a significant number of external countries experiencing a variety of emergencies. The EU Solidarity Fund (Council Regulation (EC) No 2012/2002) likewise provides emergency funding and rapid Community response potential for situations classified as 'major disasters' (e.g., in 2003 Portugal received €48.5 million, while Spain received €1.3 million).

Attention to adaptation strategies however is typically lacking in the EU policy sphere. As noted above, there is no single piece of EU legislation that deals explicitly with adaptation. Though the EU has introduced a number of directives with direct or indirect relevance to adaptation, no current Directive can really be seen as an explicit and direct response to climate change adaptation. Thus for example Directives on flood management, on forest fire protection strategies, EU biodiversity commitments, the Natura 2000 program, the Water Framework Directive (WFD) and possibly the Nitrates Directive all have potential relevance for the EU's Adaptation agenda. Moreover, a number of upcoming Directives – such as attempts to introduce an EU Soil Directive and the Invasive Alien Species Act – are likewise of direct or indirect relevance to adaptation.

Though Directives such as the EU Floods Directive and occasionally the water framework Directive (WFD) are seen as potentially direct responses to adaptation, it is important to note the distinct difficulty in determining the impact on flooding that can be specifically attributed to anthropogenic climate change. As pointed out, for example, in the Flood Directive itself, floods are considered natural phenomena that are caused by multiple factors, only one of which is climate change (EU Directive 2007/60/EC). Though the WFD too has relevance for adaptation, the primary goal has been to ensure the quality of Europe's waters and waterways. A secondary goal, however, has been to introduce water payment schemes in order to ensure a more efficient use of available water resources. As an example of the *ecosystem services approach* currently advocated by organizations like the EEA – which attempts to place a market price on the use of ecosystems and the goods and services they provide[20] – this strategy dovetails neatly with attempts to manage the potentially climate-related problem of water scarcity, in particular in the southern European states where droughts are becoming ever more prevalent phenomena.

All of these Directives, however, provide excellent opportunities for the EU to begin integrating ('mainstreaming') adaptation strategies into the EU policy framework. At the same time the general EU adaptation strategy ultimately must address a broad range of linkages across a relatively wide range of different and potentially competing policy areas. This raises at least two basic dilemmas. First, as argued below, the existing EU policy framework may only inadequately consider

---

[20]In concert with the ecosystem services approach, the TEEB Project on *the Economics of Ecosystems and Biodiversity* attempts to provide initial estimates of these costs and their potential conversion into pricing systems. See the TEEB project website at: http://ec.europa.eu/environment/nature/biodiversity/economics/teeb_en.htm.

all the competing policy linkages. It may well be necessary in this case to introduce a range of additional policy strategies in order to effectively address adaptation goals. Second, as argued below, the current strategy could ultimately be taken much further by expanding many sectoral strategies into much broader ecosystem-based approaches.

Finally, in the long run, there is no clear commitment in the White Paper to one large over-arching EU adaptation strategy. This raises important questions about how the problem of adaptation will be handled further down the line. It remains unclear, for example, whether the EU will attempt to develop a more concerted over-arching strategy – as it has for mitigation – or whether the EU will remain satisfied with its sectoral level efforts.

2.3 From Policy Linkage to Ecosystem Preservation

The monumental complexity of adaptation – in particular in some policy areas – is only just beginning to become apparent. What follows takes a detailed look at two general areas of adaptation, the thematic areas – as outlined in the Annex to the White Paper – of water and biodiversity/ecosystems. Due to time and space constraints, the third thematic element – 'land' – is not treated in this analysis.

2.3.1 Water and Policy Linkage

A good example of the complexity of adaptation is provided by a detailed look at the first thematic area: water. The effective management of water resources is in fact a highly complex issue. Well beyond the traditional set of factors likely to impact rising water demand in coming years – population and economic growth, increased demand for agricultural products, etc. – a broad range of additional considerations suggest the future governance of water will weigh heavily upon the political institutions of Europe and the Member States. Without concerted planning, the 'water wars' of Spain could easily become the water wars of Europe.[21]

A recent study from the EEA (2009a) on water resources and adaptation needs in the Alpine region points to the great complexity of pressures from and on the different sources and users of water (see Fig. 2.1). In the context of climate impacts, reduced water availability and the increased potential for droughts – in particular in the water-constrained southern EU Member States – future competition over water resources is likely to be significant and potentially fierce. Almost all relevant sectors – from households to agriculture, industry, tourism, the energy sector and river navigation – are significant users of water resources and could experience significant

[21] See e.g., 'Drought Ignites Spain's "Water War"' (Guardian.co.uk, Apr. 6th 2008). Though this single citation might give the impression that droughts in Spain are unusual circumstances, a quick news search will reveal repeated versions of the same basic water war issue on an annual basis since at least 2005.

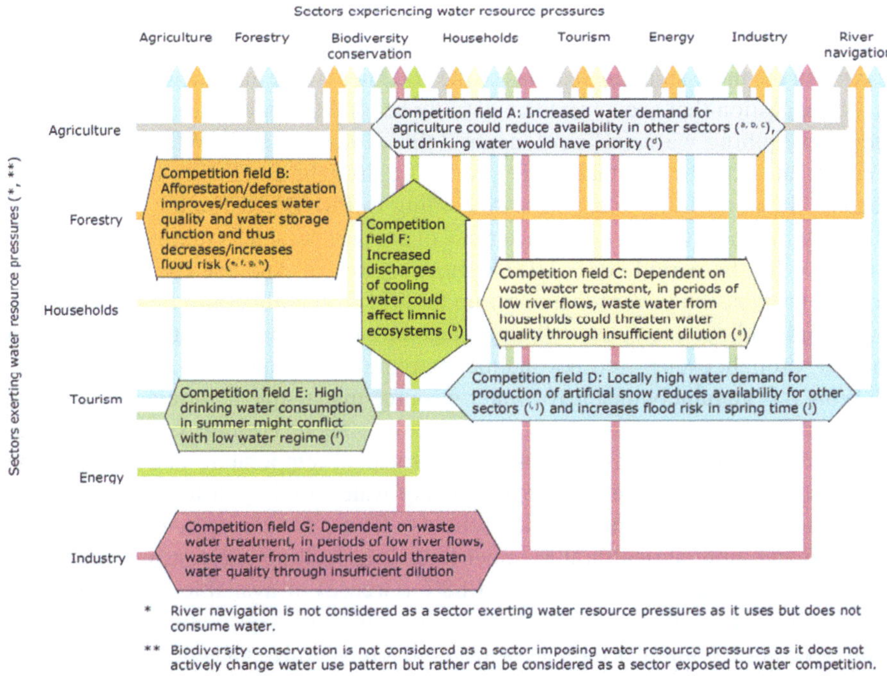

Fig. 2.1 Policy overlap in water resource management, reproduced from EEA (2009a)

constraints on water demand. As a result of climate change, water availability – in particular in summer – is likely to decrease substantially in some countries and geographical regions (see e.g. EEA, 2009b). If one couples these observations with predictions of the potential impact on agriculture and the energy sector – typically the two biggest water users – one begins to get a sense of the enormity of the problems facing water resource management.

In France, a government report on the drought conditions in the summer of 2003 suggests it was the hottest on record in the last 100 years (see Ministère de L'Économie des Finances et de L'Industrie, 2003). As the report notes, reductions in water supply in the rivers led to a 19% reduction of available hydroelectric power (10% of total power in France) and a 4% reduction in available nuclear power (84% of the total). At the same time, compared to the previous year, demand rose by 4.2%. These events led to a series of actions intended to help France keep pace with consumer demand, including appeals to consumer and industrial users to reduce energy use, a reduction of electricity exports and an increase in imports. In addition, several power plants took advantage of environmental derogations that allow warmed water releases above normally acceptable temperatures (four nuclear and two traditional power plants took advantage of these derogations). The situation was seen as critical and the state of demand in the wider European marketplace – also suffering from similar problems – was seen as quite tense. Some power outages occurred, but remained limited. In all, the 'exceptional measures' introduced to cope with these problems cost the French electricity company (EDF) approx. €300 million. The 2003 drought likewise impacted other countries in Europe (see below).

One of the more stunning features of the 2003 heat wave and the general response is the suggestion that such events lie entirely beyond the norm and are not likely to recur. Everything we now know suggests this is not the case. What we currently consider an extreme weather event producing drought conditions is predicted to become the norm in coming years. According to PRUDENCE study findings, the 2003 summer heat wave was characteristic of average seasonal (summer) conditions simulated for the period 2071–2100 in those regions.[22] And one can expect both a gradual progression toward that norm up to 2071, foretelling both more frequently warm and arid summers, as well as occasional and more frequent occurrences of 'outlier' drought events well beyond the 'extreme' 2003 conditions.

What this genuinely means for water resource management remains to be sorted out in individual countries and at the EU level. There is quite significant variation across countries both in terms of the projected temperature and precipitation changes, as well as in terms of the amounts of water used by different sources (i.e. agriculture, energy, industry and households). Thus, for example, in many of the southern and more agricultural Member States, agriculture can consume anywhere between 59 (Portugal), 72 (Spain) and 88% (Greece) of available water resources (see Italian–French Report 2006: Annex 1, 6). Moreover, agriculture consumes far more water in the summer months, when both temperatures and energy demand are also likely to be at their highest points and water availability at its lowest point. On the other hand, as further specified in the Italian-French report, several countries use a very large share of their available water resources for cooling in electricity generation: France (64%), Germany (64%) and the Netherlands (55%). While most of the water used for cooling is eventually returned to its original source at higher temperature (94% in the case of France), these countries are still dependent on water source availability (and thus can be affected by droughts and the related 'low water events'). Northern Member States, on the other hand, are typically not 'water-constrained' and tend to use significant amounts of water in industry – in particular in water-intensive cellulose and paper production.

Several competing tendencies arise from climate change and its related impacts, all of which are likely to have significant impacts on water demand and availability. For one, as part of the EU's 2020 Climate Change Package extending the Kyoto Protocol, EU Member States will be obliged to find ways to continue reducing emissions by 20% based on 1990 levels (30% in the case of an international post-Kyoto agreement), reduce energy use by 20% and raise the share of renewable energy to 20% (the share of renewable transport fuels by 10%) by the year 2020. Requirements for further emission reductions can be expected up to 2050 and perhaps beyond.

In the attempt to keep up with rising energy demand under low carbon constraints, a significant number of countries are planning to introduce more nuclear

[22]See the findings from the PRUDENCE project: http://prudence.dmi.dk/public/beniston.html.

power plants. Some (though not all) of these are southern countries, highly exposed to potential droughts. Though 'planned' projects can always be cancelled (they depend on shifting coalitions and political power arrangements), in February 2009 Italy announced a joint venture between the French EDF and the Italian ENEL to build four new nuclear reactors in Italy, as well as plans to build five nuclear reactors in France.[23] And many other EU Member States have been considering nuclear power as an option (including Germany, Poland, Hungary, Finland, Sweden and several others).

Renewable energy sources typically perform dramatically better with respect to water demand. Wind power and household solar PV systems, for example, do not consume or use water. Thus the current emphasis on renewable energy sources is a plus for most water-constrained countries. However, the advantages of renewable energy sources are not consistent across all renewable energy types. Concentrated solar power plants that use 'wet-cooling' instead of 'dry-cooling' still use massive amounts of water (the water quantity/kWh is comparable to thermoelectric systems).[24] Moreover, concentrated solar power plants are most productive where sunlight is plentiful but water scarce and wet-cooling systems tend to be cheaper (often phrased 'more efficient') than dry-cooling systems.

Carrillo and Frei (2009) find that two further renewable energy sources are likewise linked to problems of excessive water demand. For one, the shift to biomass power generation – strongly promoted by the EU and some individual countries – does not by itself solve the problem of water use, since the cooling requirements that exist with fossil fuel-based electricity generation remain. However, where dry-cooling systems are introduced, such problems can be greatly alleviated. For another, the shift to biofuels also represents a very significant increase in water demand leading Carrillo and Frei to suggest that water-constrained countries instead import biofuels from more water-rich countries.

Like France above and Switzerland (see below), Germany also experienced drought-related reductions in electricity generation in the summer of 2003. Many traditional fossil fuel-based power plants were required to either dramatically reduce electricity generation or shut down completely. However, researchers argue that the recent retrofitting of several traditional fossil fuel-based power plants in Germany with dry-cooling systems (not dependent on water) resulted in fewer power plants being cut back or shut down as a result of diminishing water availability during the 2003 heat wave (BfG, 2006, p. 185).

---

[23]See 'Italy-France Deal Sparks Nuclear Revival' (*Euractiv.com*, Feb. 25th 2009). As an indication of the likelihood this goal will be fulfilled, one might look at the progress of a similar EDF plant being built in Finland. The project has experienced significant delays and is reportedly already three and a half years behind schedule. Currently it is schedule for completion in 2012.

[24]Dry-cooling systems can also be used in traditional fossil fuel-based power plants. Where they are used, dry-cooling systems tend to reduce water use by some 80–90% or more. However, dry-cooling systems require more energy to power fan systems (cf. Sovacool & Sovacool, 2009; Feeley et al., 2008; US DOE, 2006).

The extent to which energy demand constrains water resource availability depends to a great extent on the efficient use of available resources. Agricultural interests will also place significantly increased demand on water resources in the coming years. Rising temperatures pose very immediate problems for agriculture and forestry where these lead to reduced precipitation and water availability. Many of the southern EU Member States have thus far suggested they will adapt to water constraints by building further irrigation channels and water storage facilities (see AEA, 2007). Hungary, for example, in its 3rd UNFCCC National Communication predicted it would be able to handle the climatic changes and that agriculture would not be substantially affected (Hungary, 2002). Based on global climate models, early PESETA study findings on the climate impact on agriculture seemed to concur with this finding.[25] The map-based projections suggest that by 2080, Hungary could potentially experience increased agricultural potential between 10 and 30%, while Spain and Portugal and large parts of France and Italy should witness dramatic agricultural decline (–10 to –30% in most regions).

In addition to the global scale of the PESETA study, two additional problems weaken the findings on Hungarian agriculture. For one, in the data projections discussed above and reported on the PESETA website, there is no distinction between summer and winter temperatures. For another, as noted by the PESETA researchers, the findings ignore water constraints and assume that farmers will be able to 'use as much additional irrigation water and fertilizer as desired' (see PESETA website). Later studies of Hungary, suggest that future constraints on water availability could be more severe. The 2005 Fourth National Communication to the UNFCCC draws attention to the fact that even relatively small temperature increases of 0.7°C can lead to a 60% decline in surface water availability in the Tisza catchment region, an 80% decline in subsurface water supply and a 74% decline in water available for irrigation (Hungary, 2005, pp. 93–95).

Considerable uncertainties are attached to any of the projected findings on temperature and particularly on precipitation change. And, as suggested by other data from the PRUDENCE project (cf. Christensen & Christensen, 2002), some degree of adaptation may be possible. For example, even with drought-like conditions in summer, heavier than average precipitation from more extreme weather events could be corralled in additional water storage systems for later use. Moreover, most of the model projections from the various studies cited above concur in that annual precipitation is not projected to change significantly.

Later studies however suggest that future constraints on water availability in Hungary could be severe. These projections have tended to emphasise warming temperatures, declining precipitation and the increased frequency of droughts and other extreme weather events. The most recent predictions stem from two parallel studies focused on climate impacts in Central and Eastern Europe; the CECILIA and

[25] These results are published on the PESETA website (http://peseta.jrc.ec.europa.eu/docs/Agriculture.html).

the CLAVIER projects.[26] Temperature and precipitation changes in the Tisza basin region of Hungary – the principal agricultural region – are likely to be the most dramatic. Keeping in mind the 2005 Fourth National Communication findings on the strong relationship between temperature changes and declining water availability in the Tisza region, more recent nationally and regionally generated temperature predictions from the CECILIA study suggest change in summer temperatures in this region could reach +3°C between 2021 and 2050 and as much as +5°C between 2071 and 2100. Potential changes in summer precipitation are comparably stark.[27]

Hydrological estimates of the impact on water flow in the Tisza River (see Fig. 2.2) do not currently suggest there will be dramatic changes. Annual flows for the period 2021–2050 are projected to fall by only about 2% in the lower Tisza region (and are projected to rise slightly in the upper Tisza region). Moreover, the seasonal impact is likewise projected to be relatively small – approximately 10% in the spring and 3.5% in the summer months (Pfeifer et al., 2009; Matreata et al., 2009). However, a number of differences between the CECILIA and CLAVIER studies are worth noting. For one, the ALADIN model used in the CECILIA study predicts higher temperature and precipitation changes in the Tisza region than the Clavier study REMO model. For another, the hydrological study does not provide estimates beyond 2050, though of course the CECILIA study does project temperature and precipitation changes for the period 2071–2100. Though such differences in modelling outcomes beg the question of the potential basis for comparison between studies of this type, a strong foundation for comparison is currently lacking.

Further, very little is currently known about potential future sources of change in water demand such as agriculture or the energy sector to the East of Hungary. The Tisza originates in Ukraine and meanders along the border region of Ukraine, Romania, Hungary and Slovakia before passing through the heart of the Hungarian agricultural region on its way to merge with the Danube in Serbia. Thus changing water use in these upstream countries could also influence future downstream water availability.

What this means for agricultural production in Hungary depends significantly on the degree of preparation for more extreme drought events and decreased water availability more generally. To date, though Hungary has a plan in place to build several water storage and flood management reservoirs in the Tisza region, only one of them has so far been built. The basic point is that agriculture in the southern states of Europe may ultimately be somewhat more endangered than currently predicted. As is, studies already predict that agricultural production will generally have to shift to the North and East of Europe as well as across traditional seasonal scheduling frameworks. Yet remarkably little discussion occurs about how this might be achieved – in particular with regard to the geographic shift in agricultural

[26]For information on the CECILIA project, see the project website at: http://www.cecilia-eu.org/. Information on the CLAVIER study is available at: http://www.clavier-eu.org.

[27]These findings are based on personal communications with the Hungarian Meteorological Association (OMSZ) and the graphs included below. For more general findings from the project, see Csima and Horanyi (2008).

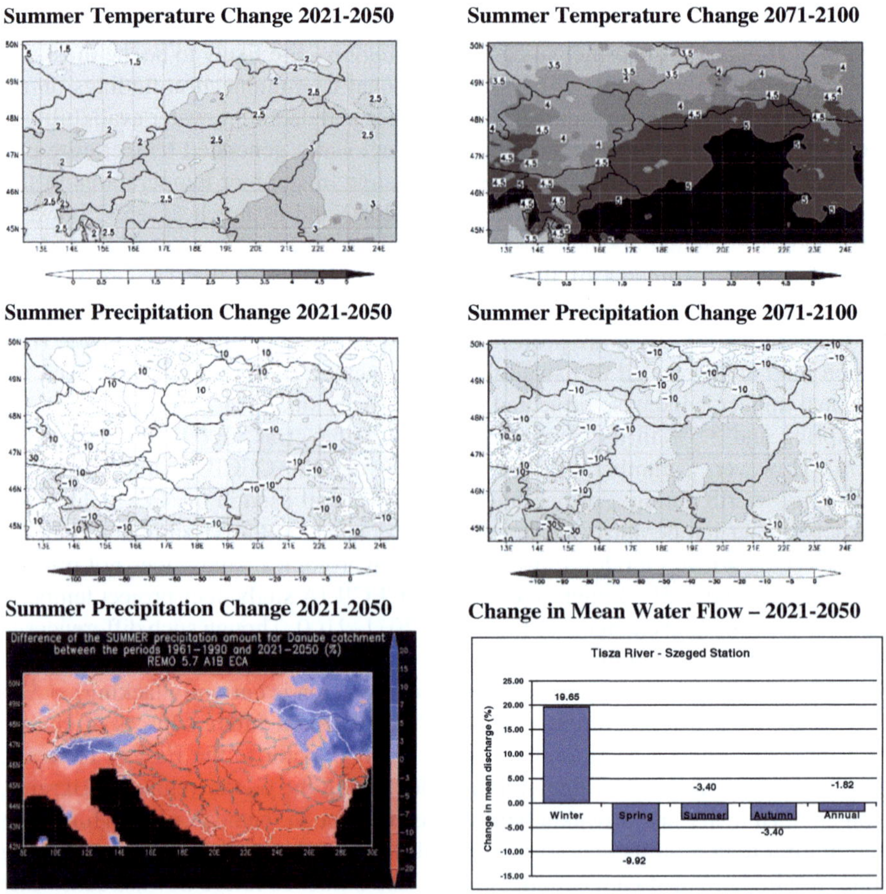

Fig. 2.2 Change in summer temperatures, precipitation and river flow compared to 1961–1990 reference period (The first four graphs in Fig. 2.2 have been graciously provided by the Hungarian Meteorological Institute (OMSZ) and the last two by the Hungarian Water Services (VITUKI))

production. Much as in economic general equilibrium models, where capital and labour are simply assumed to move to more profitable locations when economic systems careen out of balance (thereby seamlessly restoring the former balance), the same is essentially assumed about agricultural production with little or no discussion of the relative mobility of agricultural factors of production, or even of the consequences of massive land conversions (and their related carbon cost) in order to make available adequate supplies of arable land. However some authors have begun to investigate future land requirements in Europe (Rounsevell, Ewert, Reginster, Leemans, & Carter, 2005).

Water management of the alpine regions of Europe provides a particularly revealing look at the potential risks to agriculture, energy and water resource use more generally –in particular in the southern regions of Europe. The Alps currently supply some 40% of Europe's freshwater resources (EEA, 2009b). As the EEA points out:

> Spanning the centre of continental Europe, the Alps play a crucial role in accumulating and supplying water to the continent. Recognised as the 'water towers of Europe', the mountains host most of the headwaters of the rivers Danube, Rhine, Po and Rhone; as such, they deliver vital ecosystem services both within and beyond the region, underpinning social and economic wellbeing in vast lowland areas. (EEA, 2009a, p. 9)

Thus for the rest of Europe, the Alps are of great significance due to the invaluable water resources they provide. Moreover, the 2003 drought affected water resource availability very far afield. Thus due to the falling water level in the Danube, the Cernavoda nuclear power plant in Romania was required to shut down for just shy of a month and other hydropower plants along the Danube experienced reduced electricity output (EEA, 2009a). Thus declining precipitation in and water flow from the Alps can have significant impacts on water availability in quite distant regions.

Switzerland was already strongly impacted during the 2003 drought, with agricultural water demand leading to conflicts with water protection authorities. In the longer term, Swiss farmers were severely affected by the drought, which caused a reduction in harvests and a net decline in income of approximately 11% (EEA, 2009a). Though the supply of available electricity was not strongly affected, this was largely the result of an increase in electricity generation from alpine pump storage facilities which, in turn, were strongly advantaged by significant runoff from melting snow and glaciers (BUWAL, 2004). The performance of nuclear power plants in Switzerland had to be curbed by 25% for a period of two months, reducing annual electricity production by 4% (OcCC-ProClim, 2007). Whether Switzerland or other alpine regions can depend on glacial runoff in the future is unlikely. Current projections suggest that glacier coverage in Switzerland will decline by 50–90% by the year 2050 (OcCC-Proclim, 2007).

One of the more stunning findings to emerge out of Switzerland concerns the potential future demand for water from the agricultural sector. Fuhrer and Jasper (2009) note that the total amount of agricultural area requiring irrigation in Switzerland is likely to increase some eight-fold in the near future due to declining availability of water resources. Based on an analysis of the period 1980–2006, the authors find that agricultural areas totalling some eight to ten times the size of the currently irrigated area are already in need of increased irrigation. The authors further note that for the 1980–2006 period many agricultural areas in Switzerland remained dry for several weeks or even months and the longest dry spells surpassed periods of 60 days (in the more extreme cases from 150 to 200 days). Moreover, the average length of dry spells over the period 1980–2006 was rising in Central and Eastern Switzerland (with no obvious trend in areas already strongly affected by dry conditions in Western Switzerland). As the authors note, over time the types of agricultural products affected will range from the more water-intensive crops (potatoes and vegetables) to corn and grains, and ultimately to more exposed grasslands.

The relative importance of these findings extends well beyond the future challenges facing Swiss agriculture. Like the Hungarian example, they demonstrate both the broad geographic range of problems affecting agricultural production in Europe as well as their potential severity. Yet even countries that are comparatively water-rich already feel the effects of progressive climate change. With time these changes will have even more radical impacts on agricultural production in Switzerland and other countries. Irrigation will most likely be extended to a very significant share of agriculture in Switzerland, in other alpine regions and in other downstream regions in the coming decades and the type of agricultural produce planted will also change. The second major implication is that changing water resource use and management in the broader alpine region will ultimately affect water availability in downstream countries as well. If more water is used upstream in order to maintain Swiss and Austrian agricultural production or to service electricity generation, less water will be available downstream for agriculture, energy, household consumption and other purposes.

Generally missing from country-level analyses is a detailed discussion of the potential cross-border upstream and downstream impacts of changing demand for water resources and the potential rise in reduced water availability. While projections of potential water supply are available, these typically do not consider how water use habits will change in other countries. Thus, based for example on the Swiss case, rising temperatures are likely to give rise to decreased water availability, continually increasing irrigation in agriculture and increased stress on the water needs of power producers. All of this ultimately means reduced water availability further downstream. Yet, in the water-receiving countries, no attempts are made to measure the extent to which available water supply will be altered by changing water demand in the upstream countries. In the event of future average temperatures similar to or even more severe than the 2003 drought, it remains unclear what the overall impact on water availability will be in most European countries. Moreover, the potential for droughts to occur in multiple years and what this might mean for water availability is likewise not known.[28] Yet these are both likely outcomes of future rising temperatures. There may be considerable room for future work on water availability for major river catchments globally and in Europe using the WATERGAP model (Alcamo et al., 2007).

2.3.2 The Water Management Policy Framework

The European Commission and other organisations such as the EEA tout the potential advantages of the WFD as a general framework for effectively dealing with adaptation in water resource management. For example, the EEA (2009a) notes:

> Existing European legislation, particularly the Water Framework Directive (WFD), is a good basis for cross-border water coordination and adaptive management. It paves the way

[28] This potential is recognised, for example, by the Commission's JRC: 'the tendency for dry years to form clusters increases the magnitude of the drought threat. For example, any repetition of the sustained rainfall deficiencies that were a feature of a 25-year period beginning in the 1880s would, with present demand levels, represent a very severe challenge to water management throughout much of Europe' (JRC, 2005, p. 130).

> towards further integrating climate change adaptation into European policies and implementing adaptation measures, also at a river basin scale where uncoordinated actions should be avoided. Within this context River Basin Management Plans (RBMPs), a key instrument of the WFD … must be coordinated with other sectoral policies (e.g., the Common Agricultural Policy) and secure broad public participation. … The Water Framework Directive is complemented by the Floods Directive and the policy on water scarcity and droughts, which provide a more specific framework for adapting to the key water-related impacts of climate change (e.g., droughts management plans, water scarcity and droughts information system). (p. 14)

Though the WFD was initially designed as a strategy for responding to water quality issues, the additional requirement that Member States also introduce water-pricing schemes by the end of 2010 has clear implications for the management of water quantity, scarcity and availability issues. The requirement that Member States jointly develop RBMPs as part of both the WFD and the Floods Directive is a further advantage, though there is no requirement that RBMPs necessarily address water supply issues. First draft RBMPs were to be submitted by 2008 and finalised in 2009. A second RBMP round is planned for 2015.[29]

Since the WFD is a very loosely structured instrument allowing individual countries to come up with relevant strategies for managing water resources on their own and only dictating a schedule for addressing very general framework issues, it was likely to be greeted with relatively strong support from EU Member States. For the most part, these seem to welcome the WFD framework as a suitable means for attempting to get a handle on water quality while at the same time not significantly threatening national sovereignty. However, where the WFD may potentially fall short is in the successful foresight over and coordination of all the competing water use interests – cross-border/transboundary, national, sub-national and sectoral.

Ultimately the biggest problem with the development of country analyses – as opposed to regional or other international river basin approaches – is that countries highly dependent on water flows from upstream countries have only limited knowledge of the range of potential change in upstream water use and demand. However, for adequate assessments of future water quantity, these are crucial. Given the transboundary nature of many water resource management problems this may well lead to serious problems and potential future conflicts. By way of example, greater than 95% of the water in Hungary originates in other countries to the West and East.

The crisscrossing of water-use and related policy interests at the sub-national, national and transboundary levels is thus a cause for concern. The WFD framework may simply be inadequate to handling all individual Member state and EU water needs. Perhaps the biggest drawback of the WFD framework is its emphasis on 'water quality'. While water quality is certainly an important objective and one that should remain at the forefront of EU and Member state concerns, the downside is that other issues – in particular water quantity, scarcity and drought preparedness might be neglected or inadequately addressed.

[29]The timetable for WFD-related submissions is available at: http://ec.europa.eu/environment/water/water-framework/info/timetable_en.htm.

Discussion at the 2nd European Water Conference, organised by DG Environment in April 2009 suggests such concerns are warranted. For one, most of the reported discussion on the WFD surrounds water quality issues. Very little discussion even raises the issue of water scarcity and drought management (see DG Environment, 2009). In fact, based on the findings of one NGO present at the debate, only five of the RBMP surveyed by the organisation (out of a total of 17 submitted) set goals for reducing water use and only 2 of those were targeted at individual sectors (DG Environment, 2009). While the potential for introducing water payment schemes could significantly impact water use efficiency, many complain that current proposals lack adequate transparency. Moreover, most assessments of water stress are ultimately based only on precipitation trends and fail to consider 'local level characteristics' (DG Environment, 2009, p. 37). Though at least one current European study hopes to correct for some of these problems, the WATCH study focuses on global water use patterns.[30] In order to adequately inform national-level RBMP, such measurement issues need, for one, to focus on water use patterns in Europe and for another, to find their way into country level analyses. However, the SCENES project, which aims to develop and analyse a set of comprehensive scenarios of Europe's freshwater futures up to 2025, may point the way for future study.[31]

While agricultural issues were strongly discussed at the 2nd European Water Conference, there was essentially no discussion of the impact of energy demand on water use, nor were any of the major energy sector stakeholders represented at the conference. To some extent, this deficit appears to be recognised in the first two key messages to emerge from the conference:

1. Water management is affected by many other policies. Therefore, it is important to look at the impact of economic activities on water and to coordinate with a number of sectors including agriculture, industry, energy production, tourism etc. The Water Framework Directive (WFD) and its focus on integrated river basin management plans is one approach to enhance sectoral coordination.
2. Since many important water decisions are not made by water managers, it is important to involve all stakeholders in the WFD management process in order to provide the proper answers to water related issues. (DG Environment, 2009, p. 4).

[30]See the website of the WATCH project on water and global change: http://eu-watch.org/nl/25222705-Home.html. It is furthermore unclear from the project description how much focus will be placed on the two primary sources of water demand (energy and agriculture). From the project deliverables produced as of this writing, only the global evolution domestic household water demand has been considered. Though future deliverables in Work Block 2 intend to consider other aspects of water demand (in particular agriculture and industry), the energy sector or power producers are not listed in this section.

[31]For more on the SCENES project, see the website: http://www.environment.fi/default.asp?contentid=342422&lan=EN.

However, it remains unclear how the WFD framework will be revised in order to encourage Member States to address these issues.

Finally, the currently poor record of implementation in the southern EU Member States raises alarm bells. According to DG Environment's assessment:

> There is a north-south divide in Europe when it comes to the individual national river basin management plans. In northern Europe most plans have been published, while southern Europe is lagging behind. This is an issue of concern, since southern Europe is an area with more visible and multiple water problems and one would expect efforts there to be more intensive to address them. (DG Environment, 2009, p. 40)

In particular, given the problems outlined above – heavy agricultural water demand in the southern EU Member States, the rising potential for higher temperatures and extreme drought conditions, as well as the potential impact on electricity generation and planning – the need for consequent and extensive planning of future water use management is clear. However, even assuming the southern countries submit RBMP in the near future and all Member States successfully develop water pricing strategies, it remains unclear whether these will prove capable of setting an adequate framework for future adaptation planning. The basic problem is that the goal and related guidelines of the WFD were established well in advance of the current emphasis on climate impacts and adaptation requirements. This is likely to pose specific problems for individual countries since the legal framework for water pricing will most likely be passed through political systems well before it has been adequately adjusted to adaptation strategies.

The fact that water use is a cross-border, transboundary issue strongly supports the view that the EU can and presumably should play an important role connecting both water resource needs and the vulnerabilities of individual states. Moreover, given the increasingly complex awareness of the basic problems countries face, it is now necessary to connect these into a framework that can help individual Member States plan and coordinate future adaptation efforts in a far more decisive and comprehensive way. Certainly one key feature of such a strategy is the outlining of clear guidelines for integrating adaptation into WFD goals and strategies – the Commission has committed to developing guidelines for ensuring that climate goals are integrated into RBMP by the end of 2009 (European Commission, 2009a). But in the long run the EU could and presumably should go much further than this.

Though not currently under discussion, one strategy for immediate consideration is an EU Directive on Water Use Reduction (raising water use efficiency). Although EU-level climate negotiations have already focused on reducing energy use,[32] this effort needs to be extended to water use efficiency. Apart from the single basic problem of increasing water scarcity – in particular in Southern Europe – one

---

[32]EU level efforts to reduce energy use are governed by a number of different Directives and commitments. Most recently, as part of the Energy and Climate Change Package approved in December 2008, the EU committed to reducing energy use 20% by the year 2020. However, there is no single directive that lays down the specific strategy by which the EU and individual Member States will achieve this goal. That said, the Commission is currently focusing on revising and updating the 2002 Directive on the Energy Performance of Buildings (Directive 2002/91/EC). Other relevant

of the strongest justifications for this is the simple fact that the water-energy nexus has important implications for both climate mitigation and climate adaptation goals. Energy is required to pump water from the source of supply to end-users and water is required to produce energy (primarily for cooling purposes). In the United States, for example, some 4–5% of electricity use provides the necessary energy for distributing and treating water and wastewater (see US DOE, 2006). Similar amounts are likely used in Europe.[33] Thus improving water efficiency not only guards against the threat of water scarcity in the more water stressed regions of Europe, it also helps reduce energy use and thus greenhouse gas (GHG) emissions.

The European Commission now argues that improvements in water use efficiency in Europe could effectively reduce overall water use by approximately 40% and agricultural water use by 43% (see Ecologic, 2007). The magnitude of these numbers alone indicates tremendous unexploited potentials. Though which direction water demand is expected to move under a business as usual scenario remains somewhat controversial,[34] with concerted action water use efficiency could be dramatically improved. Under future drought-like conditions in Southern Europe, promoting the development and use of water-saving technologies appears more than essential. Given all of the above considerations, there is no justifiable logic why the EU should regulate energy use and not water use.

The failure to introduce a European Directive on Reducing Water Use could potentially lead to the awkward consequence that upstream countries benefiting from greater water availability face fewer incentives to introduce strong water-saving measures. This of course has the potentially ugly consequence that downstream countries depending on the flow of water resources from neighbouring countries have little impact on the upstream behaviour of fellow Member States.

Further, one could rapidly develop an argument for why the EU should require all Member States to develop Drought Management and Water Scarcity Strategies (or Plans). The WFD does in fact encourage Member States to develop Drought Management Plans (DMP) 'when and where they are needed' and the Commission produced guidelines for interested Member States on how to develop them (European Commission, 2008a). However, as with water-saving legislation, Member States that experience less immediate threats are far less likely to pursue such strategies. Nor is there any immediate requirement that cross border plans will

EU legislation on energy efficiency includes Directive 2005/32/EC on the eco-design of energy-using products, Directive 2006/32/EC on energy end-use efficiency and energy services, the EU Action Plan for Energy Efficiency (COM(2006)545 final) and finally Directive 2004/8/EC on the combined generation of heat and electricity.

[33]Data for Europe is limited. However some European reports point to the importance of energy in water distribution and treatment. See for example Carrillo and Frei (2009) and WBCSD (2009: esp. 14–15).

[34]Compare for example the results of Flörke and Alcamo (2004) who project an 11% decline in water use between 2000 and 2030 with the EEA's (2009b) European Water Resources report, which suggests that water demand will likely increase in the coming years.

be developed, posing potential threats for downstream Member States more susceptible to droughts. Presumably the most immediate reason why the Commission has so far resisted is that not all Member States are clearly affected by droughts and thus many Member States may have little need for such plans. However, as with forest fire management in the EU, this should not discourage the European Commission and the EU more broadly from pursuing a more unified strategy and taking advantage of the potential benefits of a 'Community' approach.

The emphasis on water use efficiency arising out of the WFD and potentially also the emphasis on water containment (and overflow options) arising out of the EU Floods Directive may further be inadequate when it comes to promoting additional water storage. Water storage should presumably be placed much higher on the Member state and EU agenda, since water scarce countries in particular need to be able to store sufficient water to make it through periods of significant and potentially extended drought. Though many countries have greatly extended their potential for water storage in the past few decades (among the European countries, Spain and Italy top the scale of newly constructed water storage facilities; cf. EEA, 2009b), presumably far more water storage will be necessary, in particular the southern EU Member States due to significantly increasing temperatures and declining rates of precipitation. Since at least 1990, most of the more agricultural EU Member States, foremost among them Italy, Spain, France and Greece have been rapidly extending their crop irrigation potential (EEA, 2009b).

There are at the same time potential limitations to continuously extending water storage and irrigation potential. Excessive storage is simultaneously noted as an ecological problem (EEA, 2009b), and, in the long run, little resolution of the potential conflict between increased water scarcity and the potential to overextend water storage solutions is discussed.

In all of this discussion, the role of forests is severely neglected. Though the potential role of forests does receive some discussion in the EEA's (2009a) study of alpine water resources and in another study from the Institute for European Environmental Policy (IEEP) (Anderson et al., 2008: esp. vi, 46–50), forests – in particular in the context of their impact on the water balance, water purity and flood control potential – are not even mentioned in the either the WFD or the Floods Directive. According to the EEA:

> Forest soils, which have a higher water storage capacity than non-forest soils, reduce run-off peaks and local flooding. Moreover, forest vegetation stores water and delays soil saturation. Evapotranspiration from mature forests can remove a considerable proportion of storm rainfall. ... Surface runoff can therefore be prevented or slowed to some extent, even in high precipitation events. At the local level the effect of flood reduction is particularly relevant for small watersheds and minor meteorological events. (EEA, 2009a, p. 43)

While a number of studies point to the potentially negative effects of forests on the overall water balance (Zhang et al., 2007; Jackson et al., 2005), such studies may fail to adequately comprehend or measure the role of forest ecosystems and their impact in particular on the water balance, water supply, storage and cooling effects arising from forests' ability both to retain water and to promote evapotranspiration.[35] Rising awareness that forests could or may play a more important role emerges in a Commission report on the Implementation of Forestry Measures:

> Forests and forest management have an important role in the protection of water resources. The Fifth Ministerial Conference on the Protection of Forest in Europe (MCPFE, 5–7 November 2007, Warsaw, Poland: 'Warsaw Resolution 2 Forests and Water') stressed the role of forests and forest management in protecting water quality, managing water resources for the quantity of all waters, flood alleviation, combating desertification and soil protection as well as the importance of mountain forests in the reduction of landslides, erosion and effects of avalanches. (European Commission, 2009c, p. 10)

The 2007 Warsaw Resolution on Forests and Water makes a number of broad commitments on the part of signatory countries to further investigate the relationships between forests and water and to improve the sustainable management of forests in relation to water (MCPFE, 2007). In its first follow-up to the Warsaw resolution, the MCPFE held a conference on Forests and Water in Antalya, Turkey (May 12th–14th, 2009).[36] A similar approach is also being stressed in the framework of the UNECE Water Convention. Finally, the Finnish Forest Research Institute in collaboration with the European Forest Institute (EFI) in Joensuu, Finland is also organising a number of conferences on the issue of Forestry–Water interactions.[37]

Though some awareness of potential strategies that might be implemented to improve water storage (and potentially moderate floods) through more natural techniques – in particular by exploiting the natural advantages of forests – is beginning to emerge, far more could be done. Despite the growing awareness that forests can play a positive role, there is virtually no information currently available on how and where to plants forests in order to have the greatest effect on future water

[35]Though research on the water balance is cutting edge, it is also highly controversial. Zhang, Vertessy, Walker, Gilfedder, and Hairsine (2007) argue, for example, that the impact of increased forestation on water supply is negative. However, these authors fail to consider the impact of forests on groundwater supply. Moreover, how one should think of the impact of evapotranspiration on local precipitation events is likewise controversial. Though most authors simply view evapotranspiration as a loss to the local water cycle, others view the forest-water balance on a very broad scale in its broader ecosystem context. Other authors however attempt to view the forest-water balance in a more holistic fashion, i.e. in their broader ecosystem context. These authors tend to find more support for the view that forests support increased water supply and aid significantly in improving water balance (cf. Schwärzel et al., 2009; Sheil & Murdiyarso, 2009; IUCN, 2009). Maes, Heuvelmans, and Muys (2009) likewise suggest that evapotranspiration should potentially be seen as a contribution to the water cycle, with potentially positive implications for the water balance. However, the relationship between forests and the water balance remains disputed. For a recent overview of this debate, see Ellison (2010).

[36]Much of the documentation discussed at this conference is available at the conference website: http://www.mcpfe.org/www-mcpfe/forests_and_water.

[37]The first of these was held in Joensuu in September 2009 (conference website: http://www.metla.fi/tapahtumat/2009/koli/index.htm). The second will be held in May 2010 (conference website: http://www.efi.int/portal/news___events/events/extra/2010/JFNW2010/).

supply and flood moderation or control. Though the concept of forested buffer zones between agricultural fields and lakes and rivers as a means of purifying water resources has gained some purchase, the strategy is still not widespread and receives no mention, for example, in important EU-level documents like the WFD and the Floods Directive.

The EU's Common Agricultural Policy and Rural Development Strategy is far from dealing adequately with many or most of the adaptation challenges noted above. Though DG Agriculture's European Agricultural Fund for Rural Development (EAFRD) addresses afforestation as a general strategy, no real connection is made between forests and potentially beneficial forest-water interactions. Further, although the connection between forestry and the larger context of biodiversity preservation is recognised in the EAFRD strategy (this connection is discussed in more detail below), as spending on Natura 2000 areas is permitted as part of Axis 2 spending on 'improving the environment and the countryside' (European Commission, 2009c, p. 44), few countries have thus far taken advantage of available resources.

In two further areas, discussions have not even really begun with respect to the EU's Common Agricultural Policy (CAP). The first of these – and perhaps the most urgent – concerns agriculture and water use. As noted above, in particular in the more heavily agricultural countries, agriculture is the largest user of water resources. In addition, water-saving potential in agriculture is estimated at some 43% of current use. Yet the CAP does not currently have a strategy in place for promoting efficient water use. Moreover, the current EU policy emphasis on biofuels is likely to raise water demand in agriculture. The second area has to do with the future likelihood that there will be tremendous upheaval in the agricultural sector in Europe as a result of climate change, rising temperatures and the likely shifting of agricultural zones. While potential strategies have not really been discussed at the European level, this needs to be done. How will agriculture be organised in the future and where will various agricultural products be produced? Is or will adequate agricultural land be available in Europe once the shifting of agricultural zones has taken hold? And most importantly perhaps, how will potential strategies be formulated in order to facilitate the shift of agricultural production to these new agricultural zones?

Given the above energy-related discussion, the EU clearly needs to develop a more water conscious energy strategy. This idea has been around for some time. By way of example, a 2004 report on water use recommended, among other things, that tower cooling systems be required for electricity generation in order to reduce water use (Flörke & Alcamo, 2004). Though dry-cooling systems have since upstaged and outdated tower-cooling systems, the general strategic approach is and should remain the same: attention to water use efficiency should be a requirement for all newly-installed electricity generation capacity. Dry-cooling or comparably water-use efficient cooling systems should become the rule and should be mandated in all fossil fuel and biomass-based thermoelectric power plants and concentrated solar power systems. Finally, far more consideration should be dedicated to the added advantages of other renewable energy sources (i.e. wind, solar, geothermal,[38] and tidal) that are not big water users.

The relatively severe energy problems that emerged during the 2003 heat wave will be repeated and eventually surpassed. Thus initiatives should be afoot to capitalise on the advantages and widespread availability of less water-intensive energy production and more water-efficient cooling systems.

Finally, policy efforts in general and work on the WFD in particular need to address a much broader context than is currently the case. The 2007–2009 Work Plan of the Convention on the Protection and Use of Transboundary Watercourses and International Lakes (under the United Nations Economic Commission for Europe) has opted to focus on 'Water and Climate adaptation in transboundary basins, including flood and drought risk management'.[39] In other words, this highly integrated, revised focus tacitly recognises the limitations of a more narrow focus on water quality, floods and flood risk management. The revised focus is far broader and attempts to address the broad range of climate and adaptation-related needs in the management of transboundary basins. This approach may suggest that specific advantages can be gained from a more comprehensive ecosystem-based or 'catchment basin' type approach. Moreover, the international framework tends to provide a stronger emphasis on cross-border cooperation, as demonstrated, for example, by recent Hungarian efforts to extend Ramsar wetland protection cooperation to Slovenia (Croatia and more recently Austria already participate). Though there is some indication that these more international cooperative arrangements are not always as successful as they could potentially be (see e.g., Czako & Mnatsakanian, 2008), the general direction of such efforts is both exemplary and potentially crucial to the future success of adaptation efforts.

In contrast, some evidence suggests that the WFD approach may push in the opposite direction. In particular with respect to water and river basin management, the EU WFD first encouraged countries to create several water district agencies and then delegate these with the task of developing and managing plans to meet the requirements of the WFD. In this sense, the WFD framework is poorly articulated at the cross-regional, national and/or cross-border levels. While this may work well in some regions that are relatively self-contained, the effect could be far less beneficial in other parts of Europe where river or catchment basins cross multiple borders. In such cases, ideal plans for river or catchment basin management could and presumably should involve explicit cross-regional and/or international cooperation, including well-thought out estimates of potential future water demand, extensive analysis of future water availability/quantity (based on temperature and precipitation changes) and potentially also cooperative efforts to increase or improve available supply.

---

[38]Geothermal power plants that do not use water re-injection systems can also use larger amounts of water. Thus re-injection, though more costly because it involves additional drilling in order to be able to re-inject water back into the cycle, should also be mandated.

[39]See the UNECE convention website: http://www.unece.org/env/water/cooperation/area422.htm.

Concerted action on river basin management is presumably a requirement for the successful definition of strategies to manage not only water quality but also water quantity. Though the EU has not yet taken similar steps, as argued in the following section, there are many reasons to think it provides a more solid footing for future policy efforts. Moreover, as suggested in the approach taken to water in the above section, the role of ecosystems and the services they provide is of great importance. Not only do ecosystems produce and manage water balance and supply, they also help regulate water flow, thereby diminishing the frequency and severity of flood events. Such an analysis suggests the role of ecosystems – in addition perhaps to the services they provide – should ultimately be placed at the centre of the analysis.

2.3.3 Biodiversity and Ecosystem Preservation

The treatment of biodiversity has gained new and added significance with the recent findings of the Stockholm Resilience Centre and the publication of an article on the Earth's planetary boundaries (Rockström et al., 2009). The authors identify nine planetary boundaries that reportedly define a so-called 'safe-operating space for humanity'. Among these nine planetary boundaries, climate change represents only one of three boundaries that have been overstepped. Apart from surpassing Hansen's recommended 350 ppm CO_2 atmospheric concentration barrier (Hansen et al., 2008), the world has even more substantially surpassed the biodiversity loss indicator established by the authors. This barrier is set somewhat higher than the assumed natural rate of extinction, at 10 species extinctions per million per year. Though the barrier is 10–100 times the natural rate of extinction, current rates are well over 100 species extinctions per million per year (100–1000 times the natural rate). Finally, the authors note that the nitrogen cycle barrier has likewise been significantly surpassed. Humanity, according to the authors through the production of fertilizer and crop cultivation, now exceeds the combined natural rate of all natural forms of nitrogen production.

Though in principle the connections between biodiversity, ecosystem services and the larger concepts of ecosystem resilience and adaptation remain underdeveloped, there should ultimately be little quarrel with the basic concept that ecosystems and the services they provide build the foundation for humanity's social and economic well-being (Louman et al., 2009; MEA, 2005). Nor should there be significant debate over the fact that ecosystems in general, and ecosystem services in particular – such as the provision of a clean and bountiful water supply, clean air and carbon sequestration – are threatened by changing the climatic conditions occasioned by global warming.

The strategies necessary to protect ecosystems –such as the EU's Biodiversity Action Plan and the goals of the Natura 2000 project – are often not well recognised and their relative importance is not always readily accepted. One of the best examples of this fact is the difficulty the EU has experienced with the protection of biodiversity. As parties to the Convention on Biological Diversity, the EU has declared a commitment to halting the loss of biodiversity. Yet despite the goals of the 2006 Biodiversity Action Plan to halt the loss of EU biodiversity by the year 2010, EU member states have not made significant progress in establishing special protection areas (SPA) and special areas of conservation (SAC) and thus have generally failed to implement the Natura 2000 program of the Habitats and Birds Directives.

Recently European Environment ministers, based in part on an internal assessment of implementation performance in the establishment of Natura 2000 protected areas, expressed 'deep concern' about the current state of EU biodiversity loss and argued the EU was unlikely to be able to fulfil its 2010 commitment. The Commission's internal assessment pointed out, for example, that '50% of all species and 80% of habitat types in need of protection in Europe have "unfavourable conservation" status'.[40] Based on the first evaluation of the EEA's SEBI indicators published in May 2009, the overall assessment was not very encouraging (EEA, 2009a, 2009b, 2009c).

While the 2007–2013 Framework perspective witnessed the firm integration of Natura 2000 goals and funding mechanisms into the EAFRD, NGOs such as BirdLife International and FERN continue to argue that farm lobbies are favoured over biodiversity and environmental concerns. Though these organisations note the potential advantages presented by the EU's rural development framework both FERN (2008) and BirdLife International (2009a) remain strong critics. The principal criticism concerns the failure to address biodiversity issues and to spend adequate resources on the development of Natura 2000 sites (cf. BirdLife International, 2009b).

Though one can argue EU Member States have made significant progress toward improving the quality and degree of biodiversity protection.[41] Most EU Member States are still quite far from achieving the ultimate goal of 'halting biodiversity loss by 2010'. In the words of Jacqueline McGlade, executive director of the EEA, designating relevant areas across Europe for the goal of habitat and species protection 'is only the first step'. McGlade points out that only a small share of Europe's habitats and species are currently in acceptable condition. Most are in 'unfavourable conservation status' and are potentially in need of 'ecological restoration' – in particular agricultural habitats.[42]

In this context, the EEA argues that Europe has not yet fully grasped the importance of biodiversity. In order to maintain biodiversity and ecosystems, these must

[40] See; 'Ministers 'Deeply Concerned' by Biodiversity Loss (*Euractiv.com*, June 26th 2009).

[41] The EEA's Core Set of Indicators Data (CSI data) suggests many EU Member States have done relatively well in fulfilling some of their basic commitments to the Habitat and Birds Directives. For example, a broad set of EU countries have at least 'proposed' sites that would be sufficient to protect habitats and species. Moreover, the total amount of surface area dedicated to species and habitat protection in Europe has multiplied some 6- or 7-fold between approximately 1996 and the present (cf. the CSI 008 Assessments, various years).

[42] See McGlade's speech on the Status of European Biodiversity at the Athens conference on Biodiversity Protection—Beyond 2010, (Apr. 27th 2009).

be more fully integrated into key sectors – in particular into agriculture, forestry and fisheries.[43] The EEA and European environmental ministers are currently promoting an ecosystem services approach to handling biodiversity that is likewise promoted by a relatively broad range of other European and international actors. The International Union for Conservation of Nature (IUCN), the International Union of Forest Research Organisations (IUFRC) and the Ministerial Conference on the Protection of Forests in Europe (MCPFE) are all integrating the ecosystem services approach into their core strategies. The concept of the ecosystem approach was first introduced in the framework of the Convention on Biological Diversity (2002) and the United Nations Forum on Forests (UNFF, 2003). The MCPFE has been one of the principal organisations attempting to integrate the ecosystem approach into sustainable forest management practices in Europe. The EEA and European environmental ministers are pushing for 'ecosystem goods and services' to be better integrated into the national and EU-level frameworks, seeing this as one strategy for improving the degree of biodiversity protection in the EU. The ecosystem services approach essentially attempts to place a market price on the use of ecosystems and the goods and services they provide.

The protection of ecosystems and the services they provide in the context of climate change is a complex problem and one that is not easily reduced to pricing systems for individual ecosystem services. While, as noted above, a dramatic increase in water use efficiency would certainly be a welcome evolution – efficient pricing mechanisms are one important strategy for achieving this goal – ecosystems themselves and the biodiversity they contain are likewise coming under increasing pressure from climate change. In this sense, prices on ecosystem services are ultimately only half the battle. In the long run, EU strategies could be far more broadly focused. For example, unless the revenues from such pricing mechanisms can somehow be funnelled back into what should perhaps remain the primary target of such strategies – the maintenance, preservation and even creation of ecosystems – such strategies are likely to fall far short of their goal.

How the EU, its Member States and other countries will manage this challenge remains unclear. The challenge of ecosystem maintenance and preservation is often not well understood – in particular in the context of climate change. What we know and think of as ecosystems today are likely to change substantially in the coming years as climate change progresses. Whether current EU strategies are up to the task depends significantly on how they are revised in order to respond to the climate challenge. As elaborated below, perhaps the two greatest challenges in this context are (1) coming to grips with current biodiversity emphases on 'species permanence' and the need for flexible arrangements, and (2) elaborating strategies for moving from piecemeal, patchwork strategies (afforestation, biodiversity protection and species preservation, water quality, etc.) to more grand-scale ecosystem-based strategies. Moreover, the costs associated with elaborating more adequate strategies are likely to be greater than previously estimated – though the rewards may also be

---

[43]See: Europe Must Grasp the True Value of Biodiversity (EEA Highlight, Apr. 27th 2009).

higher – raising important questions about how the EU and Member States will manage to support an already significantly underfunded objective.

2.3.4 Permanence vs. Flexibility

Though less frequently discussed in the literature on adaptation and climate change, 'biome shift' – the concept that the biological spheres in which flora and fauna thrive will migrate as the climate becomes warmer – is a real and increasingly visible problem raising all kinds of important questions (see Loarie et al., 2009). What is the potential range of biome shift over the next century? What is the potential for migration of flora (and fauna) across the shifting geography and borders of biomes? What does biome shift imply about tree-planting practices in forestry and what is the potential adaptation and/or migration potential of the existing forest stock? How much do current planting strategies need to change in order to keep pace with shifting biomes and what is the general magnitude and scope of necessary interventions? What does the concept of biome shift imply about conservation strategies – in particular since these are typically based on the concept of permanence rather than unstable and shifting bio-spheres? To what extent is the public sphere required to take action – either with regard to forestry or to nature conservation – in particular given the potential magnitude and scope of the necessary interventions?

The potential for biome shift is a direct outcome of global warming and climate change. Rising temperatures, changing precipitation patterns, shorter winters with less snow melt and longer summers with increased evaporation potential all affect the nature, character and location of existing biomes. Though the forecasting of the potential extent of biome shift is still in its infancy and the related uncertainties are high, it is already possible to predict the potential range of biome shift with some accuracy. Current predictions suggest that Europe's biomes could shift anywhere between 100 and 500 km to the Northeast by the year 2100.[44]

Just how responsive flora and fauna may be to biome shift remains open to question. Birds may provide a relatively good example. The migration patterns of many types of birds have already shifted along with historical temperature changes. Tingley, Monahan, Beissinger, and Moritz (2009), for example, find that of the 53 bird species they studied in the US, 48 exhibited tendencies to adjust their migratory patterns based on species-specific preferred environmental phenomena (sensitivity to either moisture or temperature). Shifting migratory patterns are also prevalent in Europe. The EEA's climate impact report, for example, notes the migratory patterns of birds could shift some 550 km to the Northeast by the year 2100. Moreover,

[44]Not surprisingly, there is considerable disagreement over the potential range of biome shift. The Commission's White Paper, for example, notes a range of 500–1000 km (European Commission, 2009b). Other estimates (on which the Commission's report draws) mention somewhat smaller ranges. For example, DG Agriculture's report suggest a potential range from 100 to 400 km (EFI-BOKU-INRA-IAFS, 2008) and the EEA's Adaptation study suggests the potential may shift as much as 550 km for birds and some 100–500 km for forests (EEA, 2008).

the study notes significant potential difficulties in making such migratory shifts for many species, in particular due to the rate of climate change, habitat fragmentation and the like (EEA, 2008).

Such observations immediately raise questions about the suitability of existing biodiversity protection strategies. Concepts of species permanence and the preservation of existing biodiversity typically seem to guide policies addressing the goals of nature conservation and species protection. The concept of biome shift, however, ultimately requires significant re-thinking of such approaches. Where species are likely to migrate based on the movement of biomes, it becomes impossible to preserve existing biodiversity when wildlife protection areas are established in fixed locations. On the other hand, it is relatively difficult – given the relative population density of most European Member States – to shift the location of special protection areas.

Similar problems arise when considering attempts to protect and preserve the existing range of flora. Again, shifting biomes ultimately mean that the future range of flora (and fauna) will not be the same in the coming decades. Changing temperatures and precipitation rates will lead to change in the existing flora (and fauna) and to a north-easterly movement of existing species. More compelling still is the occasional observation that not all flora will be able to keep pace with the rate of climate change, suggesting that human intervention will be required in order to preserve many species and to aid the shift of biomes. Current EU and Member state practice has been to define so-called 'special protection areas', 'sites of community interest', and 'special areas of conservation' in fixed locations. Moreover, the designation of such fixed sites has typically been pursued with the intent of preserving existing biodiversity and safeguarding species permanence. This approach presumably results primarily from the fact that many of the ideas regarding species protection were conceived quite some time ago. The EU's Birds Directive, for example, was introduced in 1979, long before all of the discussion on global warming and climate change. Later subsumed in the Habitats Directive (1992) – all 'special protection areas' under the Birds Directive were to be included as 'habitats' – this directive too still precedes much of the global warming and climate change discussion, in particular the context of adaptation.

Thus ultimately, the EU has much to do in order to rework existing legislation on the basis of what we now know and are still learning about the impacts of climate change. Current perceptions of the consequences of climate change – at least where biodiversity is concerned – often seem mired in notions of lost permanence rather than biodiversity migration. This point is tremendously important, since the key question now is what biodiversity migration really means – how many species will it affect, what will be its range – and what are the consequences for conservation practices – should we follow existing species with the establishment of new protection areas, or help new species to migrate to existing protection areas?

Finally, the basic approach of creating fixed protection or conservation areas may itself ultimately be open to question – in particular where these have been created to protect and preserve certain types of species. However, the potential for re-locating conservation areas is likewise highly problematic. For one, establishing conservation areas is a relatively complex political, economic and social problem involving considerable time and negotiation. For another, existing spatial geography exhibits many limitations. Few would currently envision tearing down existing urban settlements, for example, for the sake of conservation goals.

The EU has however made some progress in attempting to respond to these challenges. First, Policy Area No.3 (Objective 9) of the EU's Biodiversity Action Plan focuses on biodiversity and climate change. The EU strategy recognises the threat to biodiversity posed by climate change and encourages states to take action to protect Europe's biodiversity, pointing out the importance of the relationship between biodiversity and adaptation. In particular, the Biodiversity Action Plan encourages states to 'improve the resilience and connectivity of protected area networks' and to 'assist those species and habitats most at risk' (European Commission, 2006b, p. 11; see also European Commission, 2008b, p. 23). Second, several large research projects have recently attempted to assess the importance of climate change and adaptation requirements in the context of protecting the EU's biodiversity. In particular, two large projects (BRANCH and MACIS)[45] have attempted to analyse potential climate impacts on biodiversity and to provide policy recommendations. In general, these studies find that concepts of species protection need to be more flexible and adaptable to the dynamics of biome shift and species migration.

Both studies likewise point to the problem of species mobility across existing biomes and argue that ways need to be found to improve the adaptive potential of existing species. Recommendations include creating greater interconnectedness across existing conservation areas (Natura 2000 networks) in order to aid mobility, extending existing areas and/or creating new ones in order to help some species adapt. Moreover, analysts propose the concept of 'mobile sites' in areas where natural conservation areas can be eroded by natural processes (such as coastal sites). These studies also point variously to the concepts of 'migration', 'ecological' or 'biodiversity corridors'. Much like the concept of promoting interconnectedness, this approach attempts to address the necessity for facilitating species movement and thus adaptation. The corridor concept has in fact already been introduced in some Member States. In particular the Danish NAS proposes this strategy.

Finally, both of these studies point to the potential role Strategic Environmental Assessments (SEA) and Environmental Impact Assessments (EIA) could play in promoting broad scale and horizontal attention to adaptation issues if these goals were firmly integrated into the SEA and EIA Directives. And the Commission has likewise committed to developing guidelines 'to ensure that climate impacts [and adaptation needs] are taken into account in the EIA and SEA Directives' (European Commission, 2009a, p. 13). Though the phrase 'adaptation needs' does not appear in the original, it should presumably be worked into the text.

As a further illustration of the problems raised by the biome shift phenomenon, invasive species and how they will be treated in future EU policy is potentially

---

[45]On the BRANCH project see; http://www.branchproject.org/. On the MACIS project see; http://macis-project.net/index.html.

highly problematic. For one, invasive species have typically been considered pests that enter into an environment in which they do not belong. Climate change and biome shift require that we radically redefine the way we think about pests in general and invasive species in particular. Two basic problems occur with climate change. On the one hand, the survival and competition potential of some species are changing as a result of the changing climate. As a result, some species that previously posed no risk at all to the environment, forestry or agriculture have now become pests. However, these species are not 'invasive' in the true sense of the word, since they have always been present. Climate changes have simply altered their competition potential and thus upset the previously existing natural balance. A good example of a newly competitive species in the North American continent is the mountain pine beetle, which has laid waste to an area more than twice the size of Ireland.[46]

A second category of problems concerns species that are by definition 'invasive' but are not per se 'pests'. Biome shift ultimately means that significant shares of new species will come to inhabit regions where they were previously unheard of, and currently well-known species will move further along with the shifting biomes. Current legislative efforts with regard to Invasive species demonstrate some of the difficulties of coming to terms with these issues. EU legislative proposals, for example, still bear the title the *Invasive Species Act*. Yet clearly the term 'invasive', as suggested above, misrepresents the reality of the problem to be addressed.

The DAISIE project has thus far catalogued more than 11,000 species that are alien to but present in Europe. Only approximately 10–15% of these species represent potential threats in economic and/or ecological terms. Though the emphasis of the EU's efforts on invasive alien species focuses primarily on protecting European 'biodiversity', invasive species are potential threats to both flora and fauna throughout Europe. Forests too can be laid waste by the invasion of alien species as in the example of the mountain pine beetle.

To date, although the EU and individual countries have legislation in place intended to protect national and European level biodiversity, thus far no harmonised EU level approach has been put into place for monitoring and controlling invasive species and the effect of these on European biodiversity. Current efforts from the European Commission have thus far focused on the development of a European-wide 'early warning and information system' focused on reporting new and emerging species. Additionally, the Commission has proposed expanding the current list of banned species to cover other newly discovered invasive species. Other considerations include the development of legal measures for handling invasive species and the potential establishment of an independent agency. Most recently, the Environmental Council of Europe issued a statement on invasive species and drew attention to the potential importance of Natura 2000 regions for the preservation of European biodiversity (see Council of the European Union,

---

[46]See, Beetles, Wildfire: Double Threat in Warming World, (*Associated Press*, Aug. 23rd 2009).

2009). The Commission has been called upon to develop a strategy to respond to the issue of invasive species by 2010 and to continue to develop inventories on invasive species.

2.3.5 The Biodiversity Policy Framework

A quick review of two EU policy areas (Natura 2000 goals and the EU's Common Agricultural and Rural Development strategy) reveals some of the difficulties of responding adequately within the current EU policy framework. For one, the integration of the adaptation dimension in the Natura 2000 framework is likely to confront serious complications. The EU and the Member States (like many other countries)[47] exhibit considerable difficulty moving ahead with even the comparatively simple project of protecting existing levels of biodiversity by establishing Natura 2000 regions. Given current experience with the basic project, adding the adaptation dimension to the Natura 2000 goals is ultimately likely to be immensely complicating.

The principle complication – as outlined by the discussion of the consequences of biome shift – is that too little is currently known about how best to adapt Natura 2000 objectives to climate change. One possible strategy, of course, is to loosen up the original habitat objectives and to address issues of permanence and flexibility in ways that more clearly define both what can be achieved with Natura 2000 protection areas, as well as what the ultimate goals really are. In the longer term, species permanence, for example, within currently designated Natura 2000 habitats will presumably not be possible in many (perhaps even most) cases.

Second, the legal framework for species protection in Europe will require relatively radical change and reform. By way of example, most EU Member States strictly control the type of tree and to some extent plant species that can be planted in any specific country. Thus, for example, sections 6–9 of the 2008 Swedish Forestry Act provide very significant restrictions on the types of trees that can be planted and typically prohibit the planting of foreign species, including those from other EU Member States (Ellison, Pettersson, & Keskitalo, 2009).

With respect to the EU Natura 2000 policy framework, some – including the Commission's White Paper – have suggested that adaptation to climate change can be easily integrated into the management of Natura 2000 sites. Further, some have argued that Articles 3(3), 5 and 10 of the Habitats Directive provide a framework for integrating adaptation (MACIS 2008, p. 10). For example, Article 3(3) allows Member States to improve the 'ecological coherence of Natura 2000' sites, while Article 5 allows the Commission, in cooperation with the Member state, to play a caretaker role in situations where 'priority habitats' or 'priority species' are not adequately protected (by unanimous vote of the Council). Article 10 allows Member

[47] See e.g.; Efforts to sustain biodiversity fall short (*Nature*, Nov. 19th 2009).

States to intervene in potential migration corridors in order to improve the potential for species movement.

Such proposals however seem to ignore the enormity of the basic problem of biome shift and what this ultimately means for species preservation. Very few questions have been raised about the potential for Natura 2000 regions to provide an adequate basis for biodiversity protection and the bio-diverse resilience of future forests, flora and fauna. From the total land area in Europe, Natura 2000 habitats represent only a very small share. Though the Natura 2000 networks were originally conceived as a strategy for protecting Europe's biodiversity by providing areas where the broad variety of Europe's species would be guaranteed to survive, it is far less clear what will happen in the context of climate change and biome shift. In some senses, researchers and analysts appear to be looking at Natura 2000 habitats as a means of protecting and ensuring the resilience of Europe's environment. This however, is a much more extended goal and there is little certainty that the comparatively small area of terrain currently covered is adequate to the task.

Presumably a complete re-evaluation of Natura 2000 site designations should be undertaken in the context of climate change and, where necessary, Member States should consider the potential designation of new and/or additional sites in order to meet the goals originally set out by the Natura 2000 project. Though this is ultimately a much larger project, thus far no mention has been made of such a strategy. However, in order to create an adequate framework for species protection and potentially preservation in Europe, it is presumably necessary. The EU's Common Agricultural Policy and Rural Development Strategy, on the other hand, has yet to integrate or even adequately consider all of the potential linkages between forests and the goal of ecosystem protection. Though DG Agriculture's EAFRD addresses afforestation as a general strategy, only two basic goals are currently served: (1) the planting of forests in order to promote future bioenergy potential, and (2) the planting of forests in order to sequester carbon.

As a mitigation tool, the CAP's afforestation strategy has been a particularly slow moving vehicle. In a relatively long period of time, the EU has only managed to re- or afforest a very small area. Adequate data on actual afforestation and reforestation is difficult to come by. The Commission's White Paper on Adaptation, for example, notes that over the past 15 years, forest area in Europe has increased by some 13 million hectares (European Commission, 2009b, p. 81). While afforestation efforts account for a share of this increased forest cover, it is not exactly clear how much. The ECCP working group on forest sinks notes that:

> Between 1990 and 2000, afforestation and reforestation activities have extended the total EU forest area of 113 Mha by 340,000 ha/year or 3%, resulting from nearly equal surfaces of planted forests (in many cases through support from the 2080/92 afforestation scheme and the rural development regulation 1257/99) and natural forest expansion. The Group estimates that, if this process continues at the same rate during the present decade, it may result in a sequestration potential of approximately 3.84 Mt C/yr. (14 Mt CO_2 eq/year) during the first commitment period. In case of a sustained afforestation trend and taking into account an extended EU of 25 Member States, a technical sequestration potential of 34 Mt C/year (125 Mt CO_2 eq) may be reached in the long term. (ECCP WG FS, 2008, p. 4)

Roundly criticised by the European Court of Auditors (ECA, 2004) and a number of NGOs (cf. Birdlife International, 2009a; FERN, 2008), the EAFRD strategy has done relatively little to promote either the goals of biodiversity or ecosystem protection, nor has the program been very efficient at extending forest cover and sequestering carbon.

However, while in the 2007–2013 framework perspective EAFRD funding was also made available for Natura 2000 areas, to date most EU Member States have thus far not made extensive use of the available funding opportunities (European Commission, 2009c). BirdLife International (2009b) recently criticised both the EU level funding mechanisms available for Natura 2000 areas as hopelessly inadequate (giving rise to significant funding shortfalls), as well as criticising individual Member States for failing to take adequate initiative to ensure that available EU funding is actually allocated for biodiversity preservation: 'Natura 2000 and biodiversity conservation is simply not identified as a priority for EU funds in most countries' (BirdLife International, 2009b, p. 7).

These points are significant for at least two important reasons. First, the fact that the EU has only been able to re- or afforest 3% of EU forest area in a 10 year period[48] raises compelling questions about how and even whether the EU and individual Member States will be able to handle the problem of biome shift. Though it is not immediately clear from current studies just how intensive a strategy is necessary in order to respond to biome shift – we do not currently know to what extent trees, for example, will be able to keep pace with this shift – this shift will ultimately affect very broad expanses of European forests. At the very least, very large swathes of forest and wooded areas will be at increasing risk. At the worst, large areas in European forests could suffer from significant dieback – especially in areas where forests are composed of reduced numbers of species. The potential size of these areas far surpasses the capacity of current EU-level afforestation strategies, though a good share of these areas likewise falls under the forest management practices of public and private owners in individual Member States.

2.4 From the Water Towers of Europe to Ecosystem Preservation

While adaptation of course needs to be integrated into the framework of individual sectors, one of the greatest weaknesses of current efforts and approaches suggested by the above discussion of water, biodiversity and ecosystem services is the overall interconnectedness of issues across sectors. A thorough discussion of water and potential adaptation strategies requires a relatively thorough discussion of climate

[48]If we take the larger figure of 13 million hectares of increased forest cover noted above, which includes re- and afforestation under the EAFRD as well as other afforestation projects, this sum amounts to just over 11.5% of the total EU forest cover in a 15-year period, a considerably larger figure.

impacts, agricultural practices, energy use, changing water demand structure in different geographic regions of an individual river basin – in particular across the borders of individual countries – and finally, consideration of the ecosystem services which both produce and manage water flow.

In this sense, a more holistic and integrated approach to adaptation in and across individual sectors may ultimately be a requirement for policy success. Though at least some of the discussion in the accompanying documentation to the White Paper recognises the potential value of such concepts as 'Green Structure' approaches (European Commission, 2009b), as ways of improving ecosystem resilience (by improving biodiversity) and exploiting ecosystem services (such as cleaner water, air and cooler temperatures), such ideas are not well integrated into the larger EU policy context. Nor does the White Paper provide much of a model for achieving this goal.

Adequate responses to the threat of climate change, biome shift, biodiversity loss and the weakening or disappearance of ecosystems and ecosystem services are difficult to craft in the EU policy context. The general problem is not made any simpler by the fact that though there are a number of different contexts in which it is possible to begin to address these issues, the strategies that emerge are, for the most part, only partial, piecemeal or even sectorally-dependent responses to issues that should potentially be viewed in a much larger overarching context. Thus, for example, Natura 2000 and biodiversity commitments are not quite the same thing as the preservation of ecosystems and ecosystem services. Similarly, as noted above, while afforestation strategies tend to target carbon sequestration and/or future bioenergy production, they generally fail to consider such issues as the impact of forests on the water cycle, their potential value as flood management tools, their larger impact on cooling, their relative impact on biodiversity needs or – and perhaps most importantly – their potential to help create or extend important ecosystems.

As illustrated in Table 2.2, most current EU policy strategies only target single or possibly dual policy goals. Similar claims can presumably be made about national level strategies (such as National Parks or Forest preserves). With a more integrated and holistic approach, many of these policy tools could be adapted to address a much broader range of policy interests. Moreover, the benefits from doing so are presumably far greater than if policy strategies only target single or dual policy goals. Individual policy efforts may not always be able to address all of the potential targets raised in Table 2.2 above – for example, flood management or increased precipitation may not always prove meaningful depending on the ecosystem in question. At the same time, the broadening of potential policy goals has the capacity to greatly improve the quality of the outcome with respect to a broad range of adaptation-related needs and interests. Thus, for example, forests can be employed to pursue multiple goals, not just those of carbon sequestration and/or bioenergy generation.

While the EEA's current strategy of trying to raise the value of ecosystem services by imposing prices and charging for them may go some way toward strengthening support for ecosystems, there are some limitations of this approach that should also be taken into account. In the long run, the Economics of Ecosystems

Table 2.2 Potential and actual targets of individual sectoral policies

<table>
<tr><th>Sectoral policy</th><th>Actual targets</th><th>Potential targets</th></tr>
<tr><td>Afforestation</td><td>• Bioenergy
• Carbon sequestration</td><td rowspan="6">• Ecosystem creation/preservation
• Biodiversity protection
• Species preservation
• Water cycle management
o Quality/Purity
o Supply/Quantity/Balance
• Flood management
• Precipitation
• Soil Retention
• Cooling
• Air purification
• Combating Desertification
• Carbon Mitigation</td></tr>
<tr><td>Natura 2000</td><td>• Biodiversity protection
• Species preservation</td></tr>
<tr><td>WFD/RRBMPs</td><td>• Water quality
• Water quantity?</td></tr>
<tr><td>Floods Directive</td><td>• Flood management</td></tr>
<tr><td>Ramsar areas (wetlands)</td><td>• Wetlands preservation
• Water purity</td></tr>
<tr><td>Ecosystem Services</td><td>• Economically-driven use of ecosystem services</td></tr>
</table>

and Biodiversity (TEEB) strategy is more of an end-of-pipe strategy: it places a price on the 'outputs' of ecosystems, but it does not per se strengthen the commitment to the actual ecosystem that produces those outputs.[49] Thus, for example, with respect to water-pricing strategies, the TEEB approach raises the 'value' of water and will thereby likely have a substantial impact on water use efficiency (assuming of course that individual Member States introduce adequate plans). But TEEB as such does not strengthen protections for the ecosystem that produces the water, though other policy features, such as the water purity requirements of the WFD may indirectly achieve some of these goals. This should not be seen as an argument against TEEB, but rather as a defence of the notion that more is really necessary in order to adequately protect and improve the quality of existing ecosystems and/or to create new ones.

Likewise, the proposal to broaden and strengthen use of SEA and EIA assessments certainly seems advisable and on a relatively broad scale – e.g., in the assessment of structural and cohesion fund projects, large scale infrastructure projects and large scale agricultural or energy-related projects. However, it does not represent a sufficient strategy. The principal difficulty is that while afforestation strategies, for example, might benefit from some degree of climate-proofing, this in itself is no guarantee that the goals of biodiversity preservation or ecosystem maintenance and improvement will be addressed. Afforestation, as a strategy,

[49] The TEEB project attempts to estimate the costs of ecosystem services and propose relevant pricing strategies. A similar strategy is already being elaborated under the Water Framework Directive, which requires Member States to introduce water pricing systems for national water resources by 2010.

already fulfils a number of climate-proofing requirements – it provides future potential bioenergy production and longer term carbon sequestration. In this sense, though afforestation strategies may easily comply with a number of important climate goals, SEA or EIA assessments do not guarantee the goals of biodiversity preservation or improvements in ecosystem service production will be served.

More generally, the EU has not been very successful at coordinating strategies across competing issue domains. Thus, for example, though the re- and afforestation strategies under the EAFRD and biodiversity goals under the EU Biodiversity Action Plan are theoretically compatible due to the potential for biodiversity goals to strengthen future forest resilience, they appear to generate little cross-institutional cooperation and coordination. If the goals of biodiversity and forest resilience cannot even be adequately integrated into the comparatively small scale EU afforestation strategy – where it is presumably/potentially beneficial – then we should also be raising questions about the ability of EU policy to successfully integrate the far more comprehensive adaptation and ecosystem services strategies into the existing EU policy framework.

The extent to which the climate-related goals of forest and ecosystem resilience are utterly disconnected and disassociated from the more marketised (or commodified) aspects of forest system management (in particular the pursuit of bioenergy strategies and, to a lesser extent, carbon sequestration, cf. Ellison & Keskitalo, 2009) is potentially suggestive of a broader dilemma lying at the heart of EU (and potentially also national-level) governance mechanisms that may itself need to be addressed. The following section focuses on the concept of institutional constraints and adaptation strategies.

Though land use, the third thematic area defined in the White Paper, is not discussed in this chapter, land use and land use change represent perhaps some of the largest potential impacts on ecosystems and their related services. In the long run, land use and land use change cannot (or at least should not) be discussed without reference to this general adaptation-related context. Moreover, land use and land use change clearly overlap significantly with the first two areas, since decisions to build dams or extend residential areas have immediate implications for water-related issues, biodiversity and ecosystem services.

2.5 On the Governance of Adaptation: Goal Conflict And Institutional Confusion

As demonstrated in Ellison and Keskitalo (2009), EU strategies are frequently subject to a form of goal conflict that presumably results from relatively significant divisions across different horizontal and vertical levels of competence. Thus for example the introduction or pursuit of policies in one institution – for example DG Agriculture or DG Environment – can potentially conflict with policies pursued by another DG (horizontal). And of course goal conflict can also occur between EU and national level policies (vertical). Understanding the impact of the existing institutional structure is crucial to building the foundation for improved strategies for addressing both climate change mitigation and adaptation. To the extent that the existing institutional structure both determines and reinforces the fabrication of sectorally-defined, piecemeal policy solutions, models for reform gain increasing significance. Moreover, as argued in what follows, without significant institutional reform, it is unlikely the EU will be able to develop and introduce more holistic models for climate change mitigation and adaptation.

The problems of such horizontal and vertical forms of multi-level governance (MLG) are well illustrated with the example of forestry policy. With regard to forestry and forestry policy, policy outcomes exhibit strong coordination of interests around two poles: on the one hand the agricultural, energy and industry oriented Commissions/Ministries appear to favour strategies related to bioenergy, biomass and afforestation, while on the other hand environmental ministries, the EEA and environmental agencies such as the Swedish Environmental Protection Agency (SEPA) tend to favour more environmentally oriented goals such as biodiversity, the promotion of Natura 2000 natural conservation areas and a more general emphasis on ecosystems and their related services.

Some preliminary conclusions as well as identification of potential evolutionary development paths can be drawn from this very brief synopsis of institutional divisions and policy fragmentation. One is the persistence of institutional rivalry and competition across different decision-making bodies at the EU and also national levels. These divisions are reinforced by variation in the relative resource endowment of individual institutions. The result of these divisions is policy fragmentation. The continued emphasis on business-as-usual decision-making pathways would continue these problems and potentially impede the development of more coherent forest policy. However, decision-making approaches that would support more coherent policy could also be developed. The institutional approach and logic developed in Ellison and Keskitalo (2009) can easily be extended to the question of integrating adaptation into EU policy more generally. Similar goal-conflict related problems are likely to occur where attempts to address adaptation come up against competing goals – either as a result of ongoing climate mitigation efforts or as a result of other existing policy strategies.

Problems in forestry are of course only one example of such potential goal conflict. Where adaptation is concerned, the potential for such goal conflict is also high. As one important example of this, climate mitigation options pursued in particular by DG Energy and Transport strongly favour bioenergy strategies that promote the use of wood (and other forms of renewable energy use) in order to substitute for fossil fuels. While potentially not contradictory to afforestation interests – one logic for afforestation is to build the potential for future forest use in areas such as bioenergy – bioenergy goals may ultimately conflict with biodiversity goals. Moreover, the combination of afforestation and bioenergy interests – in particular where these favour the use of fertilizers, monocultures and rapid rotation energy crops – may be doubly detrimental to biodiversity goals. Adaptation interests on the other hand are far more likely to emphasise the benefits of environmental features like biodiversity – in particular for future forest resilience and potentially also for future forest regeneration. And, as suggested above, adaptation strategies could potentially go much further to emphasise the role and importance of ecosystems and the services they provide.

One alternative for the more effective coordination of interests involves the promotion of improved communication across the different EU Commission units and between the EU and other national and local levels of policy-making. At least two recent articles promote models along these lines. Mickwitz et al. (2009) recommend a number of instruments to bring about greater policy integration (communicative, organisational and procedural instruments that ultimately attempt to give greater place and prominence to the climate debate in national agendas, institutional arrangements and assessment and consultation procedures). On the other hand, Glück et al. (2009) highlight and promote the advantages of multilevel governance, decentralisation and participatory decision-making processes.

Institutional mechanisms do exist at the EU-level that are supposed to help reconcile the competing claims with respect to forestry and the use of forest-based resources. The *Inter-service Group on Forestry*, established in 2002 'to facilitate cooperation and coordination of forestry-related work between relevant Commission services' (EP 2006, p. 3), is technically responsible for insuring that forestry policy is coordinated across some 11–13 EU-level Directorates General (DG). Chaired by DG Agriculture, this body has two main purposes: to ensure the flow of information and to seek agreement across departments. There is also an Inter-service Group on International Forestry Issues responsible for the preparation of Commission positions on international issues. To what extent the general Inter-service coordination strategy is successful is controversial. Birdlife International argues, for example, that the work of the Inter-Service Group on Forestry as well as DG Agriculture's Standing Forestry Committee (SFC) should ultimately be opened up to NGOs. In addition, the power and position of DG Environment should be elevated in order to more successfully introduce forestry issues (see e.g., BirdLife International, 2006).

The European Economic and Social Committee (EESC) argues that forestry and its potential role in climate mitigation and adaptation could be utilised to far greater and more significant effect than is currently the case. Moreover, the EESC argues that far more needs to be done with regard to developing responses to the need for adaptation – in particular in forestry (EESC, 2009). Whether the failure to emphasise and improve forestry policy is explicitly the result of institutional divisions is not discussed in the EESC Opinion. However, other EU-level organisations have explicitly criticised this point in the past (ECA, 2004, p.10). The Commission, on the other hand, demonstrates considerable resistance to the idea of institutional reform and argues that a stronger legal footing for forestry policy in the EU is not feasible without greater interest from Member States.

Ellison and Keskitalo (2009) ultimately ask whether the governance structure surrounding this complex of issues – both at the EU and the national level – is well equipped to handle the increasing proliferation of actors, their related interests and the increasing potential for goal conflict. The coherent coordination of interests around a defining agenda needs to be able to overcome divisions created by the vested interests of competing actors – in particular where institutional divisions reinforce these divisions and vested interests become institutionally segregated. In such cases, more radical solutions may be necessary.

Perhaps the most important conclusion to arise from the goal conflict analysis is that the introduction of additional consultation procedures or proposals to extend and further promote decentralisation in decision-making processes may be entirely inadequate (necessary but not sufficient) to resolving such deeply-seated and broadly-situated institutional divisions and policy fragmentation. Two observations are central in this regard: first, institutional divisions that reinforce and further segregate the representation of vested interests in policy outcomes are unlikely to be overcome by the promotion of attempts to simply increase coordination across different actors. Second, vested interests are present at all levels – including the local level. Thus an emphasis on decentralisation likewise may not be able to overcome such divisions.

Assuming the basic problem regarding the coordination of adaptation strategies is the fact that there is too much institutional division across relevant policy domains and thus ultimately poor coordination and fragmentation of policy output, then at least one relevant proposal might be to coordinate adaptation policy at the EU level under one single Commission. Thus a more compelling alternative may be to create a Climate Change Commission and place principal competence for adaptation within that framework. This would have the advantage of correcting the current degree of decentralisation and fragmentation of forestry policy. Further, this would place the principal focus on forestry firmly within the context of climate change.

Creating a Climate Change Commission may help to resolve some of these dilemmas. The specific role of a Climate Change Commission should ultimately be defining strategies for responding to the challenges of global warming and climate change. Since these of course involve both mitigation AND adaptation, the creation of a Climate Change Commission might also heighten the degree of attention dedicated to the adaptation side of the debate, both in the context of forestry and in other adaptation-related policy areas.

While it is difficult for the European Commission to proceed with substantial reform without the support of the Member States, such a strategy would seem to make sense on a number of levels. For one, the EU's role in the pursuit of climate policy has been tremendously important in the international arena. Without the leadership role played by the EU, it is unlikely that the Kyoto Protocol covering the period up to 2012 or ongoing negotiations over a new agreement to cover the period 2013–2020 would have gone as far. The most important commitments to emission reductions under the Kyoto Protocol have been made by EU Member States.

Elevating EU climate strategy to Commission status would simply recognise and reinforce the EU's current leadership role in the climate debate. Moreover, it would make it possible to further mobilise both expertise and resources on a single climate agenda. Though there is currently discussion at the EU level about creating a Climate and Energy Commission, this strategy may not be the most advisable goal.

This proposal has been strongly criticised by some, in particular for attempting to shift policy competence at a strategically difficult time (just prior to the Copenhagen negotiations).[50]

The logic of the 'goal conflict' argument ultimately finds fault with this proposal. The principal issue raised in this context concerns the concept that policy choices are strongly influenced both by the institutional location in which they are developed and by the prominence of interests embedded within that context. The institutional structure promoted with the potential introduction of a Climate and Energy Commission would lend far too much weight to the Energy sector. In fact, far too much of the EU climate strategy has tended to focus attention on the energy sector and has given far too little consideration to other potential policy areas (in particular building-related energy use, forestry, transport and ultimately adaptation itself). On the other hand, the potential elevation of the climate agenda to Commission status – with a mandate for focusing on both mitigation AND adaptation – is potentially far more appealing.

Many advantages potentially arise out of the centralisation and control of the mitigation and adaptation agenda in a specifically Climate Change Commission. For one, centralised coordination provides a framework in which competing interests can more easily be made to confront each other and potentially find resolution. For another, it would heighten the sense of commitment to the issue over which the Commission has been granted competence and raises the sense of fiduciary responsibility. Third, centralising coordination would likely reduce the degree of policy fragmentation or potential for goal conflict. Moreover, it could potentially provide a more successful arena for broadening the scope of commitment from many of the currently targeted options (afforestation, flood protection, biodiversity preservation, etc.) to the maintenance, preservation and creation of ecosystems (as outlined in Table 2.2 above). Finally, centralisation of climate policy at the higher EU level can further require and promote greater cross-border coordination and planning, thus potentially discouraging free-rider behaviour where cross-border issues are concerned.

Centralising control for mitigation and adaptation strategies at the EU level in a separate Climate Change Commission may pose potential threats to local and even national level interests. Some are clearly concerned about the potential consequences of increasing centralisation.[51] However, the call for greater centralisation

[50]See e.g., MEPs Angry at Plans for Energy Shake-up (European Voice, May 14th 2009); FERN's *EU Forest Watch* newsletter (June 2009, Issue No. 139) and the letter from EU GLOBE members (May 18th 2009).

[51]Pekka Pesonen, a former state secretary at the Finnish Ministry of Agriculture and Forestry warned against the dangers of an overly aggressive degree of centralisation in forestry policy and argued that policies should remain national in character due to national-level variation in approaches to forest policy. In his view, the adoption of common rules could potentially lead to contradictions with national-level policy strategies. See Call for more EU co-ordination on forest policy (*Euractiv.com* 2007). Though forestry policy may seem a skewed example since it is not firmly integrated into the EU legislative framework, it is highly illustrative of the way in

of the EU climate strategy in a Climate Change Commission should not be seen as contradictory to the parallel calls for greater cross-sectoral and horizontal coordination, decentralisation and participatory decision-making processes. In important ways, these are complementary and not contradictory strategies.

Ultimately a significant amount of flexibility and foresight must be written into any EU-level adaptation agreement in such a way that potential conflicts over policy strategies and competence at the national and local levels are not significantly compromised. In many ways, the general structure of the EU water framework directive provides a good example of how this can be done. The WFD sets a very general framework with general guidelines and principles which individual countries are then required to implement and adopt in concert with national and local level needs and interests. While the WFD is certainly not perfect and many still wish it had gone much further (examples of how it might be improved have also been discussed herein), the general concept of providing a very general EU-level framework with considerable room for national and local level flexibility is well-represented in this directive.

2.6 Conclusions

Typically the strongest argument in favour of the development of adaptation strategies has been the observation that even if states are able to stabilise or begin to reduce world CO_2 and other GHG emissions, some degree of climate change has already been built into the system. The concept of 'committed greenhouse gas warming' or 'committed climate change' has begun to find a common currency in the larger global warming and climate change literature (cf. Parry, Lowe, & Hanson, 2009; Ramanathan & Feng, 2008; Meehl et al., 2005, Parry, Arnell, Hulme, Nicholls, & Livermore, 1998) and essentially begins to define a lower bound for climate impacts that will (or are extremely likely) to occur given current atmospheric concentrations. Though estimates of such lower bounds have typically not begun to work their way into estimates of future climate impacts, these should ultimately begin to define a minimum level of preparatory action required by all EU Member States.

Climate change and the twin goals of mitigation and adaptation are THE challenge of the 21st century. As it has for mitigation, the EU must define a more unified, cohesive and overarching agenda for approaching adaptation. The relative complexity of defining strategies for adaptation, however, provides a strong foundation for arguing that the general commitment to adaptation should be expressed in very general and broad terms and should presumably be aimed at protecting the stability of ecosystems and the services they provide. Though the details of adaptation strategies must be worked out within the relevant sectors, their overall agenda should

---

which vested interests have come to inhabit sectoral-institutional frameworks and the difficulties in sharing policies cross-institutionally (cf. Ellison & Keskitalo, 2009).

potentially be expanded in order to target a broader range of potential adaptation outcomes (as proposed in Table 2.2 above). Further, regional variation in climate impacts, vulnerability and adaptation requirements argues that a significant degree of flexibility will be required across EU-level legislative efforts and national-level Member state and local implementation. Sensitivity both to general adaptation needs in different sectors and to national and local-level requirements of adaptation must be worked into EU-level strategies.

At the same time, one should not be afraid of the potential advantages of more EU-centralised action on adaptation. The fact that failure to act on adaptation-related issues in one country has potential consequences for other neighbouring countries provides a strong foundation and motivation for the EU to intervene in a far more concerted fashion than it has to date. But perhaps more importantly, the potential for goal conflict across multiple sectors and across EU- and national-level decision-making bodies argues that responsibility for adaptation (and also mitigation) strategy should be housed and highly centralised in one EU-level institution – presumably a Climate Change Commission. The goal of centralising responsibility for adaptation strategy in one institution should ideally favour the coordination of policy goals in two important ways: (1) across issues areas (e.g., energy, agriculture, water and land use) and (2) across individual Member States.

At the very least, this chapter has outlined a number of alternative solutions to the current mainstreaming model proposed by the European Commission. These range from the potential introduction of new Directives – such as a Directive on Water Use Reduction, an EU mandate for Member States to Develop Drought Management and Water Scarcity Issues and the development of a water conscious energy strategy – to increased emphasis on forest-water interactions and the development of ecosystems in Afforestation strategies. In addition, this chapter has emphasized the importance of developing long run strategies that strongly consider interaction effects across different cross-sectoral adaptation and other policy efforts. Though this discussion has been far from exhaustive and has failed to address many issues – in particular the third theme in the White Paper on land use – it has strongly underlined the importance of ecosystems and their preservation for human survival. Favouring the centrality of ecosystems with a more holistic approach to climate change mitigation and adaptation may be the key to a successful strategy. If it is necessary to adopt new institutional structures in the EU and elsewhere in order to achieve these goals, they should be given due consideration.

References

AEA (2007). Adaptation to climate change in the agricultural sector. Report to European Commission Directorate General for Agriculture and Rural Development, AEA Energy and Environment and the Universidad de Politécnica de Madrid.

Alcamo, J., Moreno, J. M., Nováky, B., Bindi, M., Corobov, R., Devoy, R. J. N., et al. (2007). Europe. Climate Change 2007: Impacts, adaptation and vulnerability. contribution of Working Group II to the Fourth Assessment Report of the Intergovernmental Panel on Climate Change (pp. 541–580). In M. L. Parry, O. F. Canziani, J. P. Palutikof, P. J. van der Linden & C. E. Hanson (Eds.), Cambridge, UK: Cambridge University Press.

Anderson, J., Arblaster, K., Bartley, J., Cooper, T., Kettunen, M., Kaphengst, T., et al. (2008). *Climate change-induced water stress and its impact on natural and managed ecosystems*. Study prepared for the European Parliament's Temporary Committee on Climate Change. Brussels: European Parliament.

Begg, I. (2009). *EU expenditure to support transitions to a low carbon economy. EU Consent EU Budget* (Working Paper No. 9). Budapest-Mannheim: EU Consent.

BfG. (2006, April 7). Niedrigwasserperiode 2003 in Deutschland: Ursachen – Wirkungen – Folgen, Mitteilung Nr. 27. Koblenz: Bundesanstalt für Gewässerkunde (BfG).

BFH-EFI (2007). Feasibility Study on means of combating forest dieback in the European Union. Background Report. Federal Research Center for Forestry and Forest Products (BFH) and the European Forestry Institute (EFI), October.

BirdLife International (2006). *Commission draft on EU forest action plan*. Czech Republic: BirdLife International in association with Friends of the Earth.

BirdLife International (2009a). *Could do better: How is EU rural development policy delivering for biodiversity*. UK: BirdLife International.

BirdLife International (2009b). *Financing Natura 2000: Assessment of funding needs and availability of funding from EU funds*. UK: BirdLife International.

BRANCH (2007). Planning for biodiversity and climate changes. Branch Project final report. Natural England, UK.

BUWAL (2004). Auswirkungen des hitzesommers 2003 auf die Gewässer. Schriftenreihe Umwelt Nr. 369. Bern: Bundesamt für Umwelt, Wald und Landschaft.

Carrillo, A. M. C., & Frei, C. (2009). Water: A key resource in energy production. *Energy Policy, 37*(11), 4303–4312.

Christensen, J. H., & Christensen, O. B. (2002). Severe summer time flooding in Europe. *Nature, 421*, 805–806.

Christensen, J. H., Carter, T. R., Rummukainen, M., & Amanatidis, G. (2007). Evaluating the performance and utility of regional climate models: the PRUDENCE project. *Climatic Change, 81*(Suppl. 1), 1–6.

Ciscar, J.-C. (Ed.). (2009). Climate Change Impacts in Europe: Final Report of the PESETA research project. JRC Scientific and Technical Reports. Joint Research Centre, European Commission. Luxembourg: Office for Official Publications of the European Communities.

Council of the European Union (2009, June 25). A mid-term assessment of implementing the EU Biodiversity Action Plan and Towards an EU Strategy on Invasive Alien Species – Council Conclusions. Brussels.

Csima, G., & Horányi, A. (2008). Validation of the ALADIN-Climate regional climate model at the Hungarian Meteorological Service. *Időjárás, 112*(3–4), 155–177.

Czako, V., & Mnatsakanian, R. (2008). Impacts of and adaptation to climate change in the Danube-Carpathian Region. Unpublished manuscript, Department of Environmental Sciences and Policy, Central European University, Budapest.

DG Environment (2009). *Active involvement in river basin management: Plunge into the debate*. Conference Report. Brussels: DG Environment.

DG Research (2009). *European Research Framework Programme: Research on climate change*. Luxembourg: Office for Official Publications of the European Communities.

ECA (2004). Forestry measures within Rural Development Policy. European Court of Auditors Special Report No 9/2004.

ECCP WG FS (2008). Conclusions and recommendations regarding forest related sinks and climate change mitigation. European Climate Change Programme, Working Group on Forest Sinks.

ECCP WG II (2006). Impacts and adaptation: Agriculture and forestry sectoral report. European Climate Change Programme, Working Group II Impacts and Adaptation.

ECCP WG II (2007). Building National Adaptation Strategies, Sectoral Report. European Climate Change Programme, Working Group II Impacts and Adaptation.

Ecologic (2007). *EU water saving potential*. Berlin: Ecologic.

EEA (2009a). Regional climate change and adaptation: The Alps facing the challenges of changing water resources. EEA Technical Report No. 8/2009. Copenhagen: European Environment Agency.

EEA (2009b). Water resources across Europe – Confronting water scarcity and drought. Report No 2/2009. Copenhagen: European Environment Agency.

EEA (2009c). Progress towards the European 2010 Biodiversity Target. Report 4/2009. Copenhagen: European Environment Agency.

EEA (2008). Impacts of Europe's changing climate – 2008 indicator-based assessment. EEA Report No. 4/2008. Copenhagen: European Environment Agency.

EESC (2009). The role of forests and the forest-based sector in meeting the EU's climate commitments. Opinion No. 626. Brussels: European Economic and Social Committee.

EFI-BOKU-INRA-IAFS (2008). Impacts of climate change on European forests and options for adaptation. Report to the European Commission Directorate-General for Agriculture and Rural Development, AGRI-2007-G4-06.

Ellison, D. (2010). Forest-Water Interactions and Adaptation to Climate Change. Manuscript, Umea University.

Ellison, D., & Keskitalo, C. (2009). Climate politics and forestry: On the multi-level governance of Swedish forests. Drivers Working Paper for the MISTRA program, Future Forests: Sustainable Strategies under Uncertainty and Risk.

Ellison, D., Pettersson, M., & Keskitalo, C. (2009). Forest governance: International, EU and national-level frameworks. Drivers Working Paper for the MISTRA program, Future Forests: Sustainable Strategies under Uncertainty and Change.

EP (2006, February 16). *European Parliament Resolution on the implementation of a European Union forestry strategy*. Rapporteur: Heinz Kindermann. EP No. A6-0015/2006/P6_TA-PROV(2006)0068.

European Commission (2009a). *White Paper. Adapting to Climate Change: Towards a European Framework for Action*. Brussels: Commission of the European Communities [COM(2009) 147 final].

European Commission (2009b). *Commission staff working document accompanying the White Paper: Adapting to climate change*. Brussels: Commission of the European Communities.

European Commission (2009c). *Report on implementation of forestry measures under the Rural Development Regulation 1698/2005 for the period 2007–2013*. Brussels: Commission of the European Communities.

European Commission (2009d). *Stepping up International Climate Finance: A European Blueprint for the Copenhagen Deal*. Brussels: Commission of the European Communities.

European Commission (2008a). *Drought Management Plan Report*. Technical Report 2008-023. Brussels: Commission of the European Communities.

European Commission (2008b). *The European Union's Biodiversity Action Plan: Halting the loss of biodiversity by 2010 – And beyond*. Luxembourg: Office for Official Publications of the European Communities.

European Commission (2007a). *Adapting to climate change in Europe – Options for EU action*. Brussels: Commission of the European Communities.

European Commission (2007b). *Commission staff working document accompanying the Green Paper on adapting to climate change in Europe*. Brussels: Commission of the European Communities.

European Commission (2007c). Mid Term review of Industrial Policy: A Contribution to the EU's Growth and Jobs Strategy. Brussels: Commission of the European Communities.

European Commission (2006a). Annex to the Fourth national communication from the European Community under the UN Framework Convention on Climate Change (UNFCCC). Brussels: Commission of the European Communities.

European Commission (2006b). Biodiversity Action Plan. Brussels: Commission of the European Communities.

Feeley, T. J., III, Skone, T. J., Stiegel, G. J., Jr., McNemar, A., Nemeth, M., Schimmoller, B., et al. (2008). Water: A critical resource in the thermoelectric power industry. *Energy, 33*(1), 1–11.

FERN (2008). Funding forests into the future? How the European Fund for Rural Development affects Europe's forests. Moreton in Marsh: FERN.

Flörke, M., & Alcamo, J. (2004). European outlook on water use. Report commissioned by the EEA. Centre for Environmental Systems Research, University of Kassel.

Fuhrer, J., & Jasper, K. (2009). Bewässerungsbedürftigkeit von Acker- und Grasland im heutigen klima. *AgrarForschung, 16*(10), 396–401.

Glück, P., Rayner, J. Berghäll, O., Braatz, S., Robledo, C., & Wreford, A. (2009). Governance and policies for adaptation. In R. Seppälä, A. Buck & P. Katila (Eds.), *Adaptation of forests and people to climate change – A global assessment report* (pp. 187–210). Vienna: International Union of Forest Research Organizations.

Hansen, J., Sato, M., Kharecha, P., Beerling, D., Berner, R., Masson-Delmotte, V., et al. (2008). Target CO_2: Where should humanity aim? *The Open Atmospheric Science Journal, 2*, 217–231.

Hansen, J., Sato, M., Conway, T., Dlugokencky, E., Dutton, G., Elkins, J., et al. (2009, March 11). Air pollutant climate forcings within the big climate picture. Presented at the Climate Change Congress Global Risks, Challenges and Decisions, Copenhagen, Denmark.

Hungary (2005). Fourth National Communication to the UNFCCC. Budapest: Ministry for Environment and Water.

Hungary (2002). Third National Communication to the UNFCCC. Budapest: Ministry for Environment and Water.

Italian-French Report (2006). Water scarcity management in the context of WFD. Technical Report. MED Joint Process WFD/EUWI – Water Scarcity Drafting Group.

IUCN (2009). Final report study on the economic value of groundwater and biodiversity in European forests. Brussels: IUCN Regional Office for Europe.

Jackson, R. B., Jobbágy, E. B., Avissar, R., Roy, S. B., Barrett, D. J., Cook, C. W., et al. (2005). Trading water for carbon and with biological carbon sequestration. *Science, 310*(5756), 1944–1947.

JRC (2005). Climate change and the European water dimension: A report to the European Water Directors, EU Report No. 21553. Joint Research Centre, ISPRA, Italy.

Kunzewicz, Z. W., Parry, M. L., Cramer, W., Holten, J. I., Kaczmarek, Z., Martens, P., et al. (2001). Europe, in McCarthy, J. J., O. F. Canziani, N. A. Leary, D. J. Dokken & K. S. White (Eds.), *Climate Change 2001: Impacts, adaptation and vulnerability* (Chap. 3). Cambridge: Cambridge University Press.

Lenton, T. M., Held, H., Kriegler, E., Hall, J. W., Lucht, W., Rahmstorf, S., & Schellnhuber, H. J. (2008). Tipping elements in the Earth's climate system. *PNAS 105*(6): 1786–1793.

Loarie, S. R., Duffy, P. B., Hamilton, H., Asner, G. P., Field, C. B., & Ackerly, D. D. (2009, December 24). The Velocity of Climate Change. *Nature, 462*, 1052–1055.

Louman, B., Fischlin, A., Glück, P., Innes, J., Lucier, A., Parrotta, J., et al. (2009). Forest Ecosystem services: A cornerstone for human well-being, in Seppälä, Risto, Alexander Buck and Pia Katila (Eds.), Adaptation of Forests and People to Climate Change – A Global Assessment Report (pp. 15–27). Vienna: International Union of Forest Research Organizations.

MACIS (2008). Policy options to prevent/minimize negative impacts on biodiversity. MACIS project deliverable, October 2008.

Maes, W. H., Heuvelmans, G., & Muys, B. (2009). Assessment of land use impact on water- related ecosystem services: Capturing the integrated terrestrial-aquatic system. *Environmental Science and Technology, 43*(16), 7324–7330.

Massey, E. E. (2009). Adaptation policy and procedures in Central and Eastern Europe. Report R- 09/012. Amsterdam: Institute for Environmental Studies (IVM).

Massey, E., & Bergsma, E. (2008). Assessing adaptation in 29 European Countries. Report W- 08/20. Amsterdam: Institute for Environmental Studies (IVM).

Matreata, M., Corbus, C., Csik, A., Gauzer, B., Gnandt, B., Mattányi, Z., et al. (2009). Preliminary results of hydrological impact studies for catchments of central and lower Danube basin – project CLAVIER. Geophysical Research Abstracts, Vol. 11, EGU2009-11722.

McMullen, C. P., & Jabbour, J. (2009). Climate Change Science Compendium 2009, United Nations Environment Programme. Nairobi: EarthPrint.

MCPFE (2007). Warsaw Resolution 2: Forests and Water. Ministerial Conference on the Protection of Forests in Europe.

Millenium Ecosystem Assessment (2005). Ecosystem and human well-being. Synthesis. Washington: Island Press.

Meehl, G. A., Washington, W. M., Collins, W. D., Arblaster, J. M., Hu, A., Buja, L. E., et al. (2005). How much more global warming and sea level rise? *Science, 307*(5716), 1769–1772.

Mickwitz, P., Aix, F., Beck, S., Carss, D., Ferrand, N., Görg, C., et al. (2009). *Climate policy integration, coherence and governance*. Helsinki: Partnership for European Environmental Research (PEER).

Ministère de L'Économie des Finances et de L'Industrie (2003, November 28). Une canicule exceptionnelle pendant l'Été 2003.

OcCC-ProClim (2007). *Climate change and Switzerland 2050*. Bern: OcCC-ProClim.

Parry, M. L. (Ed.). (2000). Assessment of potential effects and adaptations for climate change in Europe: The Europe ACACIA Project. Jackson Environment Institute, University of East Anglia, Norwich, UK.

Parry, M., Arnell, N., Hulme, M., Nicholls, R. J., & Livermore, M. (1998, October 2). Adapting to the inevitable. *Nature, 395*, 741.

Parry, M., Lowe, J., & Hanson, C. (2009, April 30). Overshoot, adapt and recover. *Nature, 458*, 1102–1103.

Pfeifer, S., Jacob, D., Balint, G., Balteanu, D., Gobiet, A., Horanyi, A., et al. (2009). Climate change impact assessment in central and eastern Europe: The CLAVIER project. 2nd International Lund RCM Workshop, 4–8 May, Lund University, Sweden.

Ramanathan, V., & Feng, Y. (2008). On avoiding dangerous anthropogenic interference with the climate system: Formidable challenges ahead. *PNAS, 105*(38), 14245–14250.

Roberts, G., Parrotta, J., & Wreford, A. (2009). Current adaptation measures and policies. In R. Seppälä, A. Buck & P. Katila (Eds.), *Adaptation of forests and people to climate change – A global assessment report (and policy brief)* (pp. 123–133). Vienna: International Union of Forest Research Organisations.

Rockström, J., Steffen, W., Noone, K., Persson, Å., Chapin, F. S., III, Lambin, E. F., et al. (2009). A safe operating space for humanity. *Nature, 461*(124), 472–475.

Rotmans, J., Hulme, M., & Downing, T. E. (1994). Climate change implications for Europe: An application of the ESCAPE model. *Global Environmental Change, 4*(2), 97–124.

Rounsevell, M. D. A., Ewert, F., Reginster, I., Leemans, R., & Carter, T. R. (2005). Future scenarios of agricultural land use in Europe. II: Estimating changes in land use and regional allocation. *Agriculture, Ecosystems & Environment, 107*(2–3), 117–135.

Schwärzel, K., Feger, K.-H., Häntzschel, J., Menzer, A., Spank, U., Clausnitzer, F., et al. (2009). A novel approach in mode-based mapping of soil water conditions at forest sites. *Forest Ecology and Management, 258*(10): 2163–2174.

Sheil, D., & Murdiyarso, D. (2009). How forests attract rain: An examination of a new hypothesis. *BioScience, 59*(4), 341–347.

Sokolov, A. P., Stone, P. H., Forest, C. E., Prinn, R., Sarofim, M. C., Webster, M., et al. (2009). Probabilistic forecast for 21st century climate based on uncertainties in emissions (without policy) and climate parameters. *Journal of Climate, 22*(19), 5175–5204.

Sovacool, B. K., & Sovacool, K. E. (2009). Identifying future electricity-water tradeoffs in the United States. *Energy Policy, 37*(7), 2763–2773.

Swart, R., Biesbroek, R., Binnerup, S., Carter, T. R., Cowan, C., Henrichs, T., et al. (2009). Europe adapts to climate change: Comparing National Adaptation Strategies. PEER Report No 1. Helsinki: Partnership for European Environmental Research (PEER).

Tingley, M. W., Monahan, W. B., Beissinger, S. R., & Moritz, C. (2009). Birds track their Grinnellian niche through a century of climate change. *PNAS, 106*(Suppl. 2), 19637–19643.
University of Oxford (2008). Meta-analysis of adaptation and mitigation measures across the EU25 and their impacts and recommendations how negative impacts can be avoided, Deliverables 2.2 and 2.3, MACIS project study.
US DOE (2006). Energy demands on water resources: Report to Congress on the interdependency of energy and water. Washington, DC: U. S. Department of Energy.
Van der Linden, P., & Mitchell, J. F. B. (Eds.). (2009). ENSEMBLES: Climate Change and its Impacts: Summary of research and results from the ENSEMBLES project. Met Office Hadley Centre, Exeter, UK.
WBCSD (2009). Water, energy and climate change: A contribution from the business community. Conches-Geneva: World Business Council for Sustainable Development.
Zhang, L., Vertessy, R., Walker, G., Gilfedder, M., & Hairsine, P. (2007). Afforestation in a catchment context: Understanding the impacts of water yield and salinity. Industry Report 01/07, CSIRO Land and Water Science Report Number 01/07, Canberra.

Chapter 3
Climate Change Adaptation in the United Kingdom: England and South-East England

E. Carina H. Keskitalo

Abstract The UK has been one of the early actors in developing adaptation to climate change, and today has a comprehensive legislative and regulative framework for including climate change effects in planning. This chapter reviews the development of the UK approach, drawing on a literature study and semi-structured interviews conducted with several actors, the majority of whom are from public administration at the national, regional and local levels. The study focuses on England and the South-East England region in particular, one of the areas most at risk of flooding and sea level rise in the UK. In addition to discussing the national and regional levels, the chapter describes how adaptation has been integrated in a number of counties, cities and boroughs in the area. All in all, the study reveals a relatively developed approach to adaptation, made possible in part as a result of both the recognised sensitivity of selected areas to climate change and the centralised nature of the political system. Centralised as well as network capacities of the central government have made it possible for the national level to both include adaptation criteria in the performance assessment framework for local authorities, and for these to be broadly accepted among affected actors.

Keywords Adaptation · Adaptive capacity · Climate change · UK · South East England

3.1 Introduction

The UK has often been seen as a follower in European environmental policy (Börzel, 2002), as well as a relatively centralised and strongly regulatory state in the European context (Christensen and Laegreid, 2005; Jordan, 2002a, 2002b). However, the UK is also characterised by extensive privatisation and the use of

---

E.C.H. Keskitalo (✉)
Political Science/Department of Social and Economic Geography, Umeå University, Umeå 90187, Sweden
e-mail: Carina.Keskitalo@geography.umu.se

E.C.H. Keskitalo (ed.), *Developing Adaptation Policy and Practice in Europe: Multi-level Governance of Climate Change*, DOI 10.1007/978-90-481-9325-7_3,

new regulatory measures. Public-private partnerships have been advocated by the UK government since the mid-1980s, with the Blair Labour Government from 1997 explicitly promoting partnerships as a 'third way' between hierarchical and market governance (Graziano, Ferrera, Vesan, Bassoli, & Sparano, 2007). Some agreement exists over the idea that the British state has gradually transformed from the Westminster model of a unitary state with a strong executive into what Rhodes calls a 'differentiated polity' with policy networks as a characteristic feature, or even a 'disUnited Kingdom' (Rhodes, 2007).

The UK can also be considered one of the leading industrialised nations in adapting to climate change. Climate projections for the UK indicate higher year-round temperatures, increased winter precipitation and reduced summer precipitation, with greater seasonal variation and an increase in average annual temperatures of 2–4.5°C by 2080. It is predicted that weather extremes will increase, both in frequency and magnitude, while sea levels rise will contribute to adaptation needs for coastal areas, especially in the southern UK (UKCIP, 2009; Woking Borough Council, 2008; cf. Wade, Hossell, Hough, & Fenn, 1999). Several adaptation measures have been implemented at the national scale, including a Climate Change Bill and its subsequent Act, as well as a dedicated UK Climate Impacts Programme (UKCIP), created in 1997. The focus on adaptation has recently increased at the state level as mitigation and adaptation have been separated into different departments. Most recently, the implementation of a new performance assessment system, including the 'planning to adapt' to climate change indicator, has provided requirements for local level adaptation to flooding and other climate-related events and phenomena. As one interviewee noted:

> The government had declared not that long ago that they saw that there were three issues that they got to address nationally: the one was terrorism, the other one was pandemics, and the ... [third] one was flooding. And I think it took everybody by surprise when flooding was put at the same level as those ... threats ... I think that reflects ... how serious the governments – it is governments, not this particular government – have listened to the information that has been coming in. (Environment Agency, interview)

This chapter describes adaptation to climate change at the national, regional and local levels in the UK, based in a focus on the present extent of policy development. The geographical focus is placed on the South East region of England (see Fig. 3.1), which is both the second-most densely populated area of the UK (after London) and a low-lying area vulnerable to climate-related phenomena. Despite major flooding events in the area, the region faces considerable development pressure, one goal being to construct 1,098,000 new residences by 2016 (Howe and White, 2004; DEFRA, interview). The South East was also among the first regions in the country to develop a Regional Climate Change Partnership, which includes not only local authorities, but industries, research interests and NGOs in the South East (Climate South East, 2009).

Examples at the local level in this study are drawn from different types of local authorities, which each cooperates with a broader array of stakeholders in so-called Local Strategic Partnerships (LSP). The chapter includes Hampshire County Council, one of the early actors in climate change adaptation, as well as its lower-tier

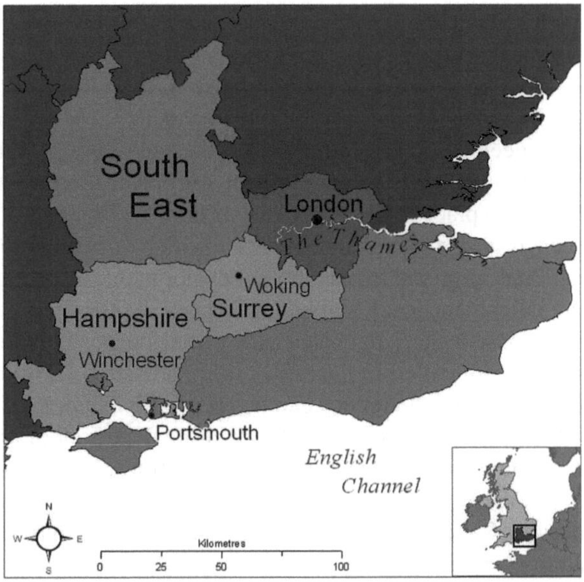

Fig. 3.1 Map of the South-East England case study region and the selected case study local authorities in the region

local authority, Winchester City Council. The chapter also reviews the unitary authority of Portsmouth City Council, England's only island local authority and an area highly vulnerable to a rise in sea level. Finally, the study includes Surrey County Council and its lower-tier local authority, Woking Borough Council, of which the latter has an extensively developed environmental policy profile. (For more detailed information on the methodology and common theoretical framework of the study, see Chapter 1 of this volume).

The chapter is structured as follows: first, environmental policy, as well as national regulation and regulatory restructuring relevant to adaptation, are described. Measures and actions at the regional level, including state actions at the regional scale with a focus on the South East are then presented, followed by actions taken by local authorities, including both state regulation that plays out at the local level and specific actions taken by the select local authorities. Finally, the relevance of the EU level, aspects of integration between scales, and the issue of transferability of 'good practices' are discussed.

3.2 National Level

3.2.1 Environmental Policy in the UK and the Establishment of Adaptation as a Priority

In the UK, the EU level has played a significant role in environmental policy. Jordan (2002a, 2002b) notes that while the UK has played a limited and even 'back-marker' (2002a, p. 47) role with regard to environmental policy in the EU, EU environmental requirements have come to play an increasingly greater role in the UK. Environmental policy has also been relatively well-integrated in the organisational change that has taken place in the UK (Jordan, 2002a), particularly since the Labour Government took power in 1997. Many environmental responsibilities have been delegated to the European scale through European environmental action programmes and the adoption of European directives. This shift in responsibility has taken place at the same time as the development of new parastatal organisations such as the Environmental Agency, which implements national regulation at the sub-national level and assists in the development of local environmental action plans at the catchment level (Gibbs and Jonas, 2000). Since 1988, a total of over 100 so-called 'Next Step' agencies and special-purpose bodies have been developed in the UK, accompanied by increases in privatisation (Smith, 2000). The development of the Environment Agency can be seen as an example of this agency structure (Gains, 2003) with its express aim of separating the development of policy frameworks from operational management (McLean, Clifford, & McMillan, 2000; Hogwood, Judge, & McVicar, 2000). Agencies are now assessed by using such techniques as 'bench-marking' scores, a mixture of process, output and outcome results, and financial audits (James, 2001).

Environmental policy has also been integrated into the UK performance assessment and audit system. In 1997, the UK Parliament established the bipartisan Environmental Audit Committee to audit government departments and public bodies to ascertain the extent to which their policies and programmes contribute to environmental protection and sustainable development (Darkin, 2006). The Committee can formally hold the Government to account, and may call individual ministers and officials to report on their activities. However, while the creation of such a body indicated the strong environment focus of Blair's Labour Government, no guarantee exists that Committee's recommendations will be adopted (Darkin, 2006). Performance assessment has also been more broadly adopted in the UK through a focus on the assessment of local government performance on a periodic basis using nationally determined criteria (discussed under 3.2.2.2).

On department level, the organisation of environmental policy has also changed over time, reflecting an attempt to integrate environmental issues across sectoral boundaries. The Blair Government 'took the decision that environmental protection could no longer be the sole responsibility of any one department of state' (Darkin, 2006, p. 258). Sustainable development was deemed a goal for all departments, a decision that was followed by the establishment of the 'super-ministry' of the Department for Environment, Transport and the Regions (DETR), a Sustainable Development Unit, and the Environmental Audit Committee. Following the perceived failure of the Ministry of Agriculture, Fisheries and Food (MAFF) to manage foot-and-mouth disease, DETR was merged with MAFF in 2001 to form the Department of Environment, Food and Rural Affairs (DEFRA), while transport and regional issues were transferred to other ministries (Darkin, 2006).

The first comprehensive Climate Change Programme for the UK was developed during this time (DETR, 2000) and illustrates the integration between industry and policy in the UK system that has formed the basis for the current partnership approach (Hulme and Turnpenny, 2004). The programme includes a focus on good practice, to some extent echoing suggestions made by Lord Marshall, former President of the Confederation of British Industries, in a report published in 1998 (Darkin, 2006; Bulkeley, 2006). However, it may also demonstrate that industry was seen as a sector that, following conflicts over domestic fuel taxes, was easier to regulate (Darkin, 2006). A second Climate Change Programme was developed in 2006 (DEFRA, 2006a). Despite its acknowledged status as a slow mover in environmental policy in the EU, the UK has moved forward on climate change (especially mitigation) during periods such as its EU presidency in 2006 (Lorenzoni, Nicolson-Cole, & Withmarsh, 2007). The UK has been on schedule for achieving the 12.5% reduction in greenhouse gas emissions by 2010 outlined under the Kyoto Protocol, and has even added more ambitious targets of –20% by 2010 and –60% by 2050 (Lorenzoni et al., 2007), illustrating the UK's progress from a laggard in EU environmental policy to a mover in climate change policy.

With regard to adaptation, significant progress was made in the 2006 Climate Change Programme which, in accordance with DEFRA's five-year strategy from 2004, also advertised the development of a climate change Adaptation Policy Framework to develop responsibilities by sector (DEFRA, 2006a). Explicitly multi-level, the Adaptation Policy Framework is intended to 'provide the structure in which adaptation strategies can be integrated into policies developed by organisations at every level of decision making' (DEFRA, 2006a, p. 132). Developed through consultations and other processes between 2006 and 2008 (e.g. Secretary of State for the Environment, Food and Rural Affairs, 2006, 2007), movement on the Framework was subsequently channelled into the development of the Climate Change Act (described below). Reflecting the strategic focus on adaptation at DEFRA, also the Environment Agency has had a Climate Change Adaptation Strategy since 2005, which aims to embed climate change into 'core business planning' with regard to the environment and environmental monitoring (DEFRA, 2008).

Issues of water and flooding, the particular focuses of this chapter, are further well institutionalised. Flooding has historically been a major concern in the UK, with measures based in legislation on drainage in the 1930s and a coastal protection act in 1949. Two reports in particular have focused national attention on flooding in the present: the Bye Report, commissioned by the government following fluvial flooding in 1998 (Bye & Horner, 1998), and the *Making Space for Water* report and consultation procedure initiated in 2004 (cf. DEFRA, 2004). Following floods in the autumn of 2000, the government also commissioned civil engineers to produce a report entitled *Learning to Live With Rivers* (in which climate change was included to some extent, cf. ICE, 2001). The commissioning of reports was considered by one interviewee to be a common reaction to such events and phenomena in the UK, and a way to incorporate lessons into policy:

> After ... major flood events, the government's natural reaction is to commission a report that really stands back from it all and tries to provide an honest overview of what could be done, what lessons really should be learned and what we should do differently. (Environment Agency, interview)

More recently, the final report from the *Pitt Review* led by Sir Michael Pitt on lessons from the 2007 flood was developed. This report advocated relatively far-ranging policy changes, including the strengthening of Planning Policy Statement (PPS) 25 on flood risk planning and the enhancement of the role of local authorities. The report also underlined the need for DEFRA to work with the Environment Agency, Natural England and partners to establish a programme for flood prevention through Catchment Flood Management Plans and Shoreline Management Plans (Cabinet Office, 2007).

Beyond the policy impact of flood events, many interviewees also noted that other past and current events such as droughts, as well as strong media coverage and relevant reports, have supported the establishment of adaptation as an issue in the South East and throughout the UK. In addition to the major floods in 2007, these events include extreme weather phenomena such as the heat waves in 2003 and 2006 that led to the generation of a heat wave plan, and the Stern report on the economic consequences of climate change that led to climate increasingly being perceived as a relevant issue (Stern, 2006). Interviewees further noted that the Fourth IPCC Assessment Report, a television series by David Attenborough, and Al Gore's *An Inconvenient Truth* have supported the development of climate change adaptation in the UK. Significant media attention to climate change and the integration of adaptation into legislation and the performance assessment framework are also noted as having quelled controversy over the existence and relevance of climate change. As one interviewee described, '[t]he IPCC and the Stern report have given us the kind of [leverage that we can] now ... move away from that debate [on climate science] and focus on the policy that is needed to support it' (LGA, interview).

The same interviewee at the local Government Association noted that the ability to develop policy on adaptation could be seen as a result of the confluence of several factors:

> At the same time as us focusing on adaptation, government was realigning itself, taking on the Stern argument, taking on the flooding, the Pitt report and so on ... it was a happy, perfect storm in that way, that these elements met. (LGA, interview)

Among the factors that potentially influenced recent adaptation policy development are changes in the governmental structure for dealing with climate change. In 2008, the responsibilities for different climate change issues were shifted as the Department for Energy and Climate Change (DECC) was created in order to combine issues of energy and climate change mitigation (the latter previously housed within DEFRA). The existing cross-government programme on adaptation (Adapting to Climate Change, ACC) was maintained at DEFRA (Shaw, 2008). This shift resulted in both the creation of a new lead minister for climate change adaptation (Shaw, 2008) and a change in focus on adaptation at DEFRA:

> To lose ... the mitigation bit came as a real shock to the departmental ministers ... so I think in the end it probably was a very political decision to keep [adaptation] here ... we [who work in adaptation] are still at the stage that we need to keep some distance with mitigation otherwise we would just get swallowed up, so in that sense, us being here is enabling us to go right to the top of the ministerial agenda. (DEFRA, interview)

DEFRA is now responsible for the majority of the government's climate change adaptation actions, and at a level '[several] civil servant grades higher than the [previous] evidence team' (UKCIP, interview), thereby increasing the departmental focus on adaptation.

3.2.2 National Legislative, Policy and Planning Frameworks on Adaptation

This and the following sections describe the national framework for climate change adaptation, which includes the Climate Change Act, the performance assessment framework and indicator on adaptation, implications of the planning system for adaptation activities, and the role and development of the UK Climate Impacts Programme. These different forms of national initiatives are summarised in Table 3.1.

3.2.2.1 The Climate Change Act

In March 2007, the UK government presented the *Climate Change Act*, an extensive document on climate change (including adaptation) that came into force in November 2008 (UK Government, 2008, cf. UK Government, 2007). The Climate Change Act requires the UK Government to produce five-year reports on adaptation in the UK beginning in 2012. The Act also requires the development of a national climate change risk assessment and a cost-benefit analysis, as well as statutory guidance for public bodies and statutory undertakers (private sector organisations that have a quasi-public sector role and are critical to national infrastructure, e.g., water

Table 3.1 National adaptation policy initiatives in the UK

Types of national adaptation initiatives	Initiatives	Year
Framework (strategies and framework legislation)	Climate Bill and Act	2007–2008
	Performance Assessment including NI188	2008/2009
	ACC Programme (England)	2008
Planning system	PPS1 (climate change included)	2005
	PPG25 (flooding)	2001
	Building standards, Treasury Green Book procurement standards	2009
Awareness-raising and stakeholder integration measures	UKCIP	1997
	Regional climate change partnerships	1999–onwards
	LRAP	2009

companies) (DEFRA, 2008). Government must also publish and update a national adaptation programme, the first of which is expected in 2012.

Among other things, this programme will draw upon work completed under DEFRA's *Adapting to Climate Change (ACC) Programme* established in November 2007. The ACC programme launched an Adaptation Policy Framework in 2008 (following recommendations in the Climate Strategy, 2006, cf. DEFRA, 2006a), focused on England but with some UK-wide elements. The programme is to be undertaken in two stages. Phase One (2008–2011) aims to gather evidence, raise awareness and embed adaptation priorities across government (among other things, into economic and performance assessment). One outcome is that guidance was published in June 2009 as part of the Treasury Green Book, the central guidance on spending and investment across government, to further embed climate change adaptation into national decision-making (DEFRA, 2008; DEFRA, personal communication). Phase Two (2012–onwards) is intended to produce the statutory National Adaptation Programme, as required by the Climate Change Act (DEFRA, 2008).

The Act also authorises the government to require assessments of climate change risks and responses from public authorities. It authorises the Secretary of State to ask any public-sector body or statutory undertaker to report their assessment of climate change risks and their responses, and the delivery of their objectives in a publishable form. Further, the Act establishes an independent Committee on Climate Change and an Adaptation Sub-Committee to oversee progress on climate change adaptation and provide advice on risk assessment (DEFRA, 2008).

The Climate Change Act thus establishes a multi-level framework for national, regional and local action on climate change, and establishes adaptation as a crucial part of acting on climate change. The Act is seen as having 'changed the ground laws in the UK significantly' (UKIRCCG, interview), moving the argument from whether climate change is of anthropogenic origin, to 'a target which everyone has to stand by' (LGA, interview). This is to some extent seen as a rather rapid transformation and establishment of the issue:

> At the government level, they didn't even recognise the importance of adaptation for several years. Maybe two years ago they started taking it seriously and now gradually we're finding ministers have got adaptation responsibilities. (SEEDA, interview)

Some interviewees described the inclusion of adaptation in the Climate Change Act (as well as in the performance assessment framework discussed below) as the result of lobbying by such actors as DEFRA, the Environment Agency, and UKCIP. While the first draft of the Climate Change Bill focused on mitigation, public consultation and actors in the Environment Agency (particularly the chief executive, who is also in the House of Lords) contributed to the inclusion of adaptation and a national risk assessment as well as to expanding the remit of the Climate Change Committee to cover adaptation (UKCIP and Environment Agency, interviews). Similarly, the UKCIP exercised its influence through both public consultation and through its relationship with DEFRA (UKCIP, interview). Political decision-making on adaptation may thus to some extent be seen as supported by existing institutional structures for the issue, such as the UKCIP.

At the time of this study, a number of decisions were still to be taken regarding the nature of implementation of the Act; for example, the requirement to report to the Secretary of State is expected to be decided (DEFRA, interview). Additionally, the role of the Adaptation Sub-Committee will be to provide advice and work as a committee for the scrutiny of government (DEFRA, interview), but its initial programme is still being developed. The delivery of five-year national risk assessments and cost-benefit analyses is also currently being developed. A scoping study commissioned by DEFRA on the structure and content of the risk assessments has raised the possibility of the use of departmental and agency risk assessments as well as of the power for the Secretary of State to require reports on vulnerability and adaptation. Given the need for government to provide guidance on these reports and the potentially sensitive nature of private company data, the number and nature of organisations to be asked to perform that task are also under discussion following a consultation launched in 2009. The ACC team has recently secured ministerial agreement for all government departments to produce adaptation plans covering the entirety of government policy areas, to be published during 2010 (DEFRA, personal communication). Finally, the relationship between the UK Climate Impacts Programme (responsible for developing national climate scenarios) and the Adaptation Sub-Committee is also to be clarified (UKCIP, interview).

To support work on adaptation, DEFRA has also been part of developing a new Partnership Board with key stakeholders to advise central government on the ACC programme and general stakeholder engagement (DEFRA, 2008). The ACC team itself has set up adaptation programme boards, including one for the local level through the Local and Regional Adaptation Partnership (LRAP) and one for domestic adaptation across the government. The latter board consists of senior civil servants from all departments and is designed to assess risks, actions and adaptation plans in each department to work towards a coherent national adaptation strategy (UKCIP and DEFRA, interviews). By contrast, the LRAP consists of all national agencies and stakeholder groups with high interest in adaptation steering and is chaired by the Government Office for London (Government Offices exist in all English regions and are the regional arm of the state) (LGA, interview). The LRAP is awarded roughly £150,000 per year for adaptation work, including the development of local performance indicators through regional workshops and an eventual national conference (GOL/LRAP and DEFRA, interviews). The production of a best practice guide by the ACC team is also under consideration (GOSE, interview).

3.2.2.2 Local Performance Assessment

In addition to the legislative changes at government level, the newly created national performance assessment framework for local authorities has important implications for adaptation. This assessment framework is a reworking of a more extensive and earlier version and forms a part of the government's policy framework for assessing local performance outlined to 2011 in the 2008/2009 Local Government White Paper (UK Government, 2006; DEFRA, 2008). Used to assess the performance of local authorities and determine funding distribution, the performance assessment framework constitutes a very important multi-level steering mechanism. The current Local Government Performance framework includes 198 National Indicators (NI), from which local authorities may choose up to 35 on which to be assessed and allocated funding. Implementation of the indicators is largely undertaken through Local Area Agreements (LAA) that set the targets for participatory Local Strategic Partnerships (LSP) between local government and other organisations. The implementation is monitored by the national Audit Commission through local reports and interviews. Assessments may also occur through the Government Offices and their work with local authorities, as well as the report of best practice activities (GOSE, interview).

Of the 198 indicators, NI 185 and 186 target emission reduction measures, NI 189 addresses flood risk, and NI 188 targets 'planning to adapt'.[1] The latter includes targets that move from the development of an initial baseline awareness of risks (level 0) to the implementation of adaptation measures (level 4), allowing each local authority to set its own goal for the four-year timeframe of the assessment framework (see Box 3.1) (DEFRA, 2009; LRAP, 2009).

Box 3.1 The progress levels in the NI 188 indicator with technical definitions of each level

(reproduced with permission from LRAP 2009, Version 1.7: 15 April 2009)

Level 0: Getting Started

The Authority has begun the process of assessing the potential threats and opportunities across its estate and services (for example, flood and coastal resilience plans, emergency planning, community risk registers/strategies etc) and has identified and agreed the next steps to build on that assessment in a systematic and coordinated way.

Level 1: Public commitment and prioritised risk-based assessment

The Authority has made a public commitment to identify and manage climate related risk. It has undertaken a local risk-based assessment of significant vulnerabilities and opportunities to weather and climate, both now and in the

[1]In order to target additional areas of adaptation, the Adapting to Climate Change (ACC) programme aims to develop a suite of national indicators on such issues as integrating adaptation into planning, and 'outcome-focused measurements' of awareness-raising, capacity-building, and policy and practice changes to be developed during 2010 (DEFRA 2008; DEFRA, personal communication).

future. It can demonstrate a sound understanding of those not yet addressed in existing strategies and actions (e.g., in land use planning documents, service delivery plans, flood and coastal resilience plans, emergency planning, community risk registers/strategies, etc). It has communicated these potential vulnerabilities and opportunities to department/service heads and other local partners and has set out the next steps in addressing them.

Level 2: Comprehensive risk-based assessment and prioritised action in some areas

The Authority has undertaken a comprehensive risk based assessment of vulnerabilities to weather and climate, both now and in the future, and has identified priority risks for its services. It has identified the most effective adaptive responses and has started incorporating these in council strategies, plans, partnerships and operations (such as planning, flood management, economic development, social care, services for children, transport etc). It has begun implementing appropriate adaptive responses in some priority areas. In its role as a community leader the council has started working with its LSP encouraging identification of major weather and climate vulnerabilities and opportunities that affect the delivery of the LSP's objectives.

Level 3: Comprehensive action plan and prioritised action in all priority areas

The Authority has embedded climate impacts and risks across council decision making. It has developed a comprehensive adaptation action plan to deliver the necessary steps to achieve the existing objectives set out in council strategies, plans, investment decisions and partnership arrangements in light of projected climate change and is implementing appropriate adaptive responses in all priority areas. This includes leadership and support for LSP in taking a risk based approach to managing major weather and climate vulnerabilities/opportunities across the wider local authority area.

Level 4: Implementation, monitoring and continuous review

Authority and LSP are implementing the comprehensive adaptation action plan across the local authority area, and there is a robust process for regular and continual monitoring and review to ensure progress with each measure and updating of objectives. The Authority and LSP are taking appropriate adaptive responses.

The integration of adaptation into the assessment framework was described as an expression of the perceived weight and importance of adaptation, and a move that went hand in hand with the development of the Climate Change Act. Interviewees in general viewed the process-based nature of NI 188 (as opposed to outcome-based) as relevant to the issue of adaptation. Most interviewees at different levels, including the Local Government Association (LGA), also responded positively to the simplified structure of the performance assessment new framework and the participatory nature of its development. As described by one interviewee at DEFRA, the new performance assessment framework 'has a much greater emphasis on what it's like to live [in a local community] and a more intuitive, area-based approach'.

These characteristics to some extent differ from previous frameworks and may place new demands on for instance the Audit Commission: 'it's quite a culture change for them ... it's a new mindset for local authorities as well' (DEFRA, interview). Despite an established UK focus on partnerships, interviewees noted that relationships are emphasised to an even greater extent in the new framework:

> [It] very much pushes towards not looking at the organisation of the local authority per se as a single ... organisation, but actually broadening that out and saying, "What is your relationship with the local authorities around you?" "What is your relationship with the people like health and emergency services in your area?" (DEFRA, interview)

However, some have raised concerns regarding the large role that unelected organisations may take in the LSP: 'It's quite hard because ... councils are bound only to deal with organisations that have some sort of democracy, and a lot of wildlife groups ... don't' (UKCIP, interview). Interviewees also noted that since it is a new and process-focused development, the adaptation indicator NI 188 will require additional guidance for both local authorities and other bodies: 'Over the last couple of months, what we have been doing is really debating across the national partnership how a well-adapting local authority partnership can sort of move through those levels on a provision' (LGA, interview). As a result, DEFRA is currently considering how to continue work on adaptation beyond process indicators, in preparation for the evaluation of the period utilising the 'preparing to adapt' indicator (DEFRA, interview).

3.2.2.3 Adaptation and the Planning System

Finally, the UK planning system plays a vital role in the integration of adaptation into activities at different scales, particularly as the requirements for development and protection are directly integrated in planning decisions. Howe and White (2004) note that '[t]he dual role of the planning system in reducing flood risk whilst providing housing land highlights a significant conflict in the current planning system' (Howe and White, 2004, p. 416). They further note that two of the most severe episodes of recorded flooding in 1998 and 2000 have led to critiques of the planning system. As a result, the Planning Policy Guidance 25 on flooding was introduced in 2001, taking a 'precautionary approach to managing development and flood risk ... to direct new development away from areas at highest risk of flooding and take ... account of climate change' (European Environment Agency, 2005, p. 37). The Planning Policy Statements (PPS) and their clarification in Planning Policy Guidance (PPG) constitute the main steering documents for government agencies and bodies in the planning system. Government actions of direct relevance to adaptation issues also include the development of a cross-government programme on flood and coastal erosion risk management (*Making Space for Water*), a heat wave plan, and a supplement to Planning Policy Statement 1 (PPS1) on Planning and

Climate Change (DEFRA, 2008; cf. Wilson, 2006).[2] In 2004, a practice advice document was also published by the Office of the Deputy Prime Minister, and in 2005 a reference to climate change as a material consideration in planning decisions was included in PPS1 (Wilson, 2006).[3]

In addition to these changes, much of the planning system is currently being revised to include a greater emphasis on foresight planning. An interviewee at DEFRA noted that the government is trying to change the general 'rules about how people invest ... [such as] building standards' and 'the planning system ... so that there's an obligation on anybody taking a planning decision ... to take account of changing climate' (DEFRA, interview).[4] PPS were also being revised in preparation for public consultation from 2009: 'we've been really successful in getting adaptation factors into ... planning at the local level and also at the national' (DEFRA, interview).

With regard to water issues, the Environment Agency (funded by DEFRA) is relevant as the main state implementing body with regard to both environmental policy and flood risk management. Organised both at the national level as well as by regional river catchment area, the Agency has increasingly come to define climate change (and its relationship with flood risk in particular) as a priority (Environment Agency, interview). Since its reorganisation in 2001, a climate change adaptation board and a climate change plan have been created to assist in the integration of climate change throughout the organisation's activities. At the same time, a part-time position in climate change was also created in each region (Environment Agency, interview). Currently, a national action plan is under development through which adaptation to climate change will be integrated into each of the Environment Agency's functions, as well as into national and regional strategic plans. The current corporate strategy, *Creating a Better Place*, includes nine environmental themes, of which two focus on limiting and adapting to climate change and reducing flood risk (Environment Agency, interview; cf. Environment Agency, 2006). Regional studies in the Thames and South Environment Agency (covering most of the South East England region) have also supported the development of national priorities, as

---

[2]This follows developments in 2004, when the UK government further published advice to planning authorities on ways for the planning system to respond to climate change. This advice was commissioned by the government in 2000, and resulted in a study including a focus on both mitigation and adaptation (with particular emphasis on adaptation), conducting interviews with members in climate change science and policy in the UK and abroad, and reviews of planning policy: Regional Planning Guidance (RPG) as well as Regional Sustainable Development Frameworks (RSDF) (Wilson 2006).

[3]Wilson (2006) noted that the publication of policy advice on climate change adaptation only took place in 2004 as a result of the initial aim to publish guidance for all of the UK, which turned out to be 'at odds with an increasingly differentiated system' (p. 13) where planning policy varied between administrations in different parts of the UK and policies on each detailed adaptation were too complex to deal with given the varying administrations.

[4]Interviewees noted, however, that increased regulation on materials used for building and to increase drainage could be developed on the state level; a local authority noted that without such legislation, they could only advise developers on what materials to use and how to avoid paving over gardens that may provide drainage (Portsmouth CC, interview).

regional priorities for this vulnerable and flood-prone region were developed before national strategies (Environment Agency, interview).

The Environment Agency is also currently reviewing the flood and shoreline management system with some relevance for integration of adaptation aims. At present, Catchment Flood Management Plans (CFMP) and Shoreline Management Plans are under development in cooperation with local authorities. Through the review and a new structural arrangement of co-ownership, local authorities will receive support in developing measures under national indicators, while the Environment Agency will gain a better overview of coastal defence than under the previous arrangement. An additional aim is to establish coordination between shoreline management plans and catchment management plans, as well as to increasingly involve the Environment Agency in the design of surface water programmes for water companies (Environment Agency, interview).[5] Together with flood defence committees (selected by the Secretary of State, the Environment Agency and local authorities), the Environment Agency also approves flood risk management plans (DEFRA, 2009b). The Environment Agency may also provide advice to local authorities on the development of potentially hazardous sites; if such advice is not followed, the Secretary of State may choose to hinder development, after which the issue goes to the relevant Government Office for review (Environment Agency, interview). The Environment Agency therefore plays an important role in reviewing local planning in relation to flood risk.

3.2.3 Creating Multi-Level Linkages: The UK Climate Impacts Programme (UKCIP)

One of the most focused features of the organisation on adaptation in the UK has been the Climate Impacts Programme (UKCIP). Set up in 1997 as a major government-funded organisation housed at Oxford University, UKCIP has developed user interfaces for climate change scenarios, developed decision-support tools and supported the development of Regional Climate Change Partnerships (RCCP) for the English regions, including Greater London. These partnerships have included businesses, corporations and local communities in the development of scoping studies and regional coordination and project development, and have resulted in the creation of autonomous regional networks in select regions (West and Gawith,

[5]Privatised water companies were created in the late 1980s, and are to a large extent controlled by the state with regard to the quality and provision of water. At their presentation at the 2007 South East England Regional Assembly Climate Change Summit, Southern Water noted that climate change will reduce river flows, increase the frequency of hot dry summers, and influence future resource schemes (Southern Water, 2007). Suggested adaptation strategies included moving away from single source dominant supply areas, looking for solutions that would make the most of opportunities afforded by climate change such as winter water, and looking for low-energy solutions (Southern Water, 2007). Unfortunately, Southern Water was not available for interview during the time of the study due to the revision of their water business plan.

2005). An informal UK Interregional Climate Change Group (UKIRCCG) has also developed out of the cooperation between UKCIP regional partnerships to support information-sharing between the partnerships, and is currently given some secretarial support from the UKCIP. As such, the UKCIP has directly supported climate change work on different levels, focusing to some extent on regions and localities. For instance:

> Without the UKCIP, I think it would have been extremely difficult, if not impossible, for regions to really take this forward. Very few regions would have the individuals with the necessary scientific understanding, the ability to interpret what the scientists are saying and taking that down to a local level, the ability to make the connections between different actors. (UKIRCCG, interview)

From its inception, however, the UKCIP was not intended to have such a strong focus on the regional and local levels. The role of the UKCIP has changed over time, from an original focus on integrated assessment of climate change impacts to an increasing focus on stakeholder support and adaptation (UKCIP, interview). The UKCIP's origins lie in work performed starting in 1996 by an expert Climate Change Impacts Review Group (CCIRG), commissioned by the Department of Environment to undertake national assessments of the impacts of climate change (UKCIP, interview). Observations that very little research produced until that point was integrated or directly comparable (for instance, between different time periods), or readily accessible to policy-makers, were among the factors behind the UKCIP's creation. It is also in part attributed to work by David Warrilow, a member of DEFRA's now obsolete Global Atmosphere Division who emphasised the need for a focus on research that met stakeholder needs. The group was particularly inspired by the MacKenzie River impact assessment in northern Canada (cf. Cohen, 1997), which focused on stakeholder engagement and the need to design participation approaches for a wide range of stakeholders (UKCIP, interview).

Tasked with developing stakeholder engagement to guide climate change research, the UKCIP began to support regional climate change partnerships in the development of impacts studies (sometimes commissioned by the partnerships for delivery by independent consultants). Here, the focus on the regional level was supported by the Department for the Environment (now DEFRA), which recognised the regions' manageable size of administration to work with climate change and the relevance of regional decision-makers to climate change. To some extent, the decision to focus on the regions was also supported by the fact that regionalisation was high on the political agenda when the Labour Government came into power in 1997 (UKCIP, interview). Interviewees emphasised, however, the non-politicised nature of the climate change issue with respect to UKCIP: 'The conservatives launched the UKCIP and Labour sort of ran with it when they got elected' (GOL/LRAP, interview).

While initially the focus was placed on assessing regional climate change impacts, the recognition of the difficulty of maintaining stakeholder cooperation in research prompted the transition to an emphasis on adaptation. This development was taking place both within the regions as well as in the UKCIP:

> We were faced with a number of regions where the partnership was coming to pieces because they didn't see a role for themselves; because they had been brought together to steer that scoping study . . . that research description task. So we said, "We think you should continue to look at adaptation and this is actually an additional goal for the programme" . . . We developed it from 2003 to 2004 and the people in DEFRA who were paying for us began to recognise the same thing. So it wasn't a conscious change, it was just we realised it was something we needed to do because [that was what] the stakeholders were saying. (UKCIP, interview)

As a result, negotiations for a new contract between the UKCIP and the government in 2003 led to a shift in aim, from the national assessment of impacts to the provision of stakeholder support to better understand impacts and adapt (UKCIP, interview).[6]

At the same time, the importance of the intersection between national- and local-level actions was being increasingly emphasised through other ongoing initiatives that could also support lower-level and cross-scale approaches on adaptation. One such initiative was the Nottingham Declaration Partnership on voluntary emission reduction, developed in the late 1990s by an officer at Nottingham City Council in preparation for a climate change conference:

> We wanted to make a statement [with the Nottingham Declaration] from the viewpoint of local government, because government had not recognised at that time . . . that those municipalities could have a role in . . . carbon reduction. They did not formally recognise that role until publishing the current UK climate change action plan or programme in 2006, and certainly in 2001 there was no recognition by government that municipalities had a positive role to play. (IDeA/Nottingham Declaration, interview)

This initiative came to support the development of successful forums for stakeholder engagement by the UKCIP:

> We realised that there was this thing, Nottingham Declaration . . . they wanted to re-launch . . . in 2005 . . . We worked with Local Government Association, the Development Agency and the Carbon Trust and Energy Savings Trust to re-launch the Nottingham Declaration with adaptation [which UKCIP helped write] as well as mitigation and with the promise that we will provide some tools to help to. . . actually deliver. (UKCIP, interview)

The resultant informal Nottingham Declaration is considered to have supported the development of the NI188 indicator, and is expected to support the LRAP process in the future (UKCIP, interview). Guidance on implementation of the NI 188 indicator is developed to a large extent through different agencies and partnerships, including the UKCIP, the Improvement and Development Agency for local government (IDeA) and the members of the Nottingham Declaration partnership (GOSE, interview). For example, the Nottingham Declaration Action Pack launched in July 2006 provides a structure for local councils to work with different roles (e.g., estate manager, service provider, or community leader) and to prepare and implement

[6] Realising that it was not possible to produce a final report on a national assessment of impacts and that more data was needed from diverse groups, the UKCIP also proposed that it should produce a report of available data in addition to the results of adaptation work (subsequently published as the *Measuring Progress* report, West and Gawith, 2005).

action plans (IDeA, 2007; cf. www.nottinghamdeclaration.co.uk). The package also sets up a 'performance reporting framework', including indicators on CO_2 reduction, climate change adaptation and resilience to be assessed through the stages of the Nottingham Declaration process (IDeA, 2007).

The Nottingham Declaration network is also viewed as having added legitimacy to the process of developing the NI 188 adaptation indicator itself (UKCIP, interview). As noted succinctly by one interviewee, 'Government could never have set up something like that unless a bottom-up process had pre-prepared the regions and local authorities to accept it' (UKCIP, interview).

To further support adaptation at the local level, the UKCIP has also developed tools such as the Local Climate Impacts Profile (LCLIP), designed to assist local authorities to better understand climate change risks and possible adaptations (DEFRA, 2008). The LCLIP provides a format for the description of costs and resources for responding to extreme events by monitoring impacts of past events described in the local media, and is seen as one way to support local authorities in undertaking the initial stages of a vulnerability assessment for the NI 188 indicator.

Given the current reorganisation of climate change adaptation structures at the national level, changes to the UKCIP's role are ongoing. One UKCIP interviewee noted that the institutionalisation of the Climate Change Act and the increase of DEFRA interest in adaptation transformed 'what should have been a light touch review of our work programme [to] ... a complete re-design' (UKCIP, interview). In addition to the reassessment of UKCIP's role in the national Sub-Committee on Adaptation (noted above), the UKCIP is currently increasing its focus on the local level through the selection of a local authority in each of the nine English regions to support work on issues such as the implementation of NI 188 (DEFRA, interview).

3.3 Regional Level

In the UK, the regional level has often been seen as relatively weakly developed in that it is mainly a focus for implementation of national legislation and regulation (Sandford, 2005). Tax income generated within the regions is collected by central government and redistributed according to set principles and development aims. However, despite its limited independence, the region is an organisationally dense scale at which a number of implementing actions and negotiations between national and local priorities are undertaken. The region in focus here, South East England, is one of those most vulnerable to climate change, in particular with regard to flooding and sea level rise, and has already experienced several floods in the 2000s. The South East region is also the home of one of the largest planned residential infrastructure development in England, primarily located on low-lying land:

> Water stress is quite a big issue particularly in the South East, where a good part of the government's main building programme and development programme is all centred around the South East ... If you have a map of the UK's water stress, then an overlay of ... where we are building our next two million houses, they are the same areas exactly... It is a completely unanswered question, almost impossible to answer...what you do about new

Table 3.2 Climate change focus in regional bodies

Regional level bodies	Extent of climate change adaptation focus
Climate South East Regional Climate Change Partnership (RCCP)	Main focus on adaptation
Government Office South East (GOSE)	State implementation (e.g., adaptation concerns in PPS and PPG)
South East England Development Agency (SEEDA)	Only within sustainability focus
South East England Regional Agency (SEERA)	(Disbanded)
RIEP (Regional Improvement and Efficiency Partnership): Improvement and Efficiency South East	Best practice focus, limited role for adaptation

> infrastructure? ... An even more difficult one is what you do about coastal towns that already exist and have existed for hundreds of years and that [could require] significant, millions of pounds of, regeneration? (DEFRA, interview)

At the regional level, key actors in adaptation include the principal regional bodies: the Regional Development Agencies, the Regional Assembly, and the Government Office. Cooperation on adaptation between these bodies and DEFRA is currently under development through for instance contracts on the provision of cost-benefit and regional risk assessments for continued work and their integration into strategies at the regional level (DEFRA, interview). The region is also home to one of the first regional climate change partnerships, now called Climate South East.

The following sections describe the role and development of the South East Regional Climate Change Partnership, as well as the actions relevant to adaptation performed by each of the major regional bodies (summarised in Table 3.2).

3.3.1 Regional Climate Change Partnerships: Climate South East

Climate South East (recently renamed in 2007 from the former South East Climate Change Partnership) was one of the first regional climate change partnerships in the country. Developed in response to the identified need for stakeholder involvement in relation to the UKCIP, in 1999 the Partnership produced one of the earliest scoping study reports, the *Impacts of Climate Change in the South East in the 21st Century* (Wade et al., 1999), which focused on key sectors of agriculture, utilities and infrastructure, tourism, planning and emergency planning (South East Climate Change Partnership, 2002). The Climate South East Partnership was also relatively early in its development of an adaptation approach. According to an interviewee at the South East England Development Agency (SEEDA), 'the rationale for Climate South East was ... adaptation rather than mitigation ... the initial trust came from spatial planning'. The South East was also seen by an interviewee as one of the 'strongest and most robust of the partnerships ... also probably the least political ... because it was a partnership of lots of organisations', it 'wasn't dominated by one partnership, one organisation or another' (UKIRCCG, interview). Some interviewees attribute this balanced structure to the fact that Climate South East is the only existing fee-based partnership, which has resulted in an increase in available funding as well as some assurance of the commitment of members. Interviewees note, however, that such a structure also poses a risk to the partnership if large, high-paying members should cancel their membership (UKCIP, interview).

Today, Climate South East has a membership of over 60 fee-paying members (its participants include among others the UKCIP and the regional bodies) and a commitment from DEFRA for £140,000 for adaptation over the next three years. The partnership is organised into sub-groups that work largely independently on themes such as spatial planning, tourism, business and economy, and meet with the Climate South East executive committee twice a year (SEEDA, interview). The partnership has selected a president to gain a high public profile and thereby raise the profile of the work: 'In our experience, you need to have a champion somewhere, someone who leads the way' (SEEDA, interview).

A challenge for the Climate South East partnership and for the partnerships in general has been to make the transition between their original responsibilities for the production of a scoping study to support research to their present role as a coordinating body for climate change work in the region. While the adaptation focus and fee-based nature of the partnership may have made this transition easier in the South East than in other regions, difficulties in procuring funding for a coordinator have still posed challenges. As one interviewee described, the absence of any clear measurable outcome of coordination efforts clashes with the rules set by the Treasury that facilitate the allocation of funds to a defined research project instead of individual positions (UKIRCCG, interview). The long-term discussions of funding and the role of regional partnerships have, however, to some extent prompted cooperation among the regions. In 2002, the then-limited provision of government funds for regional climate change activities resulted in regional discussion on the similarities and requirements of different regions, and in the creation of the informal UK Interregional Climate Change Group (UKIRCCG) to exchange experiences. The so-called Three Regions partnership was similarly created following the realisation in UKIRCCG discussions that the South East, East and London have some similar problems (UKIRCCG and GOL/LRAP, interviews). Cooperation regarding tourism between the South East and the South West has also developed out of UKIRCCG meetings (UKIRCCG, interview).

However, as the Regional Climate Change Partnerships (RCCP) are currently seen as key actors in the Adapting to Climate Change Programme (ACC), the nine English regional partnerships were provided a total funding amount of £450,000 in 2008–2009 with further commitments for the following two-year period (DEFRA, 2008). The present availability of funding for regional climate change coordinators is attributed to changes in the political process and prioritisation within DEFRA:

> DEFRA's understanding about the value of the regions is greater than it used to be ... five years ago even, there was a bit of "them" and "us" ... the relationship now is very different and very positive and very supportive (UKIRCCG, interview).

To some extent, funding for regional coordinators may also be the result of the decision in 2009 to disband the Regional Assemblies and the subsequent need to fund regional partnerships (UKCIP, interview).

3.3.2 Regional Planning and Administration

At the regional level, three bodies have been important for economic development, planning, coordination and implementation, and best practice aims. The Regional Development Agencies (RDA, established 1999) focus on supporting regional economic development and among other things produce Regional Economic Strategies. The Regional Development Agencies Act (1998) defines the principal tasks of RDAs as promoting economic development, employment and competitiveness of business as well as contributing to the achievement of sustainable development (DEFRA, 2005a). An additional aim of the Act was to even out the performance of the regions, particularly to support the less economically advanced northern regions (SEEDA, interview).

In the Adapting to Climate Change Programme, the RDAs are described as 'key strategic regional bodies' (DEFRA, 2005a, p. 18). The role of RDAs, examples of climate change activities and RDA commitments are set out in the document *Tackling Climate Change in the Regions* (England's Regional Development Agencies, 2007), developed by the RDA themselves during 2007 (SEEDA, 2008). Examples of RDA work on climate change are currently being collected for DEFRA and the Local and Regional Adaptation Partnership Board (IDeA, 2008) to serve as examples for best practice sharing.

The regional RDA, the South East Regional Development Agency (SEEDA) notes, however, that it has so far had limited direct experience with adaptation, as most work on climate change has functionally been delivered through Climate South East. Practical linkages between the two groups have been established through the employment of an ex-chair of Climate South East within SEEDA's sustainable development team. SEEDA also funded a research project in 2003–2004 to update the original 1999 regional scoping study (SEEDA, interview). One interviewee noted that climate work had not always been easy to integrate with SEEDA's economic focus. While sustainable development is included as a goal of RDA work,[7] its development as a priority has been a long process. For instance:

> In 1999, the regional economic strategy didn't even mention the word environment in an original draft ... In 2002, things had moved on, a member of our board was very aware

[7] In addition, Business Link, which is funded by the government but run by the RDA, audits and advises businesses and has resources to work with business in a way that Climate South East does not (SEEDA, interview).

> about sustainability ... In 2006, it was a lot more prominent and ... the ecological footprint policy has been accepted as one of three over-arching key goals of the regional economic strategy. (SEEDA, interview)

Today, the South East Regional Economic Strategy includes policy for encouraging business to plan for climate change adaptation, implemented through support to Climate South East (in the form of office space and secretarial resources).

The second group of regional bodies, the Regional Assemblies (RA), were partnership bodies between regional and local stakeholders originally created as a potential step towards elected regional government in England (cf. Sandford, 2005). However, as this transformation has largely been seen as politically unviable, RAs are currently undergoing a process of dissolution, to be replaced by a new form of regional accountability body currently under development. Until early 2009, RAs acted as the regional statutory planning bodies, scrutinising the RDA and supporting their development of Regional Economic Strategies while promoting regional strategies such as the Regional Sustainable Development Frameworks (since 2000) and integrated policy (DEFRA, 2005b). RA responsibility for the interpretation of national priorities based on regional needs and opportunities and focus on the planning system has differentiated it from national monitoring: while RA indicators may be linked to national indicators, national indicators range across the full spectrum of local authority activities and not only planning (SEERA, interview).

Until it was disbanded in March 31, 2009, the South East England Regional Assembly (SEERA) worked largely as a regional partnership body focused on planning, thus adding to the otherwise state-based implementation structure. With regard to climate change, SEERA has been responsible for developing and consulting on policies for mitigation and adaptation, including those related to the design of new buildings and water efficiency standards (SEERA, interview). In addition, SEERA developed the *Climate Change and the South East Plan* in 2007 to guide regional and local planners, as well as a climate change mitigation and adaptation implementation plan and a relevant local development framework (SEERA, interview). Both the climate change plan and the mitigation and adaptation implementation plan were partly developed with national and international experts from Germany, the Netherlands and Belgium. This expertise was made available through the region's involvement in the EU *European Spatial Planning: Adapting to Climate Events* (ESPACE) project, developed from the outset within the South East Regional Climate Change Partnership and led by regional bodies and the Hampshire County Council. The project addressed adaptation measures in relation to pressure on water resources, flood risks, and other water-related climate change impacts (SEERA, 2007).[8]

---

[8]SEERA also held a Climate Change Summit in 2007, including participation from industry and NGOs (IDeA, 2007). Aside from explicit adaptation priorities, SEERA worked to promote sustainable development across the region through a framework of 24 objectives adapted from UK national sustainable development principles and based on indicators developed by the various

Regional structural reform and the disbanding of the RAs was noted by a number of interviewees in relation to the need to replace the oversight and democratic functions the organisations had provided:

> The regional assemblies were very much seen as a stepping stone on to elected regional government but that never happened ... a lot of the responsibilities of the regional assemblies in terms of town and regional planning are given to the RDA and then what do we do about this democratic deficit? (DEFRA, interview).

Similarly, the RA provision of oversight of regional development agencies may now be replaced by a regional select committee or by local leader forums consisting of local councillors (UKCIP and DEFRA, interviews).

With the RAs disbanded, the Government Offices (GO) remain the principal regional bodies. In English regions, GOs are part of the central government and produce such items as the Regional Sustainable Development Action Plan, and support the Regional Development Agencies and other stakeholders in implementing government policy (DEFRA, 2005a). While GOSE and SEERA have worked together on implementing and negotiating state policy, SEERA has maintained a focus on the implementation of climate adaptation and other measures within the planning system.[9] Work relevant to adaptation by the Government Office South East (GOSE) includes the goal of integrating adaptation to climate change in all regional strategies (i.e. the South East Plan, the Regional Sustainability Framework and the Regional Economic Strategy)[10] as well as integrating adaptation under the Public Service Agreements (PSA). These agreements describe the objectives for UK government departments over three-year periods (including PSA 27 on climate change and PSA 28 on environmental quality). An increasing focus on climate change adaptation in GOSE work is reflected in the establishment of positions with relevance to climate change, including the position of senior climate change policy advisor since January 2007 (GOSE, interview). However, climate change and adaptation in particular were relatively small aspects of GOSEs work and of the PSA (which are currently under revision) at the time of the study.

Finally, the regional partnership structure is currently being augmented through the establishment of Regional Improvement and Efficiency Partnerships (RIEP) in each English region. The RIEPs are funded by DEFRA and CLG/Communities and Local Government departments to a total of over £4 million over the next three years, with the aim to 'draw ... out best practice examples to share regionally and nationally' (GOSE, interview). Among other things, this may support the function of bodies such as the national Improvement and Development Agency for Local Government (IDeA). In the South East region, the RIEP Improvement and

---

regional partnership organisations. SEERA has conducted annual assessments of the implementation of the plan across the region and has worked to ensure consistency between local and regional plans; results are compiled in an annual monitoring report.

[9] Here, national priorities such as the PPS (e.g., Climate Policy Statement 1, published in December 2007, and PPS25 on risk) have guided the development of the South East Plan (SEERA, interview).

[10] Other regional strategies also exist, such as Integrated Regional Strategies or Integrated Regional Frameworks (DEFRA, 2006b).

Efficiency South East (IESE) receives £350,000 in funding for its work in ten different themes, of which sustainability includes some reference to climate change; however, the priorities under IESE are very broad and include housing, education, crime and health (GOSE, interview).

One challenge has thus been the translation of climate change adaptation as an issue into a central aspect of regional work (cf. SEEDA, interview). Climate change can be seen as integrated in particular within the RCCP, which includes membership from the regional bodies and a number of local authorities in the region. Adaptation aims mainstreamed in policy areas such as planning are also implemented as a part especially of GO responsibilities.

3.4 Local Level

3.4.1 Adaptation and the Structure of Local Government

The regulatory state described in New Public Management literature, with its focus on partnership development in relation to privatisation, performance assessment and output audit, is strongly manifested at the local level in the UK. The local level receives relatively strong steering from the state level, and can only undertake actions that are statutorily allowed (as opposed to local government arrangements in many other parts of Europe, which can within legal frameworks decide which actions are in the interest of the community, so called general competence) (cf. Bache and Olsson, 2001). Some loosening of these legal constraints took place through the 2000 Local Government Act, which introduced the 'power of well-being': while not a general competence, it enables local government to promote well-being in their areas rather than only focusing on existing service delivery (Wilson and Game, 2006). In addition, the majority of local funding is redistributed from central government, only a small percentage of the local funds collected being retained locally. The UK national government is thus able to steer development at the local level to a considerable extent using among other things local funding as a tool.

An extensive privatisation of services has also taken place. Through reforms since the 1970s, the private and voluntary sectors have become responsible for several services that formerly were the responsibility of local government, including some part of provision of education, transportation and housing. During the 1990s, a range of performance indicators was introduced in conjunction with the development of local partnerships between industry and other stakeholders (Bulkeley and Betsil, 2005). In 1993, the Local Governance Programme was launched in order to formally reorient local governments toward a partnership culture (Jones, 1998). Following the enactment of the Local Government Act in 2000, Local Strategic Partnerships (LSP) were promoted to prepare local strategies such as the Sustainable Community Strategies (Regalia, 2007). Several local partnerships were the result of 'a number of government bidding programmes which required bidders to show that they were working in partnership' (Regalia, 2007, p. 20). The partnership approach was thus established as a result of government policy and funding initiatives for local governments, and has since been encouraged especially in the case of new, experimental national UK programmes (Regalia, 2007).

The creation of partnerships and their assessment is enforced through the national performance assessment framework, which includes target-setting through the Local Area Agreements (LAA) and the National Indicators described in the previous sections, within which local adaptation to climate change is being implemented. The National Indicator framework described above is now ‘the single route through which central government sets priorities for local government’ (DEFRA, 2008, p. 40). National Indicators, including NI 188 for ‘preparing to adapt’, must be implemented in cooperation between the local authority and local stakeholders under the LAA. Given this structure, additional sectors (e.g., industry) are to some extent included in local decision-making.

Locally focussed bodies relevant to climate change include the Local Government Association (LGA), created to promote and lobby for better local government in England and Wales (DEFRA, 2005a). The LGA has been active on climate change adaptation, issuing a report on the role of local government in adaptation and mitigation through their Climate Change Commission in December 2007 (LGA, 2007), which emphasised the need to raise awareness of adaptation and build adaptive capacity among local governments. The LGA has also published documents on the need for local authorities to act on climate change through the provision of duty of care (LGA, 2008), and on different local adaptation strategies issued on the anniversary of the 2007 floods (LGA, 2008b). Guidance and policy has also been provided by the LGA together with national partners through the ‘Small Change, Big Difference’ campaign (LGA, interview; cf. LGA, 2009).

Both the LGA and local authorities express a favourable view of the performance indicator framework and of the way adaptation has been treated by the government and by DEFRA. Although the new adaptation indicator has yet to be assessed for outcome or process effects, nearly one-third of all English LAAs had included the new adaptation indicator or taken it on as a local target in 2008 (DEFRA, 2008). In part, the incorporation of adaptation into the local performance assessment framework occurred in response to earlier criticisms that without a specific target or indicator, climate change would not be considered a priority by local authorities (Demeritt and Langdon, 2004). Wilson (2006) considered that ‘for many Local Planning authorities across England, [climate change] would only gain such standard as a material consideration if either a specific PPG was devoted to the issue or its status as such was made clear’ (p. 18).

Through its nature as a process indicator and its emphasis on the development of risk awareness, NI 188 requires contextualisation and is thus seen as providing the local level with a greater role than indicators that set numerical targets for given items. This greater focus on process may to some extent derive from the local specificity of the adaptation issue, where adaptation needs are defined for a given local situation; however, the performance assessment framework also reflects what the LGA sees as an increased role for the local level. Noting that the previous comprehensive performance assessment was a ‘very inspector-led .... dry, singular process which didn’t necessarily reflect the size and differences that local variances had', an interviewee from the LGA described the new NI 188 as 'the renaissance of adaptation, which coincided with a new way of measuring performance and improvement in local government'. The same interviewee described the shift from a top-down approach to a sector-led approach that allows for the regional negotiation of targets in order to reflect community interests:

> I guess it's like a culture change, not just for government but also with the way that central treats regional and local. There has to be a realisation that it can't be managed from London or Westminster. There's got to be everyone's chips in at the same time ... it is about encouraging people to take that extra mile and actually own the process a bit more. (LGA, interview)

NI 188 may thus support the integration of adaptation at the local level within a broad array of partnerships and may also serve to integrate adaptation priorities across levels in the centralised UK system. One local authority further noted that the possibilities for local prioritisation of adaptation is providing additional support for performing well on the indicator:

> When I'm going to [council] members, talking about the importance of the work, instead of saying "it's one of the national indicators" we are saying, "it's a... local area agreement priority". Which gives it that extra little bit of an incentive to them working on it, I suppose. (Winchester CC, interview)

3.4.2 Actions at Selected Local Authorities in the South East

The local authorities targeted in this study exemplify a wide range of capacities. While the local authorities described in this section range from higher-tier authorities (County Councils) to lower-tier City or Borough councils lying under these or – in the case of Portsmouth – acting as independent authorities, it seems that the distinction between lower- and upper-tier authorities plays only some role in determining Council resources to climate change. Other factors that come into play include political leadership in the Council and institutionalised environmental aims. Some of these broadly expressed strengths and limitations in terms of adaptive capacity are summarised in Table 3.3. This section describes the development of vulnerability assessments and policy and the implementation of adaptation actions for select local authorities in the South East region. Differential local vulnerabilities and adaptive capacity among different actors are described in terms of existing environmental policy tradition, available funding, general awareness and leadership, and the perceived ability to identify and undertake adaptation measures.

3.4.2.1 Hampshire County Council and Winchester City Council

Hampshire County Council

The local authority of Hampshire County Council, currently led by conservatives, has an extensive history of engagement with environmental issues. In general,

Table 3.3 Particular adaptive capacity strengths and limitations at local authority level

Local authority	Type of authority	Expressed financial capacity	Expressed political capacity (including council leadership)	Expressed institutionalisation of environmental aims
Hampshire	County Council	High	High	Well developed
Winchester	City Council	Limited	(linked to county council leadership)	(not emphasised)
Portsmouth	City Council (unitary)	Limited	Limited	(not emphasised)
Surrey	County Council	Limited	Limited	(not emphasised)
Woking	Borough Council	High	High	Well developed

Hampshire is seen as one of the focal counties on adaptation: 'I can't think of an area where government is ahead of us in thinking on adaptation' (Hampshire CC, interview). The Council was a 2001 signatory to the Nottingham Declaration, a founding member of Climate South East, a partner in the EU INTERREG *Biodiversity Requires Adaptation in Northwest Europe under a Changing Climate* (BRANCH) project 2003–2007, and a member of the international advisory group of the European Commission on adaptation to climate change (Hampshire County Council, 2009a). In addition, Hampshire also has an accord with central government on developing and contributing best practice examples with general benefit for local authorities. Hampshire was further a lead partner of the EU ESPACE project, along with the Environment Agency, the South East Climate Change Partnership, Surrey County Council, West Sussex County Council and the South East England Regional Assembly (Hampshire County Council, 2009a). Although the project ended in August 2008, it was discussed by several interviewees as having supported a focus on adaptation in the greater South East region: 'I think that it's been ... informing policy-makers in Europe as well as ourselves, and it certainly was one of the first adaptation projects in Europe' (Hampshire CC, interview).

Perhaps the most significant development in Hampshire adaptation was the establishment of the Climate change Commission of Inquiry in November 2006 (Hampshire County Council, 2009c). The Climate Change Commission was the first of its kind in the UK, instituted, funded and supported by the Council but set up as an independent body to advise on climate change adaptation. Among other things, the Commission convenes hearings, compiles evidence reports and holds awareness and evidence sessions (Hampshire CC, interview). A report from the Commission concluded that Hampshire County Council and its residents were already experiencing climate change and should work with governmental and non-governmental organisations in the UK and elsewhere to adapt to these challenges (Hampshire

Climate Change Commission of Inquiry, 2007). The report notes that nearly half of Hampshire's coasts are undefended and characterised by soft and erosion-prone mudflats and sandy cliffs. Sea level rise is noted as a potential risk to multiple interests. Coastal issues are also noted in a report following up the progress of Hampshire County Council and its partners, which states that a long-term strategy for adaptation of the Council's significant coastal land holdings to change will be developed (Hampshire County Council, 2009b).

According to one interviewee, adaptation has been mainstreamed in the County Council's decision-making as a result of these priorities and identified risks: 'climate change will be taken account of in every decision the council takes ... we have systems that do that now, every report that goes in for decision has a climate change balance attached to it' (Hampshire CC, interview). This development has occurred in response to pressures that include a call from the County Cabinet for the County Council to adopt a framework policy for incorporating climate change considerations into current and future actions (Hampshire County Council, 2009b). The Commission's 2007 report identifies two principal drivers at the County Council for the formation of a Commission of Inquiry: the Stern report published in October 2006, and the County Council leader's statement to Council in November 2006 that 'within a decade Hampshire will prosper without risking our environment' (Hampshire Climate Change Commission of Inquiry, 2007, p. 4).

The County Council has also decided that a Member of the Cabinet should be appointed to champion climate change issues across the Council's operations (Hampshire County Council, 2009b). Hampshire has further set the goal of achieving a Level 3 status (where Level 4 is the maximum) for NI 188 across Hampshire's local authorities by 2011 (Hampshire County Council, 2009b; Shaw, 2008). Partners in the LAA include the Sustainable Business Partnership and the Partnership for Urban South Hampshire (PUSH, a local authority network broadly focused on growth and sustainable development, cf. PUSH, 2009) and the Solent Forum coastal organisation (Hampshire County Council, 2009b).

Whereas economic imperatives may limit action on climate change elsewhere, adaptation has been seen in Hampshire as a way to maintain competitiveness. Hampshire was a proactive local authority in the UK with regard to the development of water and soils strategies, as well as an early actor on waste management in response to EU directives. An interviewee considered these to be factors that enabled the Council to adjust its responses prior to the official implementation of the Waste Directive in the UK. With regard to adaptation, the strategy has been 'doing the same again ... The primary reason has to be to make us more competitive and to improve the services ... [in] Hampshire' (Hampshire CC, interview). The Hampshire approach was also largely formed through its early response to the highly publicised 1998 UK climate change scenarios. When the release of the scenarios and UKCIP work prompted the South East RCCP to produce a regional scoping study, this persuaded Hampshire Council of the severity and need to devote resources to climate change (Hampshire CC, interview). An interviewee also noted the importance of both political and voter support: 'it's very powerful [for council members] ... if ... somebody is going to vote for them and that person is saying climate change is an issue' (Hampshire CC, interview). The interviewee noted further: 'I cannot overemphasise the importance of getting the councillors behind it [a commitment to adaptation]'.

Practically, the Hampshire approach has also been developed through very clear examples of the possible effects of climate change. Predictions of a 2°C increase by 2100 and changing precipitation patterns have been portrayed as a shift from Hampshire's green and rolling landscape to one such as is found in Bordeaux, a much drier environment. Hampshire County Council's promotion of a newspaper, local video, and wine-tasting event (where wine from Bordeaux and Hampshire wine were compared) both influenced councillors and gained publicity (Hampshire CC, interview).

Other explanations for Hampshire's strong policy development on climate change adaptation include its high vulnerability to both current climate events and future climate impacts. The County's large agricultural sector and dense population situated along a long coastline have already been exposed to extreme events such as storms, wind events, flooding, and drought. Strong pressure for development has obligated Hampshire to attempt to direct and plan for growth in a sustainable way:

> Because of this development agenda ... we've got about 100,000 houses to be built over the next twenty years in Hampshire alone, that's on top of what we had over the last forty years so Hampshire's grown immensely. We're used to dealing with long-term planning (Hampshire CC, interview)

Finally, Hampshire is a large and relatively wealthy council, and has been able to increase the number of staff working with climate change to a total of twelve people (Hampshire CC, interview). The large size of the council is considered by some actors as a possible explanation for its engagement with adaptation: 'Hampshire has always been a very forward-looking type of authority ... it's officially big [enough] to ... feel it has an influence on things' (Hampshire CC, interview). The proximity to national government, given the absence of elected regional administration in England, and the international orientation of the Council are also noted as possible strengths:

> We've always had an international office, for example ... We've had an accord with Normandy for decades ... we are part of the Western Europe growth triangle, London, Paris and Brussels ... Our politicians have always invested money in forward planning ... in working with researchers and lobbying ... and working with the European Commission. (Hampshire CC, interview)

Hampshire can thus be seen as a local authority that has mainstreamed adaptation relatively extensively in relation to the vulnerabilities identified that might affect infrastructure development. It has achieved this through a strong adaptive capacity, established through leadership on environmental issues, and by drawing on its size, relative wealth, and international orientation. These are, additionally, parameters of a growth region that can be seen as relating to factors discussed in the new regionalism literature (cf. Veggeland, 2000).

Winchester City Council

Winchester City Council is a local authority under Hampshire County Council that illustrates both the potential of cooperation within a larger and well-resourced county and the role of differential resource access in a smaller city council vis-à-vis the county level. In addition to its adoption of the NI 188 as one of its local indicators, Winchester City has directed its action on adaptation through the Winchester City Council framework document on climate change, which determines 'what the local community, through the Winchester District Strategic Partnership, can do in delivering action' (Winchester City Council, 2007, p. 1).

The latest restructuring of the Winchester District Strategic Partnership has shifted its organisation from a single group to an executive group with five subgroups or strategic outcome groups (Winchester CC, interview). At the same time, the Winchester Action on Climate Change was also established, a community group focused on climate change (mainly mitigation) that is now active within the strategic partnership.[11] The goal for adaptation planning is to ensure that the district is 'climate change ready' and able to maximise opportunities and minimise costs of climate change (Winchester CC, interview). This includes the development of more efficient water use, fewer problems caused by flooding, minimising impacts of emergencies arising from climate change, and the adaptation of crops to a new climate. An indicator for extreme weather events is also requested, for which complete data do not yet exist (Winchester City Council, 2007).

Despite these developments, Winchester considers itself relatively limited in terms of financial resources to act independently on climate change adaptation: 'We are a little city; on pure scale we might not rate much more than a town [although] we do enjoy city status' (Winchester CC, interview). As a result, the Winchester City Council has placed a large focus on working in partnerships. The City Council is influenced by priorities at the Hampshire level, which has resulted in a two-tiered structure in the county level network and the local strategic partnership that has created some coordination challenges (Winchester CC, interview). The authority also noted that membership in Climate South East is too costly for Winchester, but that it nevertheless benefits from the partnership through the membership of its cooperation partners (Winchester CC, interview).

3.4.2.2 Portsmouth City Council

The UK's only single island city, Portsmouth is highly vulnerable to climate change. The 2007 Hampshire Climate Change Commission of Inquiry report notes that 72% of Portsmouth's industrial areas lie within the city's coastal floodplain, and that more than 60,000 residents are expected to reside in this area by 2026, creating substantial concern for the expected rise in sea level and increased storm

[11] As Winchester Action on Climate Change is mainly focused on mitigation, they for that reason declined a request for an interview in this study.

events (Hampshire Climate Change Commission of Inquiry, 2007). The report further describes the principal threats/vulnerabilities to the city and its port from climate change in terms of the 'extra costs of defence against sea level rise, impact of national and regional transport curbs biting into cargo movements; enhanced specifications required of new port development and operations; and impact upon tourism, leisure and retail, plus information/communications technology' (Hampshire Climate Change Commission of Inquiry, 2007, p. 20). An additional survey has indicated that 61% of Portsmouth's residents are concerned about the effects of climate change (Portsmouth City Council, 2008a). According to a Portsmouth official, these results have influenced councillors' priorities where climate change is concerned (Portsmouth CC, interview).

As a result, adaptation to climate change is one of four priority areas of Portsmouth's Climate Change Strategy (Portsmouth City Council, 2008a). Portsmouth has also held a number of climate change strategy consultation events aimed at creating awareness on potential climate change impacts in Portsmouth and identifying its key priorities. The outcome of these consultations was the signing of a multi-agency climate change strategy action plan for Portsmouth (Portsmouth Sustainability Action Group, 2008; cf. Portsmouth City Council, 2008b). In order to raise awareness of the issues, the strategy noted the need for champions and identifies one of the councillors as a potential Member Champion (Portsmouth City Council, 2008b). Portsmouth City Council has also developed a strategic flood risk assessment, a coastal strategy for the city (including a specific sub-group on adaptation), and a sustainability strategy (Portsmouth CC, interview).

The Portsmouth Climate Change Action Network (PCAN), a local NGO, has been a key actor in adaptation development in Portsmouth and has successfully lobbied for a full-time Sustainability and Climate Change Coordinator position (Portsmouth Sustainability Action Group, 2007). The City Council also set up the Portsmouth Sustainability Action Group (PSAG), a partnership of public, private and voluntary sector organisations that has been active in producing a climate change strategy for the city, due to be published during 2009 (Portsmouth CC, interview). According to the Climate Change Action Network, the PSAG partnership 'would not exist without PCAN and ... [the] post as a principal officer for climate change would not happen without PCAN so there is ... a clear impact' (PCAN, interview). The PSAG partnership is now a project under the Local Strategic Partnership and is largely responsible for delivery of the climate change strategy in the city. Portsmouth City Council also works on climate change within the city's Local Strategic Partnership and the Partnership for Urban South Hampshire (PUSH), as well as PSAG (Portsmouth City Council, 2008a, 2008b; Portsmouth Sustainability Action Group, 2007).

However, the Portsmouth City Council also admitted some difficulty in engaging with climate change adaptation, noting that 'the adaptation side of the strategy is quite light on the specific actions simply because we are at a stage where we need to ... work out where we are first' (Portsmouth CC, interview). In order to advance their work on adaptation, 'we're ... working with Hampshire county council ... doing an LCLIP' (Portsmouth CC, interview). While Portsmouth's sensitivity to climate change is high, the costs for Portsmouth to adapt to climate change may also be unusually high. An interviewee at the Council noted that mitigation actions have been prioritised, since the actions for adaptation are 'a lot harder and a lot bigger', among other things as they may require investments and commitment from water companies. In addition to the high costs of adaptation, one interviewee also noted competing issues: 'Portsmouth has got quite significant social issues and you have to have priorities as to what you can do if you are a local authority' (Environment Agency, interview). An interviewee at the Portsmouth City Council noted, however, that once the climate change strategy was released, additional funding for staff and environmental awareness sessions for councillors could become available (Portsmouth CC, interview).

3.4.2.3 Surrey County Council and Woking Borough Council

Surrey County Council

Surrey County Council provides an example of a county where adaptation has not been a major focus, but within which a smaller local authority, Woking Borough, has come to provide a best-case example of responding to climate change. In Surrey, work on adaptation to climate change has recently started through the development of the Surrey Climate Change Strategy, born of a project commissioned by council members active in climate change and accepted in 2008 (Surrey CC, interview). The Strategy notes that Surrey County Council has a role to play both in mitigation and adaptation, and that the Council will undertake a local climate change impact assessment, establish a system for taking account of the inclusion of climate impacts and risks into strategies, policies and programmes, and implement, monitor and report on the strategy's progress (Surrey County Council, 2008). The strategy also directs the Council to share information and best practice with additional partners (through the LSP and LAA). Adaptation measures include a focus on risk management and resilience (e.g., reviewing and updating emergency plans and capacity), development and economy (e.g., ensuring resilient infrastructure development), water consumption (e.g., reuse of water and measures to reduce water consumption), flooding (e.g., working with planning authorities and water companies), and transport (Surrey County Council, 2008).

Surrey has also formed a Climate Change Task Group to take evidence from persons they see as key witnesses on climate change and use this as a basis for discussing potential adaptations. The Task Group is modelled on the work of the Kent County Council, a similarly Tory-led county council that carried out an LCLIP relatively early in 2007 (Surrey CC, interview). Whereas the Climate Change Task Group will focus on adaptation issues in general, the County Council has also set up a task group to look specifically at the Pitt Review and risks from flooding (Surrey CC, interview). A Surrey Climate Change Partnership was also set up in 2008, but was noted by one of the interviewees as being relatively limited due to the absence of any legislative basis and the fact that it is not led by the Council (Surrey CC, interview). However, the Partnership will commission an external Surrey-wide strategy in the longer term with the aim of developing a common LCLIP rather than having a profile conducted in the County and Boroughs separately (Surrey CC, interview).

Interviewees at Surrey County Council noted that limited awareness on climate change, a limited history of environmental work, and limited political will among executive members of the Council on adaptation has restricted action on the strategy (Surrey CC, interview). Some scepticism regarding the validity of climate change was also noted (Surrey CC, interview). Contrary to Hampshire's description of the coordinating role of the county level, interviewees at Surrey County Council noted that:

> Most of the legislative responsibilities are independent from us. It is a ... legislative and hierarchical split between us and the boroughs and districts. In climate change terms ... they are entirely independent to carry out their own [work, with their own resources] ... as Woking has over the last ten years. (Surrey CC, interview)

Where Hampshire (another upper-tier local authority) describes the county level as crucial to adaptation at lower levels, Surrey may thus represent a case in which the county may have a more limited role. Interviewees also highlighted the differences between different local authorities given that action prior to the establishment of the new performance assessment system has been largely voluntary:

> Because there hasn't been any real government intervention or any real government leadership the inconsistencies you are seeing like Nottingham, Woking ... I'm sure they are down to individual champions: political or employed individuals who have sufficient influence to make a difference. (Surrey CC, interview)

Surrey also illustrates other difficulties of integration in climate change adaptation. Despite the existence of a Surrey Climate Change Partnership that includes all boroughs and districts in the county, there is limited integration or coordination of adaptation work between them (Surrey CC, interview). Interviewees also noted that the few existing incidences of local extreme events may have limited awareness raising. In response to the newly established performance targets, the Council further noted the limited availability of funding:

> There is a big sort of cultural shift we are supposed to be making ... LCLIP results should filter through the executive down into the services and education, housing ... and should be converted into actions within transport, education ... Inevitably it is going to require new money ... and we have got no idea where that is going to come from. (Surrey CC, interview)

Other issues that were identified as potential limits on adaptation work include the relatively small size and recent establishment of the climate change team at Surrey County Council (which consists of two positions since October 2008), and recent reorganisations in the Council. In contrast to Hampshire, Surrey described limited institutional access to European funding structures and the EU context (Surrey CC, interview).[12]

---

[12]However, interviewees also noted that adaptation is not a party political issue, and that both Tory and Labour local authorities have been proactive.

Woking Borough Council

Situated in the County of Surrey, Woking Borough Council has been acknowledged as a 'best practice' example on climate change, particularly with regard to its established focus on mitigation and environmental policy. At their presentation at the South East England Regional Assembly Climate Change Summit, Woking Borough Council described the environment as one of the Council's top three priorities since 1990/1991, buttressed by associated corporate council commitment and 'cross-party political support' on environmental policy (Woking Borough Council, 2007b).

Recognition for Woking's work on sustainability has taken the form of its receipt of Beacon Awards, awarded under a government programme for the promotion of excellence in local service provision administered by IDeA. Woking Borough Council has been awarded Beacon Awards for its promotion of sustainable energy and sustainable communities through the planning process and, more recently, under the 2008–2009 theme of 'Tackling Climate Change' (Woking Borough Council, 2008). Among other things, the six councils that received the 2008–2009 awards participated in a workshop as a part of a national conference to share results and promote awareness raising. In their description of the Beacon for Woking, IDeA notes that:

> The council's Climate Change Strategy takes an integrated approach to the mitigation of and adaptation to the effects of climate change. Over forty actions have been identified across eight key themes with priorities divided into short, medium and long term timescales. These actions have been embedded corporately and their responsibilities have been assigned to officers within the appropriate council service area. (IDeA, 2008, p. 22)

The financial reward associated with a Beacon Award is around £60,000 and is often used to employ a coordinator for the year (IDeA/Nottingham Declaration, interview). Aside from national awards, Woking also joined the ICLEI Local Governments for Sustainability *Cities for Climate Protection Campaign* (CCP) in December 2006, and was granted the special status CCP City of Ambition for its work (IDeA, 2008).

While Woking's main focus has been on mitigation, engagement with adaptation began as a result of the application process for the Climate Change Beacon. In their application for the Beacon award, the Woking Borough Council described a number of practical initiatives, including the adjustment of flood defences in accordance with requirements suggested in Planning Policy Guidance Note 25 (PPG25) for development and flood risk (Woking Borough Council, 2007). The Council further noted that certain properties were flooded in 2006 during heavy rainstorms due to the inadequacy of surface water systems, in response to which an overflow area was developed.

The Woking Borough Council Climate Change Strategy was adopted in December 2002 and published in March 2003. The current strategy includes ten key themes: planning and regulation, energy, waste, transport, procurement, education and promotion, green spaces, water, working with business, and community and residents (Woking Borough Council, 2007). Although principally focused on mitigation, the risk for flash rises and falls in water levels is noted under the Green spaces theme. Case studies under this theme include Hoe Valley landscaping for flood protection, the Surrey Heathland project on fire protection, and the Woking park pond restoration scheme. Under the water theme, the strategy describes work on flood mitigation (following extensive flooding in the Hoe Valley in 2000 and a subsequent report by the Environment Agency), drought, and water efficiency. Work on the strategy is in part carried out in cooperation with a Local Agenda 21-related group, one of the few remaining LA 21 groups in the UK (Woking Borough Council, 2008).

Financing for Woking's work on sustainability derives in part from an unusual form of attracting economic resources through the Thameswey Energy Company, formed in the late 1990s. The company is owned by Woking Borough Council but attracts private financing to develop renewable and sustainable energy installations (Woking LA21 and Woking BC, interviews). Under Thameswey, the subsidiary Energy Centre for Sustainable Communities (ECSC) provides consultancy services, while profits from the subsidiary Thameswey Energy are used to fund environmental projects in the borough (Woking Borough Council, 2007b).

Despite changes in leadership over the last several years, interviewees noted that commitment to environmental issues has not changed, rather, it has remained a top priority for the last 15-year period (Woking BC, interview). Political support is also provided through formal and informal agreements on environment and climate issues:

> We have a champion in the form of our chief executive, who is very proactive on climate change and sustainability issues, and we have supporting members as well as ... councillors as well; we have a climate change Woking group that meets every quarter and that is attended by officers and councillors to promote action within the climate change strategy. (Woking BC, interview)

3.5 The Role of the EU

Although local adaptation is strongly steered from the national level (but with differences in prerequisites and priorities between different local areas), all of the above cases also describe a discernable impact from the EU level on adaptation initiatives. While to date the EU does not have a committed adaptation policy beyond the Green and White Papers (cf. Introduction and Chapter 2), interviewees at the different levels generally discussed the impact of the EU along two strands: (1) the impact of existing EU environmental policy directives and associated funding on land and water use, and (2) the possibility or relevance for the EU to act on adaptation.

With regard to the first point, interviewees stated that the Habitats, Birds and Water Framework Directives have not addressed climate change, and that the EU Natura 2000 network of protected areas has thus far been treated on the national level (rather than with regard to a network for species migration on the EU scale). Only the Floods Directive acknowledges climate change, raising concerns that other directives have been 'predicated upon a fixed state' (Hampshire CC, interview). Thus, '[it] means that we are struggling with stakeholders who have got to balance an EU requirement that assumes a static climate, and the reality that the climate is changing' (UKCIP, interview). As a result, several interviewees suggested the need for a revision of the directives (and other areas of EU regulation, such as planning) with an eye to include climate change issues. Interviewees also acknowledged the scale of effort that such a revision would require: 'one of the reasons why we were having such difficulty getting DG [Directorate-General] Environment to talk about adaptation ... [may be] because they still have got no EU 27 [member] signup to the Birds and Habitats Directives' (Hampshire CC, interview).

It is noted, however, that it is not necessarily the EU directives that cause the problem in all cases, but the UK interpretation of them. For instance:

> Government is affecting the way we do things because there are parts of the government that don't really understand and so some of their thinking is constraining us ... Things like shoreline management planning ... to some extent this is driven by the EU ... [but] the more work we've done in Europe the more we're coming to realise that this, large [parts of] of it, is driven by the UK's interpretation. (Hampshire CC, interview)

In some interviewees' understandings, the specific government interpretation of EU policy may thus impact local adaptation responses (Hampshire CC, interview). Limited funding for the implementation of EU directives has also been a concern: for example, while the UK implementation of the Water Framework Directive has no designated funding attached to it, the Habitat Creation Programme does. The result has been the need for some administrative re-shuffling: 'if we can link the two together so we are creating habitat where we also need to create habitat for Water Framework Directive, then we can meet [both] those requirements at the same time' (Environment Agency, interview). Currently, the Habitats and the Birds Directives in the UK require the Environment Agency to replace any loss of European-designated habitat through the Habitat Creation Programme. In the Environment Agency Southern Region (covering part of South East England), the Habitat Creation Programme represents the second of its kind, established voluntarily in 2004. The Programme has begun to integrate its work with other Environment Agency areas, as well as with local authorities, in order to incorporate habitat needs involving multiple actors into the Programme. Currently, the Environment Agency's head office is also establishing guidance on setting up habitat creation programmes in all regions (Environment Agency, interview).

Beyond the mainstreaming of climate change adaptation issues into existing directives, however, concerns as to the degree to which the EU can address adaptation also exist. Some interviewees shared the perspective that given the context-specific nature of adaptation, it would be better addressed through national programmes, providing the EU with examples and models from different countries that could serve as a basis for advice to individual states (UKCIP, interview). The suggested role for the EU is to ensure adaptation through programme funding: 'we and a lot of other people have said that the single thing the European Union could do is to require [that] anything they put money into should demonstrate they have taken risks of climate change into account' (UKCIP, interview).

Interviewees also noted the potential for organisational difficulties in driving adaptation issues at the EU level, both with regard to the civil service system (where established connections to certain positions may be lost as people move through their career), and with regard to the organisation of the Directorates-General. Rather than designated an issue only for DG Environment, climate change would require the involvement and commitment across the Commission (UKCIP, interview). One interviewee also noted the potential for confusion if adaptation policy is not clarified:

> In English the word "resilience" has many, many meanings, and the White Paper is full of the term "resilience". And I know in English it is misunderstood and misinterpreted; when we start translating it into 27 different languages it is going to be . . . meaningless. (UKCIP, interview)

3.6 The Participation of Voluntary and NGO Networks

Voluntary networks have also had an impact on climate change adaptation. Of these, the Local Agenda 21 (LA21) initiative for local sustainability action, established following the 1992 Rio de Janeiro UN Conference on Environment and Development, is among the most noteworthy. Following his election in 1997, Prime Minister Tony Blair (Labour) announced the need for each municipality to adopt an LA21 group by the year 2000. However, LA21 groups are now relatively rare. Some interviewees noted that LA21 ideas had been taken up by the Local Strategic Partnership and the LAA: 'To my mind community strategy is the natural successor to the LA21 strategy . . . and the LSP is what the LA21 working group was, although nowadays it's certainly strengthened' (Winchester CC, interview). Some actors also noted that the decrease in LA21 groups may to some extent be a function of the terminology. 'Once you get into the 21st century, you're kind of losing the point in the terminology . . . I know in some areas they changed the A to mean Action' (Winchester CC, interview).

With regard to the specific local case study areas assessed in this chapter, Surrey County Council noted that they never had a strong LA21 group, while in other areas the LA21 has disappeared. Even in areas where LA21 groups still exist, connections with the Rio initiative from which it sprang have been loosened, as in the case of Woking, whose LA21 has become a relatively independent (but still partly council-funded) environmental action group. Today, Woking LA21 (now known as Local *Action* 21) comprises roughly twenty active individuals and has produced, for instance, a green services listing and a 'greener homes' construction guide (Woking LA21, interview). Interestingly, one actor noted that:

> It seems like the authorities who really do specialise in climate change or who have made a point of making it a political agenda still have Local Agenda 21 as a key role within their organisation. But other than that, I haven't really seen it be a huge focus of local authorities. It's similar to the Nottingham Declaration really, those who have made the point of signing it have made the point of advertising the fact . . . but in terms of general implications with authorities I haven't really found it. (Surrey CC, interview)

Beyond the LA21, the voluntary network ICLEI Cities for Climate Protection (CCP) has gained some attention in earlier studies (Bulkeley and Betsil, 2005), but was not extensively discussed by interviewees. This may be in part due to the fact that while IDeA managed the UK pilot of CCP, the programme was only funded for 18 months, after which the national importance of the ICLEI CCP diminished. In addition, the ICLEI organisation has been more successful in obtaining funding from the European Commission to work on sustainable procurement than on CCP work (IDeA/Nottingham Declaration, interview). However, Woking Borough Council has received some support from its participation in ICLEI's *Sustainable Now* project on adaptation and mitigation (Woking BC, interview). The voluntary Aalborg initiative on sustainability principles for local authorities was additionally noted by one interviewee, but only in passing as a part of integration of climate change and environment issues throughout Council (Hampshire CC, interview).

Other networks between environmental NGOs exist, such as the Portsmouth Climate Action Network (PCAN), set up as a small informal group in November 2005 with linkages to Greenpeace, Friends of the Earth and smaller, climate-focused groups such as Stop Climate Chaos, the National Campaign for Climate Change, Climate Camp and the World Development Movement (PCAN, interview). However, these are mainly focused on mitigation and, with the exception of PCAN in Portsmouth, were not discussed by interviewees.

3.7 Adaptation as an Issue: Shifts in Thinking, Policy Transfer and Integration Across Levels

3.7.1 Policy Transfer Between Contexts?

The analysis above indicates that while no one model for adaptation to climate change exists, several of the local councils refer to each other and to known examples, as well as to state support structures such as the UKCIP. With respect to the UK context, policy transfer and good practice can be conceived in terms of several different types. One such type is transfer within the domestic context that largely focuses on general adaptive capacity-building measures (such as the development of coordination and communication). However, general principles similar to those relevant for the national context may also be relevant for international policy transfer, which could result in transferability of principles and models such as science-policy coordination or a focus on the regional and local level.

Table 3.4 indicates the types of adaptation that have been targeted in the UK context, differentiated according to the outline presented in the introduction (see Chapter 1, this volume) in terms of policy priorities, binding targets, adaptive capacity development, and organisation in terms of issue-specific dedicated organisations or mainstreaming within the existing administrative structure.

Table 3.4 Summary of main types of adaptation policy development on different levels. For the UK case, most policy priorities on the national level discussed in this chapter have also developed into binding targets or now exist in organisational form

Level	National	Regional	Local
Type			
Policy priority	Mainstreaming adaptation into policy in Green Book, planning system	(To some extent in SEEDA sustainable development priorities; SEERA climate change strategy; RIEP sustainability theme)	LGA strategies Local authority climate change strategies
Binding measures (such as legislation)	Reporting powers of the Secretary of State Performance Assessment Framework for local government Revised PPS, PPG	(To some extent in e.g., PPS targets in South East Plan)	NI 188 as priority indicator
Adaptive capacity-building measures	Risk assessment Cost-benefit analyses IDeA Beacon programme	Conferences	Residents' surveys on climate change Flood prevention EU projects Events
Dedicated organisation	Committee on Climate Change Sub-Committee on Adaptation LRAP UKCIP	Climate South East	Hampshire Climate Change Commission of Enquiry Community groups on climate change
Main-streaming in existing organisations	ACC within DEFRA Cross-departmental domestic adaptation programme board (To some extent, Environmental Agency improved coordination with relevance for adaptation)	(through strategies as above)	Council Champions on climate change County-wide partnerships

3.7.1.1 The Domestic Context

As can be noted from the above sections, adaptation policy has been institutionalised at all levels in the study (national, regional, and local) through a mix of different types of priorities, capacity-building, and organisational forms. Throughout the various scales, the potential for the transfer of best or good practices – or quite simply the sharing of examples – is emphasised in both policy and practice, and has been institutionalised within the UK context. For instance, under the heading 'Beacon Authorities are Here to Help', IDeA describes the benefits for local authorities engaged in the Beacon Scheme for Tackling Climate Change as 'being able to network, learn about best practice and get the opportunity to work alongside some of the best services in the country to close peer mentoring relationships' (IDeA, 2008, p. 4). This sentiment is echoed among interviewees, who noted that sharing good practice among local authorities is a tradition rather than (only) a result of government steering (IDeA/Nottingham Declaration, interview). DEFRA even noted that local authorities may be able to get the message across more easily than central government:

> Something we try to do in ... looking at what an individual local authority ... might do is to say, when you decide on projects and you are looking at a particular course of action, think about the national applicability of that, about how you can share that with other people in terms of best practice ... That is one of the principles behind having an accord with Hampshire ... [if] people at local government ... don't believe a word I say ... if somebody from a neighbouring authority comes along and says we have done this, then you know they really relate to that ... and so having specific Beacon authorities in this area ... it's a really good way of working really. (DEFRA, interview)

An example of transfer of adaptation approaches within the domestic context is DEFRA's development of integration between levels and sectors with regard to adaptation by employing measures such as the exchange of personnel between departments and local authorities. The goal of the exchange is to minimise conflict that could arise through distributing key issues in lead agencies with later involvement by local authorities by providing instead personnel to work on adaptation issues within a given area, for instance, in selected councils (DEFRA, interview). Such agreements currently exist with Hampshire and Kent County Councils. A similar method is used to support integrated policy development at departmental level:

> When policies develop at one department, it's normally developed in-house to quite a high degree, so your ability to influence is weak at that stage. Whereas if you can get somebody in who is working with the policy people from inception ... then you can actually influence the product while it is being developed ... [it is] quite a small amount of resource we've put in [for having a person from us at another department] ... but it has worked quite well actually. (DEFRA, interview)

In order to develop action on adaptation domestically (and at the local scale in particular), several interviewees expressed the crucial need for commitment among executive officers or other high-level positions. These actors may also be the most difficult to attract to awareness-building events, which may hamper progress on the issues. Interviewees also noted the need to allow for participation in the development of vulnerability assessments and adaptation measures, such that local actors could feel ownership of the process and thus be inclined towards its local application:

> Allow the experts in their field to come up with the answers to those questions, rather than you providing them with the answers. So if you are talking to the health service you can provide scenarios, you can ask questions about how they are currently impacted by extreme

> weather, how that might change over time. You can postulate situations, imaginary situations, and how do you respond to those situations. They know the answers to the questions if you ask the right questions. (UKIRCCG, interview)

The need for moving beyond the provision of tools and knowledge, to the provision of actual support, training and education, was also noted as a requirement for local adaptation and the transfer of adaptation approaches, as was the need for more collaborative approaches than have been used between government and local authorities in the past (UKCIP and DEFRA, interviews). To this end, DEFRA has launched online web community meetings between local officials and the department, using IT and forums such as YouTube to disseminate information and encourage participation (DEFRA, interview). Attempts have also been made to achieve stronger integration between different sectors at the local level (DEFRA, interview).

So far, challenges to integration across sectors on adaptation in the domestic context have included the perception of adaptation as an issue for environmental specialisations only. Both DEFRA and Climate South East noted that local level representation on issues of adaptation were commonly of environmental background, while other sectors or departments may not even have considered themselves relevant. Despite the LSP system, several interviewees further noted the challenge of including businesses into regional partnerships or adaptation at county level, particularly given the competitive nature of businesses in small areas. Limited incentives also exist for coordination across industries despite the potential for mutual support given similar activities in the same geographical area. Industry may also have a much shorter time horizon than local authorities or other organisations, with the exception of sectors such as water utilities and water companies. These often employ a long planning timeframe, and have thus been among the private enterprises most involved in adaptation (UKCIP, interview).

In order to amend the absence of the private sector, however, attempts by the Environment Agency as well as by the UKCIP to encourage businesses to consider their vulnerability to potential disruptions in supply chains and flooding impacts are underway (Environment Agency, interview). One actor also noted that larger or well-established companies should also be encouraged to increase their planning horizons to longer time scales (SEEDA, interview). With regard to the private sector, issues of insurance were also considered as potential ways of developing incentives and disincentives with regard to property location, although these were not emphasised by interviewees. Mortgages and insurance issues were identified as relevant to residential owners or municipalities with existing property in flood plains, where an assessment of vulnerability may decrease its value. While the Association of British Insurance has noted that such property should be able to obtain insurance, they have not yet given an open-ended commitment for this (Environment Agency, interview).

3.7.1.2 The International Context

In addition to these domestic examples, other mechanisms may apply with regard to policy transfer across national contexts. Regarding the possibility for identifying 'good practice' cases and the extent to which they may support development in different areas and contexts even outside the country, most interviewees agreed on the transferability of general adaptation structures. Possible lessons for transfer could include the process itself, including ways of engaging difficult-to-engage sectors, or the utility of a science-policy interface organisation such as the UKCIP. Specific features such as tools developed by the UKCIP could also be relevant for other countries. For countries or areas with similar problems, comparisons could also be made between approaches to common issues such as water scarcity, heat waves, erosion and flooding (GOSE, interview). One interviewee noted, for instance, that while there are cultural and political structural differences, 'the general lessons and the general process … [are] absolutely transferable' (UKIRCCG, interview).

Several comments on the transferability of lessons learned thus referred to adaptive capacity-related issues relevant for implementation of approaches:

> Most projects are transferable and do have the ability to be replicated elsewhere to varying degrees. There's always going to be that limitation that [something] might not quite work, you might not have for instance, if you take, finance is a huge obstacle, so, you may not have the mix of officer and member support, you may not have the ability to get the finance or you may not have the support of local residents … we've been lucky in terms … [of] the level of officer and member support and that continuing support and commitment … [on] the environment. (Woking BC, interview)

Developed or advanced industrial states are further noted as a particular type of case:

> [In these cases], governments are quite well sorted out, it's clear who owns almost everything, it's clear that disadvantaged parts of society ought to be and will be protected, there isn't an issue with massive corruption … So actually the problems of delivering adaptation in such a developed economy are perhaps less because of that, perhaps greater because of the increased vulnerability that some of the systems [have]. But the scale of the impacts is much more about financial cost and inconvenience … The principle [developed in the UK], I think, is transportable to any developed economy with a reasonable level of good governance. (UKCIP, interview)

Within developed countries, however, interviewees also noted major differences. The relatively large geographical area and smaller population of Finland when compared to the South East of England was suggested as an example of where a top-down organisation may be more easily implemented (UKIRCCG, interview). However, also other differences between national contexts were highlighted:

> [In the UK] the relationship between central and local level is much more direct and for all the words like devolving and freedom and flexibilities in actual fact going back to this, the government has huge control over what they do. And then the other thing … is the split between the public sector and private sector … all our utilities are in private hands … as opposed to direct control over local infrastructure. (DEFRA, interview)

3.7.2 Adaptation as a Shift in Thinking

In general, as a result of their work on adaptation, a number of interviewees commented in considerable depth on the challenges of thinking about adaptation and developing adaptation policy and practice in the future. Adaptation was seen as something that may require not only policy transfer within established systems, but also a stronger re-assessment of present planning and legislative contexts.

Many of the interviewees noted that adaptation has required a shift in thinking, akin to the shift required to implement a process- rather than outcome-based type of indicator. However, this shift was noted as only a first step in developing and institutionalising adaptation: 'very much "stage one" of a much wider paradigm shift: it is planning to adapt, it is not yet adapting' (LGA, interview). Even within the relatively highly-developed policy context for adaptation in the UK, adaptation was thus very much seen as 'under development'. Interviewees noted that adaptation would need to be developed empirically at a number of different locations in order for further approaches to be defined:

> The challenge for adaptation at this point is to understand what it means. We are still at an early stage of exploring what it means, what we mean by an adapted community or adapted local government, what we mean by adaptive capacity or how you build capacity in an organisation to make it adaptation proof. (IDeA/Nottingham Declaration, interview)

> There is no such thing as a well-adapted community; it is always going to be ongoing. (LGA, interview)

Given the context-dependent nature of adaptation, this challenge was also related to the wider diversity between organisations and how organisations relate to change. One actor noted:

> I think there is still a huge amount we don't know about adapting to climate change ... because we have not tested so many systems equally. We also know awfully little about how organisations react to change, how individuals or organisations make decisions, how organisations plan for the future. (UKCIP, interview)

To develop an understanding of adaptation, it was suggested that lessons could be learned from the process of 'preparing to adapt' at the local level, followed by the application of the adaptation framework using evidence to determine what measures need to be taken and how to apply scarce resources (LGA, interview). Lessons may also be learned from the ways organisations work, optimise or target their activities and be applied to modify policy. As a UKCIP interviewee noted: '[One person here] has a theory ... that the way organisations are judged or measured would increasingly [need to] be how they perform under extremes ... It would be extremely helpful if [this]... was written into performance targets'.

In addition to its specific impacts, adaptation may also challenge established ways of thinking that relate to the preservation of infrastructure and consider the time scale for which such preservation is feasible – something that is treated by some interviewees with a certain awe and regret. To some extent, these concerns are comparable to discussions of adaptation in terms of planned retreat or accommodation

(Nicholls and Klein, 2000). The examples given relate to the way in which increasing risks and costs may require planning decisions to abandon land or fundamentally change the way areas and infrastructure are developed:

> Especially when it comes to something like sea level rise and impacts on coastal communities and that, we are not going to be able to defend every single last part of the coast that we have defended in the past. And the coast[al infrastructure] was basically laid down in Victorian time ... [from] 1850s and up to 1900 when most of the sea defences were built. And since then the coast has been sinking. The South East is sinking ... getting that message across to people, getting people to understand what their future is, is one of the main challenges for us. (Environment Agency, interview)

> We had very bad floods in the west of England in 2007 ... almost the immediate reaction was of the Environment Agency to build more flood defences. In some places that might be the right thing to do, but in other places it's absolutely not the right thing to do. But it is only when people have got a reasonable level of knowledge that you can recognise ... the rights and wrongs in every individual case ... we have moved from a situation where 90% of the people will say build a floodplain defence ... now it's less. There's a bit more of an informed response ... but it takes a long time because certainly to be able to recognise potential solutions does take a lot of understanding. (UKIRCCG, interview)

> We can see no way economically or practically of defending Portsmouth against a two meter sea rise ... in what stage between now and 150 years [in the future] should we be making a decision that would change our investment strategy. (Hampshire CC, interview)

Adaptation may also require the development of approaches that are not purely centred on planning or infrastructure, but that target other ways of thinking. Novel situations may provide yet another example of the need to deal innovatively with problems without focusing on a set solution or solutions. One interviewee takes an example from the major storm Gudrun that impacted much of the Baltic area:

> Two or three days before [the storm] Gudrun struck, there had been a shipment of high value cars ... into Tallinn, and the compound where they stored these cars was pretty much on the dockside. Gudrun came in ... filled up the compound ... What was the adaptation response? The obvious response is to build a higher wall. Actually the simple response is to pay a few people a few euros to drive those cars half a mile inland to somewhere else. So you can use those sorts of examples to illustrate different types of responses other than the traditional response of "let's build something", which is the normal reaction. And building something is almost inevitably expensive. (UKIRCCG, interview)

One proposed solution to such an issue was the development of warning systems to facilitate communication between relevant actors should an extreme event occur. Such integration between sectors, as well as the development and implementation of diverse and potentially novel approaches, was seen as a major challenge and potentially the greatest requirement posed by adaptation. One interviewee expressed this succinctly:

> If you have some rain falling on a field, while it's in the field, it's the farmer's responsibility. The moment it flows under the gate and into the road, it's the Highway Authority. If it flows into a drainage ditch, it's the local authority; if it then flows into a river, it's the Environment Agency. So the problem is not that nobody is in overall command, the problem is those boundaries, and how do you manage the responsibility for something moving across the boundary. (UKCIP, interview)

As a result, adaptation may relate to both managing the borders between issues and developing new types of policy responses: 'a useful way to look at when you judge adaptation [may be]: have you actually managed the risk, or have you just pushed it on to someone else?' (UKCIP, interview).

3.8 Conclusion

In the UK, adaptation policy has been developed at all levels of study. The UK experience with developing policy on adaptation to climate change highlights some rather contrary characteristics of the British system. On the one hand, the possibilities for a committed centralised state to develop an issue on all levels is prominently displayed, perhaps most so in the Climate Change Act and reporting duties of all public bodies and in the Performance Assessment Framework including requirements for local authorities. On the other hand, the UK also displays features of what has been called a differentiated polity, with policy networks as a characteristic feature (a 'disUnited Kingdom' according to Rhodes, 2007). This can be seen in the extensive focus on stakeholder engagement and vertical and horizontal coordination, and in the integration of voluntary networks such as the Nottingham Declaration partnership as well as in the inherent developments in relation to policy bodies (such as the development of adaptation in the UKCIP) within the state framework.

On the whole, the development in the UK (England and its South East region in particular) can be seen to indicate the opening of a policy window through which adaptation has been institutionalised on the political agenda. Interviewees noted both an established (and historical) policy focus on flooding and the importance of recent focusing events and resulting policy development (e.g., the Pitt Review). The Stern Review's presentation of adaptation as an economic and risk-oriented issue in line with the focus in the UK on cost-benefit assessment has also resulted in a re-framing of adaptation from a more limited environmental issue, to one with extensive economic implications. As interviewees noted, the treatment of adaptation has shifted from focusing on scientific debate relating to the possibility of climate change, to it being an issue that must be addressed through legislation, policy and planning frameworks. Issue establishment has taken place relatively rapidly through a 'policy bandwagon' where no real politicised or partisan differences on the necessity of acting on climate change are currently expressed; instead, adaptation development has been carried by a number of driving actors and champions acting on multiple levels.

The multi-level organisation of adaptation has been pronounced in the UK, and has likely been supported by centralised and partnership features of the UK state. National organisation now includes both horizontal partnerships (the Whitehall cross-sectoral board) and the vertical LRAP Partnership Board and its work across the local and regional levels. The focus on integrating several levels and sectors may to some extent be a result of the express NPM characteristics of the UK system, which has placed a focus on 'best practice' transfer; however, it can also be seen as related to the regional and stakeholder focus in the UKCIP (to some extent drawing inspiration from a Canadian initiative focused on stakeholder integration). The multi-level context of adaptation in the UK is especially clear in the UKCIP's focus on creating multi-level linkages and straddling the national level (as a funded national body), regional level (through support to regional climate change partnerships) and local levels (through the provision of tools and support for integrating adaptation in local authorities). The importance of the UKCIP was emphasised by several interviewees, who noted that 'without UKCIP. . .it would have been difficult, if not impossible, for regions to really take [adaptation] forward' (UKIRCCG, interview). In total, then, while the UK's approach to climate change adaptation is under rapid development, there is a clear commitment to adaptation as well as recognition of the need for further development (for instance, in DEFRA's consideration of adaptation beyond process indicators in upcoming performance assessment periods). While explicitly implemented actions are so far relatively less developed given the focus on 'preparing to adapt', adaptation policy integration into a multi-level governance framework is well developed.

With regard to the regional level, adaptation is a focal issue particularly for the regional climate change partnerships, although planning policy statements and guidance (with relevance e.g., for government offices) also include adaptation priorities. Adaptation plays a relatively small role for other regional bodies such as SEEDA and the RIEP, while the Environment Agency manages climate change adaptation organisation, strategies and plans at both national and regional level. The region of South East England is highly vulnerable due to a combination of development pressures and it having large low-lying areas. The region has several upper-tier local authorities with an NI188 target, and has developed an early and fee-based climate change partnership focused on adaptation. A great variety of organisations work within the partnership, indicating considerable interest in the issue and internal financial commitment even prior to the development of state funding for regional coordination. The establishment of funding for regional coordination coincided with a rise of adaptation on the agenda and an increased focus on the multi-level delivery of adaptation. This change in the relationship between the central and regional scales is mirrored to some extent in the performance assessment framework, demonstrating the expansion of the top-down or centralised process to more extensively include other identified UK characteristics such as the use of network governance and partnerships. Some interviewees also noted that the disbanding of the Regional Assemblies could strengthen the importance of stakeholder bodies such as the RCCP, given the need for new bases for regional legitimacy. Thus, while the regional role is to some extent in flux, the regional responsibilities for climate change adaptation could potentially support a stronger role for the region in general and support initiatives by economically and politically prominent counties or regions to gain larger leeway in relation to the state.

At the local (including county) level, several of the same features found at regional level are prominent, including the use of partnerships (and the implementation of the NI188 in partnership) and the need for local champions to promote policy development. Although bodies such as the Local Government Association have developed policy on adaptation and supported the larger local role in relation to NI 188, the abilities of local authorities to respond to identified vulnerabilities are differentiated with regard to such factors as wealth, political leadership, environmental policy traditions and established focus on adaptation-relevant issues such as flooding. From the local case studies, it becomes clear that there are significant variations at the local level although all local cases exhibit some policy development on adaptation, often in terms of existing or forthcoming strategies or through developing practical initiatives, ranging from awareness sessions to the development of indicators for extreme weather events. The local level has also developed voluntary adaptation commitments prior to the performance assessment framework through the Nottingham Declaration, a development that may to some extent have been a prerequisite for the formulation of the adaptation indicator NI 188. However, despite the focus on partnerships, e.g., in the LAA, the involvement of actors beyond government and administration is also differentiated. Industry has been a difficult sector to fully engage (given competition and short-term focuses), and the role of NGOs is a relatively modest one in this material.

With regard to particular authorities, a number of different factors that support adaptive capacity and adaptation policy development can be discussed. For Hampshire County Council, Hampshire's strong environment profile, early development on water, soil and waste management strategies, position as a lead partner of the EU ESPACE project, and accord with DEFRA to support development of national priorities may all have supported the development of adaptation actions within the council. A council leader's decision to move on adaptation, voter interest in the issue, the use of very clear examples to influence councillors and gain publicity, and significant environmental sensitivity to climate change and exposure to events have all been identified as important features among others in the development of the Climate Change Commission of Inquiry. The Council's relative wealth and ability to dedicate staff to adaptation have also been important determinants of capacity, as well as its assumed role as a leader in long-term planning. In comparison, the lower-tier authority of Winchester City Council exhibits significant limitations on adaptive capacity in terms of its small size and limited finances, resulting in a more limited staff and difficulties in the coordination of adaptation actions. One of the most vulnerable areas, Portsmouth – the UK's only single island city – also exhibits limited adaptive capacity in terms of funding and the low prioritisation of adaptation within the context of a more limited established environment focus and competing demands.

Despite being an upper-tier authority, Surrey County Council also exhibits some of these limitations, particularly due to the relatively low occurrence of focusing events and the relatively limited focus on environmental policy. The case of Surrey also highlighted a legislative and hierarchical split between local authorities not expressed in Hampshire County, as well as the importance of individual champions for the different local councils. The lower-tier authority Woking Borough, situated in Surrey County, has nevertheless established a high cross-party priority on environmental policy since the early 1990s, an unusual funding mechanism for environmental projects (thereby raising financial capacity), and distinguished itself as a leader through the receipt of multiple Beacon Awards. In Woking, practical initiatives on adaptation were developed out of existing practices as a result of the Beacon application process. Woking's supporting members/councillors, and the fact that the climate change working group was attended by officers and councillors alike, are additionally noted as important supporting factors in the development of adaptation.

Despite these differences across cases, a relatively in-depth understanding of the complexity of the adaptation issue does seem to exist both in local examples (particularly among those authorities extensively engaged in adaptation) and among actors at regional and national levels. Drawing upon best practices was seen by many of the interviewees as a tradition that has been normalised, and attempts for lesson-drawing have to some extent been institutionalised domestically through the Beacon Scheme (as may be expected given the relatively high institutionalisation of NPM in the UK). Interviewees emphasised that the nature of the adaptation issue may require a shift in thinking towards a performance-based evaluation of actions under extreme conditions and the management of responsibility for risks across organisational boundaries. The need for such change is for instance pronounced with regard to the EU level, where existing EU directives or national implementation that fail to consider climate change may impede adaptation. With regard to broader transferability, many noted that general lessons and processes may be transferable to other industrialised states that suffer from similar vulnerabilities, despite structural differences between countries as well as differences in political, financial or public support. The use of such lesson-drawing could also be a potential contribution to the development of EU adaptation policy.

References

Interviews

Climate South East, Guildford, January 16, 2009.

DEFRA (Department for Food, Agriculture and Rural Affairs), London, February 19, 2009.

Environment Agency, Interview 1, Worthing, February 10, 2009.

Environment Agency, Interview 2, Worthing, February 10, 2009.

Environment Agency, Interview 3, Worthing, February 10, 2009.

Government Office London/Local and Regional Adaptation Partnership Board (GOL/LRAP), London, February 18, 2009.

Government Office South East (GOSE), Guildford, February 17, 2009.

Hampshire County Council, Winchester, January 21, 2009.

Improvement and Development Agency for Local Government, Nottingham Declaration (IDeA/Nottingham Declaration), telephone interview, February 27, 2009.

Local Government Association (LGA), London, February 18, 2009.

Portsmouth City Council, Portsmouth, January 20, 2009.

Portsmouth Climate Action Network (PCAN), Portsmouth, January 19, 2009.

South East England Development Agency (SEEDA), London, January 16, 2009.

South East England Regional Assembly (SEERA), Guildford, January 16, 2009.

Surrey County Council, Kingston upon Thames, January 16, 2009.

UK Climate Impacts Programme (UKCIP), Interview 1, Oxford, February 11, 2009.

UK Climate Impacts Programme (UKCIP), Interview 2, Oxford, February 11, 2009.

UK Climate Impacts Programme (UKCIP), Interview 3, Oxford, February 11, 2009.
UK Interregional Climate Change Group (UKIRCCG), Woking, February 9, 2009.
Woking Borough Council, Woking, January 15, 2009.
Woking LA21, Woking, February 9, 2009.
Winchester City Council, Winchester, February 19, 2009.

Secondary References

Bache, I., & Olsson, J. (2001). Legitimacy through partnership? EU policy diffusion in Britain and Sweden. *Scandinavian Political Studies, 24*(3), 215–237.

Börzel, T. A. (2002). Pace-setting, foot-dragging, and fence-sitting: Member State responses to Europeanisation. *JCMS, 40*(2), 193–214.

Bulkeley, H. (2006). Urban sustainability: learning from best practice? *Environment and Planning A, 38*, 1029–1044.

Bulkeley, H., & Betsil, M. (2005). *Cities and climate change: Urban sustainability and global environmental governance*. London and New York: Routledge.

Bye, P. & Horner, M. (1998). *Easter 1998 Floods. Volume I: Findings Sections*. Report by the Independent Review Team to the Board of the Enviroment Agency, Bristol.

Cabinet Office (2007). *The Pitt review: Learning lessons from the 2007 floods*. London: Cabinet Office.

Christensen, T., & Laegreid, P. (2005). Regulatory reforms and agencification. Working Paper 6 – 2005. Oslo: Stein Rokkan Centre for Social Studies.

Climate South East (2009). Partners. Retrieved November 15, 2009, from http://www.climatesoutheast.org.uk/index.php/partners.

Cohen, S. (Ed.). (1997). *MacKenzie Basin impact study: Final report*. Downsview: Atmospheric Environmental Service, Environment Canada.

Darkin, B. (2006). Pledges, policy and performance: an assessment of UK climate policy. *Climate Policy, 6*, 257–274.

DEFRA (2004). *Making space for water – Developing a new Government strategy for flood and coastal erosion risk management in England*. Consultation Document. London. 2004.

DEFRA (2005a). *Adaptation policy framework. A consultation by the Department for Environment, Food and Rural Affairs*. London: Department for Environment, Food and Rural Affairs (DEFRA).

DEFRA (2005b). *Securing the Future. The UK Government Sustainable Development Strategy*. London: Department of Environment, Food and Rural Affairs.

DEFRA (2006a). *Climate Change. The UK Programme 2006*. London: Department of Environment, Food and Rural Affairs.

DEFRA (2006b) *Securing the regions' futures. Strengthening delivery of sustainable development in the English regions*. London: Department of Environment, Food and Rural Affairs.

DEFRA (2008). *Adapting to climate change in England. A framework for action*. London: Department for Environment, Food and Rural Affairs.

DEFRA (2009). Local Government Performance Framework. NI 188 – Planning to adapt to climate change. Retrieved June 22, 2009, from http://www.defra.gov.uk/environment/localgovindicators/ni188.htm.

DEFRA (2009b). Environment Agency. Retrieved June 22, 2009, from http://www.defra.gov.uk/Environ/fcd/policy/opauthsea.htm.

DETR [Department for Environment, Transport and the Regions] (2000). *Climate change: the UK programme*. HMSO, Norwich, UK.

Demeritt, D., & Langdon, D. (2004). The UK Climate Change Programme and communication with local authorities. *Global Environmental Change, 14*, 325–336

England's Regional Development Agencies (2007). Tackling climate change in the regions. Retrieved June 22, 2009, from www.eeda.org.uk/files/tackling_Climate_Change.pdf.

Environment Agency (2006). *Creating a Better Place. Corporate Strategy 2006–2011*. Environment Agency, Bristol.

European Environment Agency (2005). *Vulnerability and adaptation to climate change in Europe. EEA Technical Report No. 7/2005*. Copenhagen: EEA.

Gains, F. (2003). Executive agencies in government: The impact of bureaucratic networks on policy outcomes. *Journal of Public Policy, 23*(1), 55–79.

Gibbs, D., & Jonas, A. E. G. (2000). Governance and regulation in local environmental policy: the utility of a regime approach. *Geoforum, 31,* 299–313.

Graziano, P., Ferrera, M., 1Vesan, P., Bassoli, M., & Sparano, V. (2007) Local partnership consolidation in Italy – Analytical framework, research strategy and case selection. NewGov Report no. 18b/D05a. European University Institute, Italy.

Hampshire Climate Change Commission of Inquiry (2007). Hampshire's Climate Change Commission of Inquiry October 2007. Hampshire County Council, Hampshire.

Hampshire County Council (2009a). Taking action. Retrieved January 9, 2009, from http://www3.hants.gov.uk/climatechange/action-policy/environment-climate-progress.

Hampshire County Council (2009b). Hampshire's Climate Change Commission of Inquiry. Report on the response and progress of Hampshire County Council and its partners since October 2007. Hampshire County Council, Hampshire.

Hampshire Country Council (2009c). Hampshire's Climate Change Commission of Inquiry. Retrieved June 23, 2009, from http://www3.hants.gov.uk/climatechange/climatechange commissionof.htm.

Hogwood, B. W., Judge, D., & McVicar, M. (2000). Agencies and accountability. In: R. A. W. Rhodes (Ed.), *Transforming British Government. Volume 1: Changing institutions* (pp. 195–222). Houndmills, Basingstoke: Macmillan Press.

Howe, J., & White, I. (2004). Like a fish out of water: the relationship between planning and flood risk management in the UK. *Planning Practice and Research, 19*(4), 415–425.

Hulme, M., & Turnpenny, J. (2004). Understanding and managing climate change: The UK experience. *Geographical Journal, 170*(2), 105–115.

ICE [Institution of Civil Engineers] (2001). *Learning to live with rivers*. Final report of the ICE Presidential Commission to review the technical aspects of flood risk management in England and Wales. London: Institution of Civil Engineers.

IDeA (2007). The Nottingham Declaration action pack. Presentation by S. Waller, Sustainability Advisor, to the South East England Regional Assembly Climate Change Summit, March 30, 2007.

IDeA (2008). The Beacon Scheme excellence in local government: tackling climate change theme guide. IDeA, London.

James, O. (2001). Evaluating executive agencies in UK Government. *Public Policy and Administration, 16*(3), 24–52.

Jones, M. (1998). Restructuring the local state: Economic governance or social regulation? *Political Geography, 17*(8), 959–988.

Jordan, A. (2002a). Efficient hardware and light green software: Environmental policy integration in the UK. In A. Lenschow (Ed.), *Environmental policy integration: Greening sectoral policies in Europe* (pp. 37–56). London: Earthscan.

Jordan, A. (2002b). *Europeanization of British environmental policy: A departmental perspective.* Gordonsville: Palgrave Macmillan.

LGA (2007). A climate of change: final report of the LGA Climate Change Commission. Retrieved June 22, 2009, from http://www.lga.gov.uk/lga/publications/publication-display.do?id=20630.

LGA (2008). Cutting through the green tape. Retrieved June 22, 2009, from http://www.lga.gov.uk/lga/publications/publication-display.do?id=874295.

LGA (2008b). Be aware, be prepared, take action. Retrieved June 22, 2009, from http://www.lga.gov.uk/lga/publications/publication-display.do?id=566301.

LGA (2009). Small change, big difference campaign. Retrieved November 15, 2009, from http://www.lga.gov.uk/lga/core/page.do?pageId=269299.

Lorenzoni, I., Nicolson-Cole, S., & Withmarsh, L. (2007). Barriers perceived to engaging with climate change among the UK public and their policy implications. *Global Environmental Change, 17*, 445–459.

LRAP (2009). *Adapting to climate change. Guidance notes for NI 188.* London: LRAP.

McLean, I., Clifford, C., & McMillan, A. (2000). The organisation of central government departments: A history, 1942–92. In: R. A. W. Rhodes (Ed.), *Transforming British Government. Volume 1: Changing institutions* (pp. 135–155). Houndmills, Basingstoke: Macmillan Press.

Portsmouth City Council (2008a). Tackling climate change together. Information and consultation leaflet directed at the public. Portsmouth City Council, Portsmouth.

Portsmouth City Council (2008b). Climate change community event. July 24, 2008. Portsmouth City Council, Portsmouth.

Portsmouth Sustainability Action Group (2007). Action minutes April 20, 2007. Portsmouth.

Portsmouth Sustainability Action Group (2008). Action minutes 29 May 2008. Portsmouth.

PUSH (2009). Partnership for Urban South Hampshire. Retrieved June 23, 2009, from http://www.push.gov.uk/intro.aspx.

Regalia, I. (2007). Territorial pacts and local level concertation in Europe. A multi-level governance perspective. NewGov Project Report No. 18b/D05b. European University Institute, Italy

Rhodes, R. A. W. (2007). Understanding governance: ten years on. *Organization Studies, 28*(8), 1243–1264.

Sandford, M. (2005). *The New Governance of the English Regions*. Palgrave Macmillan, Houndmills Basingstoke.

Secretary of State for the Environment, Food and Rural Affairs (2006) *Climate change. The UK programme 2006*. Report to Parliament, London.

Secretary of State for Environment, Food and Rural Affairs (2007). *Taking forward the UK Climate Change Bill. The government response to pre-legislative scrutiny and public consultation.* Secretary of State for Environment, Food and Rural Affairs, London.

SEEDA (2008). News – RDAs welcome pioneering Climate Change Bill- but say 'hard work starts here. Retrieved January 9, 2009, from http://www.seeda.co.uk/news_&_events/press_releases/2008/20081127.asp.

SEERA (2007). Climate change and the South East Plan. Guide for planners. Climate change mitigation and adaptation implementation plan. South East England Regional Assembly, March 2007, Guildford.

Shaw, J. (2008). Speech by Regional Minister to the Hampshire Climate Change Commission, November 4, 2008. Hampshire County Council, Hampshire.

Smith, M. J. (2000). Prime Ministers, Ministers and Civil Servants in the Core Executive. In R. A. W. Rhodes (Ed.), *Transforming British Government. Volume 1: Changing institutions* (pp. 25–25). Houndsmills, Basingstoke: Macmillan Press.

Southern Water (2007). Business and climate change. Southern Water perspective. Presentation to the South East England Regional Assembly Climate Change Summit March 30, 2007, Brighton, UK.

Stern, N. (2006). *The economics of climate change*. Investigation under the H M Treasury/Cabinet Office, London.

Surrey County Council (2008). Climate change strategy. Surrey County Council, Surrey.

UKCIP (2009). UK climate projections (UKCP 09). Retrieved August 15, 2009, from http://www.ukcip.org.uk/index.php?option=com_content&task=view&id=163&Itemid=287.

UK Government (2006). Strong and prosperous communities. The Local Government White Paper. Retrieved June 22, 2009, from http://www.communities.gov.uk/publications/localgovernment/strongprosperous.

UK Government (2007). Climate Change Bill, as amended in Public Bill Committee. Bill 129. London, UK.

UK Government (2008). Climate Change Act 2008. Retrieved June 22, 2009, from http://www.defra.gov.uk/environment/climatechange/uk/legislation/.

Veggeland, N. (2000). *Den Nye Regionalismen. Flernivåstyring og europeisk integrasjon.* Fagbokforlaget. Bergen.

Wade, S., Hossell, J., Hough, M., & Fenn, C. (Eds.) (1999). Rising to the challenge: the impacts of climate change in the South East. Technical report. Epsom: WS Atkins.

West, C., & Gawith, M. (2005). *Measuring progress: preparing for climate change through the UK Climate Impacts Programme*. Oxford: UKCIP.

Wilson, E. (2006). Developing UK spatial planning policy to respond to climate change. *Journal of Environmental Policy and Planning*, *8*(1), 9–26.

Winchester City Council (2007). Live for the future. Tackling climate change. Framework. December 2007. Winchester City Council, Winchester.

Woking Borough Council (2007). Thinking globally, acting locally. Beacon application: Tackling climate change. Information sheets for the Beacon Assessment Team, November 2007. Woking Borough Council, Woking.

Woking Borough Council (2007b). The Nottingham Declaration in practice. Presentation by S. Rendall, Woking Borough Council, to the South East England Regional Assembly Climate Change Summit, March 30, 2007, Brighton, UK.

Woking Borough Council (2008). Think globally, act locally. Climate change strategy 2008–2013. Woking Borough Council, Woking.

Chapter 4
Mainstreaming Climate Change Adaptation: The Case of Multi-Level Governance in Finland

Sirkku Juhola

Abstract Although mitigation of climate change dominates the climate change agenda in Finland, adaptation to climate change is increasingly recognised as an important policy issue across all levels of governance. Finland was an early mover on adaptation, being the first country in Europe to publish a National Adaptation Strategy to climate change in 2005. After a few years of mainstreaming of adaptation into regular planning, implementation and monitoring at the national level, adaptation has been recognised important and some measures have been implemented but that there are also sectors where hardly any measures have been taken. At sub-national level, actors are pursuing voluntary climate strategies that are not directly linked to the developments at the national level. This chapter highlights how the different levels of governance are disconnected in terms of their actions on adaptation. On the one hand, at the national level, the NAS predominantly concentrates on administrative sectors by mainstreaming adaptation. On the other hand, the lower levels of governance are pursuing their separate climate strategies that are based on voluntary initiatives with little input from the national level. Thus, despite the early action on adaptation, it can be argued that implementation of adaptation measures has been slow and fragmented across levels of governance.

Keywords Finland · Climate change adaptation · Multi-level governance · Mainstreaming

4.1 Introduction

Finland has a long background in dealing with environmental issues, partly due to its high reliance on natural resources that has contributed to the development of a strong environment administration. Climate policy in Finland has evolved from

S. Juhola (✉)
Department of Social and Economic Geography, Umeå University, Umeå 90187, Sweden;
Centre for Urban and Regional Studies, Helsinki University of Technology, Helsinki 02015, Finland
e-mail: sirkku.juhola@tkk.fi

E.C.H. Keskitalo (ed.), *Developing Adaptation Policy and Practice in Europe: Multi-level Governance of Climate Change*, DOI 10.1007/978-90-481-9325-7_4,

early emphasis on mitigation and energy policy into an early action on adaptation at the national level. The Government has pursued Climate and Energy Strategies in co-operation with all sectors of the administration and the preparation of the adaptation strategy also followed this tradition. The Finnish National Adaptation Strategy (NAS) was the first European national strategy published in 2005 and is currently being implemented through mainstreaming by different administrative sectors. The NAS relies on future scenarios to identify potential impacts across different sectors and possible adaptation measures with emphasis on the need for more research to identify possible adaptation measures. However, local action on adaptation has been slower, so far based on voluntary measures with more emphasis on mitigation, although adaptation is considered. So, far action has been hindered by lack of knowledge on precise climate change impacts and the consequent measures that need to be taken by local authorities. Furthermore, the voluntary nature of strategies means that additional resources need to be secured, and smaller municipalities have been able to do this as a result of building networks.

This chapter first discusses the background of environmental policy-making in Finland and sheds light into governance structures within which these decisions are made. Secondly, an overview of Finnish climate policy describes the development of policies that aim to reduce green house gas emissions and increase the use of renewable energy. The Finnish adaptation policy has been formulated through preparation of the NAS from 2003 onwards. At the sub-national level, voluntary initiatives have emerged across the country, mainly in the form of voluntary climate strategies that aim to find ways of addressing climate change concerns on these levels. The regional and local case considered in this chapter is the region of Uusimaa, the most populous area of the country on the south coast. The area has suffered from flooding due to storm surges as well as flooding from inland rivers as well as from occasional droughts. Overall, the chapter shows that the importance climate change adaptation has been recognised across all levels of governance in Finland. Issues that emerge from the analysis highlight the need for more vertical integration in terms of implications of the NAS to the lower levels of governance.

4.2 Background

The political style of Finland can be characterised as a multiparty democracy with a drive towards consensus-building. Finland's public sector is similar to those of the other Nordic countries, including a highly-developed sense of public responsibility for the well being of the citizens, and strong local authorities that are responsible for delivering services to citizens. The Government of Finland consists of twelve ministries that are responsible for the preparation of issues that fall within the scope of the Government and for the proper functioning of the administration. Governments are generally stable and different party coalitions across the political spectrum are formed regularly. In fact, it has been argued that in the last thirty years leadership is exercised by strong majority governments without much effective opposition (Raunio, 2004).

Given a lack of elected regional governments, influence from the national level is mainly channelled through the municipalities by joint municipal authorities of regional councils, of which importance in terms of regional governance has increased significantly since the beginning of the 1990s (Jauhiainen & Niemenmaa, 2006). Regional councils were first established in 1994 and are responsible for overseeing the interests of the region (Kuntaliitto, 2009). The 20 regions governed by Regional Councils now serve as forums of cooperation for the municipalities of a region with the municipalities as the main actors within the councils. The main tasks of the Regions are regional planning and the development of enterprise and education, and are responsible for the management of regional development work and regional planning. Through these recent arenas of cooperation, it can be argued that municipalities have acquired even more influence (Niemi-Iilahti, 2001).

Regional councils also play key international functions through their involvement in developing EU regional policy, drawing up programmes required for the granting of support from EU structural funds for their own regions, and in part, implementing these plans. Regional development plans produced by the Regional Councils outline the vision and goals for long-term development in the region, which are in turn implemented through regional development programmes (normally produced for the near future). Regional land use plans are also based on the regional development plans and serve as directives in regional land use planning. Most regions in Finland have produced these documents over the last two to four years.

As in Sweden, Finnish local government has its roots in the parish council tradition of the Lutheran church. Municipalities enjoy considerable powers, especially with regards to land use. Since the beginning of the 1990s, there has been an increase in the role that regions play in terms of land use with respect to the state and local authorities. Although compared to the rest of Europe, their role is still relatively weak (Jauhiainen & Niemenmaa, 2006). National land use guidelines, approved by the Council of State, influence the regional land use plans, drafted by the regional councils, which are further approved by the Ministry of the Environment. These regional plans then guide the development of a general plan of the municipality. Although the state influence is channelled through the national guidelines, the municipalities are the authorities that approve the general plans.

Overall, in the recent years, municipalities have been be taking part in the most comprehensive restructuring process ever (von Bergmann-Winberg, 2000). These processes have been both internally and externally driven, and have changed the sub-national governance structures. The internal restructuring process has transformed the municipalities towards independent financial entities and marketisation, while simultaneous regional administrative restructuring has created new horizontal and vertical governance networks. In addition, external processes have shaped the sub-national system through the process of European integration by extending the reach of these networks and forcing the actors towards an increasingly varied and complex governance structure. Since joining the EU, the local government level has shown significant interest in transnational cooperation (von Bergmann-Winberg, 2000).

In more recent developments, the regional level of governance is going through further structural changes that will take place beginning of 2010. The Government began a Reform Project for the Regional State Administration (ALKU) in 2007 that aims to clarify the roles, duties, steering and the regional division of all regional state administrative authorities. The Bill enters into force in January 2010, consisting of over two hundred acts and over a thousand provisions. The main structural change due to the reform is the phasing out of all state provincial offices, employment and economic centres, regional environmental centres and permit agencies, road districts and occupational health and safety districts (Ministry of Finance, 2009). In turn, two new entities emerge, the Regional State Administrative Agency (AVI) that executes all legislative implementation and steers and supervises the functions in the regions. The Centres for Economic Development, Transport and the Environment (ELY) on the other hand, foster regional development. As these reforms are only beginning to unfold, it is difficult to estimate whether the reforms will enhance the ability of the regional councils to coordinate and harmonise regional development, as claimed by the Government, nor their impact of climate change or adaptation policy.

4.2.1 Environmental Policy in Finland

Finland has been rated among the world's leading countries in many international comparisons of environmental protection standards. In the Global Economic Forum's Environmental Sustainability Index, Finland achieved the highest ranking in environmental sustainability in the world in 2001, 2002 and 2005 (Esty, Levy, Srebtnjak, & de Sherbinin, 2005). Amongst Finland's strengths are its highly effective environmental administration and legislation, and the consideration of environmental protection in all sectors of society. Wide-ranging and detailed environmental data and high levels of technological skill additionally form the basis of Finland's effective environmental protection policies.

However, there is definitely a need for efficient environmental administration when one considers Finland's large ecological footprint, high levels of material and energy consumption, and levels of greenhouse gas emissions. The ecological footprint of a Finnish citizen is on average approximately three times the size of the global average and amongst the highest in the world (WWF, 2004). It is estimated that these high figures are due to Finland's high standard of living, large-scale intensive metal and forest industries, and the high demand for energy due to the cold climate and long distances. The country's contribution to global emissions is fairly small in absolute terms but very high when measured per capita. So far, emission levels have exceeded the targets set in the Kyoto Protocol despite a few years where the trend has been encouraging. Renewable energy sources account for a quarter of all energy produced in the country; national energy policies aim to increase this percentage while encouraging energy saving measures.

Environmental administration and policy is largely responsible for Finland's relatively high rates of environmental sustainability over the last few decades, as all three levels of government participate in environmental governance according to their own mandate. The environment administration was substantially reformed in 1995, resulting in a more decentralised system (Niemi-Iilahti, 2001). At the national level, the Ministry of Environment formulates policies and retains many of the administrative duties while the Finnish Environment Institute (SYKE) has research and development duties with the aim of feeding research data and findings into the administration.

Regional Environment Centres (REC) have belonged to state administration under the Ministry of the Environment, and form the regional tier of the environmental governance system for the past decades. The underlying idea behind the decision to establish regional centres was the realisation that environment issues are better dealt with at the regional scale and are more likely to correspond to for example watershed boundaries. Overall, the goal of these Environment Centres is to promote the sustainable use of natural resources, restrict and control the load burdening the environment, as well as preserving various natural environments and the aesthetic and cultural values of the environment (Ympäristöministeriö 2009). The 13 existing centres have a considerably wide array of responsibilities, ranging from environmental protection, land use and nature conservation. RECs also oversee the use and management of water resources under the Ministry of Agriculture and Forestry.

In recent years, all REC have prepared Environment Strategies that set out the long-term visions of each regional environment administrations (e.g., Länsi-Suomen ympäristökeskus, 2007; Lounais-Suomen ympäristökeskus, 2007; Panula-Ontto-Suuronen, 2005; Pirkanmaan ympäristökeskus, 2001; Pohjois-Pohjanmaan ympäristökeskus, 2005; Uusimaa Regional Environment Centre, 2007). The Strategies are complemented by an Environment Programme that sets out concrete measures with which the goals outlined in the Strategy can be achieved within two-to-four year timeframes. In all cases, the Strategies and environmental programmes represent a product of wide participation from the different sectors of the region.

As elaborated in the previous section, the new administrative reforms taking place in January 2010 will have significant impacts to how the environment administration will function at the regional level. The regional environment centres will become part of the new Centres for Economic, Development, Transport and the Environment (ELY) (Ministry of Finance, 2009). These centres are designed to promote entrepreneurship, enhance cultural activities, ensure smooth transport and support sustainable use of natural resources and a healthy environment. The granting of environmental permits will remain the responsibility of the regional state administrative agencies (AVI).

4.3 National Climate Policy

The early years of climate policy in Finland can be characterised as being influenced by economic and energy considerations and policy (Tirkkonen, 2000). The role of different Ministries and Ministerial Working Groups in climate negotiations changed pre- and post- Rio Summit, as both the Ministerial Working Group on Economic Policy and the Ministry for Foreign Affairs have assumed the lead role in turn. Finland's first mitigation measures and goals for international negotiations were defined by the First Carbon Committee (Carbon Committee, 1991). The main tasks for early climate policy were threefold: emissions were to be actively reduced, existing carbon sinks to be maintained, and finally, developing countries were to be provided aid in their efforts to mitigate climate change. The first piece of legislation related to climate change was a carbon tax enacted in 1990, the first of its kind worldwide (Sairinen & Teittinen, 1999). The tax was introduced through the reform of 'fuel tax law and has since been revised several times (Finnish Environment Institute 2008b). In January 2008, tax rates were raised by 9.8 per cent on average, focusing on fuels used in sectors outside emission trade.

After the signing and ratification of the United Nations Framework Convention on Climate Change (UNFCCC) in 1992, the Convention entered into force in Finland in 1994. Since then, the Finnish Government has committed itself to all the international agreements alongside the Member States of the European Union. This also includes the Kyoto targets for GHG reductions to no more than 1990 levels. However, directly after signing up to the UNFCCC, there was scepticism as to whether the reduction targets could be met. Already in the 1992 National Energy Strategy, it was admitted that the tools provided by existing energy policy were not enough to reduce emissions to the 1990 levels by the year 2000 (cf. Tirkkonen, 2000).

4.3.1 National Climate and Energy Strategy 2001

As a result of these pessimistic statements, a Ministerial Working Group was appointed to prepare a national action plan to achieve the agreed targets. The National Climate and Energy Strategy was submitted to the Parliament in 2001 in the form a Government report, and outlined the principles, targets and measures that were considered necessary to achieve the national targets under the international commitments in the period 2008–2012 (The Government of Finland, 2001). Several ministries were involved in the preparation of the strategy and in producing sector-specific reports, while the overall work was co-ordinated by the Ministerial Working Group on Energy and Climate.

The Strategy outlined the main measures for mitigation for the reduction of GHG emissions, stating that emissions were approaching the 1990 level but that they would continue to increase in the long term if no effective action was taken. In order to meet the targets, the strategy outlines the implementation of two main programmes that focus on the conservation of energy and measures for promoting renewable sources of energy, respectively. It was estimated that reductions achieved in these two programmes accounted for approximately half of the emission reductions needed to reach the national target. As a result, the use of coal as an energy source must be substituted by the use of natural gas or by increasing the nuclear power capacity (The Government of Finland, 2001). According to the Strategy, the implementation will result in increased expenditure for the consumers of energy and to the national economy, and consequently in a sizeable investment by the Government.

4.3.2 Revised National Climate and Energy Strategy 2005

Only two years after the publication of the strategy, it was decided that global climate policy had significantly moved on in the last few years and that the national policy on climate change and energy needed to be amended to take these issues into account. As a result, the National Climate Strategy of 2001 was updated in November 2005 with the Government's finalisation of a revised National Climate and Energy Strategy (Valtioneuvosto, 2005). The revised National Strategy of 2005 consisted of an outline of the changes in international policy that have affected the energy and climate policy environment nationally and focuses on measures that needed to be undertaken in order to preserve and secure the diversity of the energy supply. The revisions mostly dealt with the EU Directive on emissions trading and the Kyoto mechanisms. The Strategy also estimated that the implementation of the measures outlined would result in extra costs to energy users as well as the economy as a whole.

The Revised Strategy further brought forth the municipal sector as a stakeholder with regards to energy and climate policy, as both an energy producer and user (see Box 4.1 for measures to be implemented at the municipal level). The environmental assessment conducted on the impacts of the Strategy highlighted the role played by the municipalities in reducing the impact on the environment and climate through planning, construction and refurbishment of new community structures. According to the Strategy, cooperation between the State and the municipalities in mitigating climate change through saving energy and promoting the use of renewable energy sources should be further strengthened.

Box 4.1 Measures for implementation at the Municipal level (adapted from Valtioneuvosto, 2005)

- Continuation and further development of the municipal climate protection campaign (Cities for Climate Protection)
- Promotion of creation of sound community, creating energy efficient local communities that also act to reduce traffic
- Continuation of cooperation between urban areas in order to develop energy efficient solutions
- Promotion of renewable energy sources
- Promotion of energy efficient solutions and cooperation between municipalities

- Coordination of climate policy measures at regional and municipal levels is develop so that they further the objectives of the national Energy and Climate Strategy, while taking the diversity of the municipalities into account. Adapting to climate change will be taken into account in the operation and interaction of the various municipal administrations
- In the implementation of the Strategy, local expertise and competence related to energy and climate issues is particularly needed, and such expertise must be introduced

4.3.3 Revised National Climate and Energy Strategy 2008

National climate and energy policy were revised again in late 2008 after significant changes in international processes had taken place, mainly in the form of the European Commission's Climate and Energy Package of 2007. A product of wide governmental participation, the revised strategy of 2008 outlines measures to be taken until the year 2020 (Ministry of Employment and Economy, 2008). The Strategy's new aim is to significantly increase the use of renewable energy up to 38%, as well as of production and use of domestic energy. Different administrative sectors are also to improve energy efficiency and decrease the reliance on fossil fuels. The option of further increasing nuclear energy is also highlighted as a possibility. The measures to achieve these targets include economic steering mechanisms, (i.e. taxation of energy and certificates), as well as a national plan for increased production and use of renewable energy.

Overall, Finnish climate policy has been heavily influenced by concerns of energy production, energy security and economic development at the national level. Climate concerns have become more integrated into national decision-making as the national strategies have been revised every few years. Similarly, international developments have naturally played a large role in the evolution of Finnish climate policy in terms of emissions reduction targets and targets for the use of renewable energy. The 2008 strategy brings forth the role of renewable energy to the extent that cannot be seen in the other strategies. The Finnish climate policy is likely to be further redirected by the Government Foresight Policy published at the end of 2009.

The role of different Ministries in terms of climate policy has become clearer since the 1990's when several Ministries participated in negotiations. The Ministry of the Environment continues to responsible for the international negotiation processes, whilst the Ministry of Employment and the Economy focuses on mitigation and energy policy. The Ministry of Agriculture and Forestry has taken the lead on adaptation issues. Despite one ministry responsible for coordination, many participate in the preparation of the strategy, focusing on their own sector. For example, representatives from eight Ministries make up the climate and energy policy network that took part in the revision of the 2008 strategy. The division of tasks in climate policy has also influenced the way in which adaptation has been addressed in

Finland. Adaptation emerged as a policy issue in the beginning of the century when climate change and its potential impacts were widely discussed at the national level. After the decision to revise the 2001 National Climate and Energy Strategy, the approach toward the formation of policy towards climate change adaptation begun to take shape.

4.4 National Adaptation Policy

4.4.1 Preparation of the National Strategy

Adaptation was first discussed in a Delphi exercise that was organised as a part of the first nationwide climate change research programme in 1995 that brought together stakeholders to discuss the future of Finnish climate policy (Wilenius & Tirkkonen, 1997). This exercise involved environmental researchers, civil servants, political decisions makers, representatives of the economic sector and non-governmental organisations. In the discussions, it was 'repeatedly emphasised that national mitigation efforts cannot be appreciable results globally unless climate change can be slowed significantly' (Wilenius & Tirkkonen, 1997, p. 853). It was then suggested that adaptation might be more an advisable strategy to pursue at the national level.

The fact that the 2001 National Climate and Energy Strategy did not address climate change adaptation was recognised straight away in 2001 by members of Parliament and some committee representatives. The Finnish Parliament recommended that a separate programme for adaptation to climate change be initiated as a reply to the Government. As a result, a task force coordinated by the Ministry of Agriculture and Forestry was set up in order to prepare the strategy in the latter half of 2003. Representatives from the Ministry of Trade and Industry, Ministry of the Environment, Ministry of Transport and Communications, Ministry of Social Affairs and Health, Ministry of Foreign Affairs, as well as the Finnish Meteorological Institute and the Finnish Environment Institute all took part in the preparation process. The strategy was prepared through seminars, use of existing data, expert consultations and assessments. The resultant proposal for Finland's National Strategy for Adaptation to Climate Change was presented at an open seminar in October 2004 and was also sent to several stakeholders for consultation. The general public was additionally able to comment on the draft through the Internet. All comments were taken into account when finalising Finland's first National Strategy for Adaptation to Climate Change in 2005 (Marttila et al., 2005).

The preparation of the Strategy largely reflects the process of environmental policy-making in Finland. Similar processes have been carried for other issues, for example sustainable development (The Government of Finland, 2006). The preparation of this strategy included representatives from all ministries and as well representatives of the civil society, all of which recognised the need mainstreaming of sustainable development goals to public policy. In the case of adaptation, the recognition of the need to adapt into the policy arena was seen to be result of international negotiations and the realisation that despite how successful mitigation efforts turn out to be in the long run, adaptation to changes in climate will be necessary in any case (Ministry of Agriculture and Forestry, interview 2). Policymakers and researchers working in this area shared a common understanding of the importance of developing an adaptation strategy that resulted in the redrafting the agenda on national climate policy to include adaptation from 2003 onwards.

4.4.2 Key Projected Impacts of Climate Change

Climate policy in Finland relies heavily on scientific knowledge as with other public policy issues too. The focus on climate related research in Finland has been placed on scenario work whilst relatively little research has been done in relation to identification of vulnerabilities. Reliable climate observations date back to the middle of the 19th century in Finland, with the systematic collection weather data since 1846. A study by Tuomenvirta (2004) found that the mean temperature has increased by 0.76°C degrees in the 20th century, though most of the warming has occurred in the spring, (an increase of approximately 2°C degrees), further observed in the earlier break of up ice and in the earlier onset of spring blooms. The study found no significant in nation-wide precipitation trends during the period studied, in contradiction with Sweden where a 15–20% increase in precipitation has been observed during the 20th century (Tuomenvirta, 2004).

The most recent climate scenarios used in the NAS are based on the IPCC emissions scenarios (B1, A1B and A2) and outline the likely changes in climate in Finland (Ilmatieteenlaitos, 2009). Although the models are based on the alternative emissions scenarios and thus differ significantly, warming of temperatures occurs in calculations based on all three scenarios, ranging from an optimistic 3°C to an increase of 6°C according to the most pessimistic scenario. From 2040, the speed of warming is more dependent on the cumulative amount of GHG emitted. The NAS also outlines future climate change scenarios as well as socio-economic scenarios and scenarios for natural systems, including soil, water, atmosphere, and flora and fauna (Marttila et al., 2005). The NAS further outlines the impacts of climate change on the fifteen sectors considered in the national strategy.

According to the NAS, warming of temperatures is more likely to occur during the winter than during other seasons. Precipitation is likely to increase between 12 and 22% depending on the emission scenario used, and more likely to increase in the winter rather than the summer, resulting in less snow cover during the winter. It is estimated that at the end of the century, climatic conditions in southern Finland are likely to resemble that of the current conditions of central Europe. The impacts of climate change will naturally affect all sectors of the society (Marttila et al., 2005); the NAS process identified impacts across all sectors, including the use of natural resources, industry, energy, land use, health, tourism and the insurance sector.

4.4.3 The National Adaptation Strategy

The overall goal of the National Adaptation Strategy is to mainstream adaptation measures into the Government's administrative sectors (Marttila et al., 2005). The strategy is a 281-page document that presents the latest scenarios (as described above), and identifies impacts of climate change and adaptation measures in 15 sectors. According to the NAS, adaptation measures are to be taken up by all sectors of public administration as a part of their regular planning, implementation and monitoring. All sectors are required to analyse and develop their capacities and increase the use of research information in order to adapt to climate change. Furthermore, the importance of coordination and cooperation between different branches of administration (sectoral, regional and local authorities) is underlined in the National Adaptation Strategy. General measures available to authorities include administration and planning, legislative measures and economic-technical measures. The Strategy further discusses a number of tools and systems (i.e. environmental management systems, environmental impact assessments and risk assessments) that can be used in the development of administrative capacities.

The NAS outlines different adaptation options within different sectors, taking into account the time frame and the nature of actors within each one (see Box 4.2 for the sectors considered). For each sector, the adaptive capacity of the actors, possible measures and research on adaptation within each sector are covered. In addition, indicative measures for adaptation to climate change are given; within each sector, adaptation measures are divided between those to be undertaken by the public or the private sector. Measures to be undertaken by the public sector are further divided into three categories of administration and planning, research and information, and normative (legislative) framework. Measures are additionally divided between anticipatory and reactive measures either in the immediate (2005–2010), short term (2010–2030) and long term (2030–2080).

Box 4.2 Sectors considered in the National Adaptation Strategy (adapted from Marttila et al., 2005)

- Use of natural resources
  - *Agriculture and food production*
  - *Forestry*
  - *Fisheries*
  - *Reindeer husbandry*
  - *Game management*
  - *Water resources*
- Biodiversity
- Industry
- Energy

- Transport and communications
- Land use and communities
- Buildings and construction
- Health
- Tourism and the recreational use of nature
- Insurance operations

Although it is recognised that most of the adaptation measures are to be taken at the local level, global inter-linkages between the impacts and adaptation of climate change are also acknowledged. As well as focusing on adaptation measures in Finland, the national strategy also highlights the need to adapt to changes taking place in other parts of the world (Marttila et al., 2005), by identifying preliminary estimates of how global climate change impacts will reflect on different sectors in Finland.

4.4.4 Implementation of the NAS

Mainstreaming and sectoral integration of adaptation into public policy have been generally highlighted as important and so far constitute an area with relatively few empirical studies (Kivimaa & Mickwitz, 2009; Kok & de Conick, 2007; Urwin & Jordan, 2008). The study by Urwin and Jordan explores the extent to which climate change concerns have been integrated into non-climate policy-making sectors. The study concluded that only a few existing policies explicitly encourage climate change adaptation across the sectors that were studied, and that the biggest challenges in polices arose from the mismatches of temporal and spatial scales with existing policies and objectives of adaptation policies. The Finnish case presents an interesting comparison in this regard, in that the integration of adaptation is required within each sector of administration.

According to the NAS, practical implementation primarily takes the form of various programmes and strategies, where initial plans cover immediate actions only (2005–2010). The Strategy is also implemented through planning of the ministries' operations and is subject to monitoring and changes as part of the regular follow-up carried out by administration. The Strategy further presents a preliminary set of indicators associated with the impacts of climate change and adaptation measures for the follow-up of the National Adaptation Strategy.

The Ministry of the Environment was first to establish a network on climate change adaptation in 2006. The main aim of the network is to prepare a work plan for the implementation of the NAS for the environment administration (Ympäristöministeriön työryhmä 2008). From the beginning, it was clear that the implementation of the NAS was to require specific, practical measures. The main task was to identify these within the sub-sectors and to designate a responsible body for each one (Ministry of Environment, interview 2). A network consisting members from each sub-sector was put together to identify both immediate and long-term measures related to adaptation and to promote their implementation and foster cooperation between various groups of stakeholders. The scope of the exercise expanded during the process as it became clear that more and more sub-sectors within the environment administration needed to be included in the action plan.

During the process, two main issues arose, mainly related to the issues of information and human capital (Ministry of Environment, interview 2). Firstly, most of the measures outlined were directed towards getting more information on climate change impacts on specific sectors. Design of specific measures remained relatively difficult given the unavailability of knowledge of very specific impacts. Secondly, the process of drafting a strategy was found to be taxing on human resources and at times was considered a problem when this form of strategy work was being carried outside normal duties as an extra layer. This is a similar constraint to those identified at the municipal level, where normal bureaucracy and administrative duties leave little time for developing new or innovative measures to tackle already existing or new issues.

Generally, the 2008 implementation plan underlines the importance of research into the impacts of climate change and the possible alternatives to adaptation as a starting point for any adaptation strategies. Adaptation measures are to be carried out by all actors in the environment sector: the Ministry of the Environment, regional environment centres, the Finnish Environment Centre and the Metsähallitus.[1] The adaptation measures to be undertaken are detailed in Box 4.3.

Box 4.3 Adaptation measures in the environment administration (Ympäristöministeriön työryhmä 2008)

Biological diversity and outdoor recreational use

- Administrative measures, legislative measures, research

Land use and building

- Revision of land use legislation, monitoring, research

Housing (residential and commercial)

- Building legislation, building management, research

Environmental protection

- Environmental permits, waste management, research

Water resources management

- Flooding, waste water management, dam safety, rehabilitation of waterways

---

[1] Metsähallitus is a state-owned enterprise that runs business activities while fulfilling many public administration duties, and administers more than 12 million hectares of state-owned land and water areas.

Within this sector, there are sub-sectors that have already integrated adaptation concerns into their decision-making procedures. For example, information on climate change impacts and adaptation needs within water resources management have already been integrated into work within the environment administration (Ministry of the Environment, interview 1). Within many of the other sectors, however, a lack of adequate research knowledge in terms of possible impacts or in terms of different adaptation options has reportedly slowed down the implementation of the national strategy (Ministry of Agriculture and Forestry, 2009). For example, measures within the fisheries and reindeer herding are based on current climatic variability and monitoring, since there has not yet been enough basic research into impacts and adaptation. Within the natural resource field, several research programmes are currently underway that will eventually aid the evaluation of different adaptation options.

Although the national level is crucial in steering adaptation, it is also possible that regulations and recommendations may be changed or redrawn due to climatic changes already ongoing at the sub-national level. For example, changes in water courses that can be attributed to climatic changes have been observed in the case study area and changes in regulations have already been made (Uusimaa Regional Environment Centre, interview). In addition, other pieces of legislation that relate to adaptation have been revised in terms of climate concerns. The 2000 Land Use and Building Act was revised in 2008 in terms of mitigation and adaptation, and now considers not only issues to do with flooding, but building regulations in terms mitigation and energy efficiency as well (Valtioneuvosto, 2008).

With regards to regional environment centres, the extent to which the NAS has been implemented has varied (Uusimaa Regional Environment Centre, interview). There have not been many centrally-coordinated or explicit measures to mainstream or feed the NAS into the regional centres. Instead, the uptake of the NAS within the regional environment centres depends very much on the individual capacities of the centres. In the case study area, the main obstacle in integrating NAS measures into normal day-to-day activities was considered the lack of adequate human capital (Uusimaa Regional Environment Centre, interview).

4.4.5 Evaluating the Implementation of the NAS

Progress in implementing the adaptation measures was evaluated for the first time in early 2009 (Ministry of Agriculture and Forestry, 2009). Initial measures in the NAS target 2006–2015, while another evaluation will be carried out in the period between 2011 and 2013. Different administrative sectors' success in implementing initial adaptation measures has been varied. Many of those interviewed considered it unsurprising that the environment administration had been the first to push the adaptation agenda forwards, as it was considered to be a closely-related or natural area for the Ministry to consider. This to a certain extent highlights the fact that adaptation is still considered as an environment issue in Finland, and not an issue that all sectors need act upon immediately.

In terms of other sectors, early actions by the Ministry of Agriculture and Forestry have concentrated on the natural resource sector and in 2001 prepared a background document that aided in the drafting of the National Climate Strategy (Maa- ja metsätalousministeriö, 2001). The document dealt with mitigation almost exclusively in discussing carbon emissions emanating from the agriculture and forestry sector, as well as the Ministry's role in mitigating emissions in order to reach the Kyoto targets. The main interest of the Ministry has been the role of agricultural and forestry policies in national climate strategy and policy. Particularly of interest has been the function that forests and agricultural lands have as carbon sinks (Metsäntutkimuslaitos, 2004; Regina, Lehtonen, & Esala, 2008; Uusivuori et al., 2008).

In a report on the future of the agriculture and forestry sector until 2015, the Ministry of Agriculture and Forestry set the implementation of the national climate and energy strategy as a priority (Maa- ja metsätalousministeriö, 2006) which also includes the implementation of the National Adaptation Strategy. The Ministry has also produced a report on future scenarios in the agriculture and forestry sector with regards to carbon emissions (Maa- ja metsätalousministeriö, 2008b). The National Forestry Programme 2015 prepared by the Ministry has measures to evaluate the impact of climate change and the need to build up a preparedness system for future climate related forest destruction (Maa- ja metsätalousministeriö, 2008a). More research is deemed necessary in terms of adaptation measures in the agricultural sectors, while adaptation in the management of water resources has been partly influenced by the EU Flood directive. The need for changes in water levels and damns in terms of adaptation has already been evaluated and the necessary changes in guidelines are underway. However, as of yet, there are no explicit plans to mainstream adaptation measures to the day-to-day activities within the sector.

The Ministry of Transport and Communication is also responsible for climate change mitigation and adaptation in their sector (Jalasto et al., 2007). The Ministry appointed a Committee for Climate Policy that is currently drafting a proposal for 2009–2020 in order to carry out the national climate policy objectives. The issues of concern include energy efficiency, land use, planning of transportation systems and their greenhouse gas emissions, and measures for adaptation. In May 2008, the first seminar of the Committee for Climate Policy was held in which issues related to climate change and transport sector were discussed (Järvi & Laurikko, 2008; Tynkkynen, 2008). Sub-sectors of the transport sector were also discussed, including rail transport (Niemimuukko, 2008), ship transport (Mustamäki, 2008), aviation (Salonen, 2008) and road transport (Hirvelä, 2008), among others.

The Ministry of Employment and the Economy has been responsible for preparing the national climate policy since the beginning of 2008, having received responsibility from the Ministry of Trade and Industry that ceased operations on December 1st 2007. The Ministry also heads a high-level working group of government officials, which has assisted the Ministerial Working Group on Climate Change and Energy. As the Ministry is in charge of the energy policy as well as the climate policy, much of the emphasis is placed on mitigation of climate change. Prevailing themes in the Ministry's agenda include energy security, the impacts of climate policies on the energy sector (Forsström and Lehtilä, 2005), and the effects of climate policies on the national economy (Honkatukia, Kemppi, & Kerkelä, 2005). Emissions trading, energy saving, renewable energy strategies and the implementation of the Kyoto mechanisms also fall under the remit of the Ministry. So far, the Ministry of Employment and the Economy has focused on raising awareness of adaptation within its area of responsibility (Ministry of Agriculture & Forestry, 2009). A limited amount of research on the socio-economic aspects and impacts of adaptation has been conducted thus far; furthermore, it is unlikely that there will be active involvement in adaptation until it is seen to more dramatically affect the economic performance of the sector (Ministry of Employment and the Economy, interview).

Overall, the implementation of the NAS has been progressing well in some areas and more slowly in others (Ministry of Agriculture and Forestry, 2009). Through the use of a preliminary adaptation indicator developed for the evaluation of implementation, it is estimated that overall the need for adaptation has been recognised across different sectors of administration, and overall, Finland is on the second step of the indicator, see Box 4.4 for more details.

Box 4.4 Indicator of progress on adaptation (adapted from Ministry of Agriculture and Forestry, 2009)

Step 1.
The need to adapt has been recognized among a small group of pioneers in the sector
Little research done on the impacts of or adaptation to climate change
Some adaptation measures identified but not yet implemented
Step 2.
Need for adaptation measures recognized to some extent in the sector (some decision-makers)
Impacts of climate change known to indicatively (qualitative information), taking account of the uncertainty involved in climate change scenarios
Adaptation measures identified and plans made for their implementation, some of them launched
Step 3.
Need for adaptation measures quite well recognized (majority of decision-makers) in the sector
Impacts of climate change quite well known (quantitative information), taking account of the uncertainty involved in climate change scenarios
Adaptation measures identified and their implementation launched
Cross-sectoral co-operation on adaptation measures started
Step 4.
Need for adaptation measures widely recognized and accepted within the sector

Adaptation incorporated into regular decision-making processes
Impacts of climate change well known, within the limits of the uncertainty involved in climate change scenarios
Implementation of adaptation measures widely launched and their benefits assessed at least to some extent
Cross-sectoral cooperation on adaptation measures an established practice
Step 5.
Adaptation measures under the Adaptation Strategy or recognized otherwise implemented in the sector

Preliminary knowledge of climate change impacts exists and initial adaptation measures have been identified and some are being carried out. The environment sector, traffic, land use, agriculture and forestry have made progress in implementation, while in other sectors the progress has not been as fast (Ministry of Agriculture and Forestry, 2009).

The environment administration has advanced furthest in implementation with very specific measures. The sections within the environment administration that have already integrated adaptation concerns into their day to day activities have admittedly benefited from existing knowledge and supporting policies and directives within that sector. Sectors such as water management and flooding have already considered adaptive measures and have thus benefited from the NAS process. However, the reasons for the early implementation within the environment administration were considered to be the recognition of the importance of adaptation within the sector and its consideration as an important issue that relevant actors are already engaged in (Ministry for the Environment, interview 1).

On the whole, some sectors of the Finnish administration have been successful in integrating adaptation into their existing management plans. Despite the early efforts in planning and designing adaptation strategies, it was admitted that the 'importance of adaptation will be realised more as soon as more concrete impacts of climate change will be seen' (Ministry of Agriculture and Forestry, interview 1). It was questioned whether adaptation concerns will be taken into account within the economic sector unless there are economic incentives as in emissions trading schemes in mitigation policy (Ministry of Employment and Economy, interview).

This case study here, then, argues along the lines of a recent study on the integration of climate concerns, both mitigation and adaptation, into policy in Finland (Kivimaa & Mickwitz, 2009). These authors studied the integration of climate concerns at the national level and conclude that within the key government strategies, the inclusion has been extensive, although the weighting given to climate issues differs. Furthermore, the authors argue that consistency of climate issues with other policy objectives is rarely addressed nor are questions of financial or human resources. Sufficient coordination horizontally at the national level, as also demonstrated in this chapter, does exist but no new measures or ways of tackling the climate challenge in terms of organisational structures have been put forward.

4.4.6 Role of Research in Climate Change Adaptation

Research into environmental change has always played a major role in Finnish environmental policy-making, and is also true for climate change. The environmental research field is characterised by collaborative research programmes that involve both universities and public research organisations, thus crossing the divide between basic and applied research. The role of public research organisations in providing applied research within each governance sector has been a strong and useful tool in developing knowledge-based political decision-making (Rantanen, 2008). If this argument is followed, it can be claimed that engaging the public research organisations into climate adaptation research through research programmes increases the likelihood of research findings that directly contribute to implementation of adaptation measures. In Finland, basic research on climate change has focused on climate change scenarios and socio-economic scenarios, as well as impact studies, while less research has been done on mapping vulnerability.

As indicated above, Finnish climate research has long traditions, with the earliest observations of climate through ice break up series in Lapland since the late 17th century (Tuomenvirta, 2004). One of the first concerted efforts in terms of climate change research was the Finnish Research Programme on Climate Change (SILMU), a multidisciplinary programme of the Academy of Finland that ran from 1990 to 1995. Research themes were divided into atmosphere, water, land ecosystems and human interaction, bringing together more than 80 research projects and involving over 200 researchers in seven universities and 11 research institutes. Climate scenarios were the main priority so as to make the results of different research projects comparable in order to gain a better understanding of climate impacts (Finnish Environment Institute 2008a). Scenarios were an important part of SILMU; three scenarios of temperature and precipitation change were developed based on global climate models results over Finland: a central, 'best guess' scenario, together with lower and upper estimates representing an unspecified uncertainty range.

Further climate scenarios were produced as part of the Finnish Global Change Research Programme (FIGARE) during 1999–2002. The programme consisted of projects in natural and social sciences, including technology and economics, in order to understand changes in the global environment and the resulting effects in Finland. Funding for the programme was provided by the Academy of Finland as well as different Ministries. As part of FIGARE, the FINSKEN project developed projections of changes in environmental, socio-economic and technical, sea-level and atmospheric scenarios in Finland during the 21st century (Carter, 2004). FINSKEN scenarios were further used in the Integrated assessment modelling of global change impacts and adaptation project (FINESSI). This three year project developed a computer-based evaluation framework that investigated the impacts of global change on natural and managed systems in Finland (Carter et al., 2004).

Adaptation to climate change was first discussed in a preliminary analysis conducted by the Finnish Environment Institute (Carter and Kankaapää, 2003). This preliminary analysis highlighted that adaptive capacity across different sectors was poorly understood and that further research was necessary. It was also stressed that adaptation should be an integral part of any climate change research. Since this initial assessment on adaptation, there have been several nationwide research programmes that have brought together various governmental research institutes and universities. Overall, the main emphasis in research programmes that focus specifically on adaptation has been and still is to support national and regional measures for adaptation and further link climate change impacts to possible measures for adaptation.

The first of these was a larger two-year research programme on adaptation titled *Assessing the Adaptive Capacity of the Finnish Environment and Society under a Changing Climate* (FINADAPT). This research programme was carried out to improve the understanding of how Finnish society and the environment can adapt to a changing climate (Carter and Kankaapää 2003), running parallel to the NAS though not explicitly linked to it. Several research institutes (11 in total) participated in the programme, coordinated by the Finnish Environment Institute (SYKE). Within FINADAPT, studies were carried out during 2004–2005 based on literature reviews, interactions with stakeholders, seminars, and targeted research efforts. The programme consisted of 14 different work packages that covered various different sectors. [2] FINADAPT produced a number of publications based on the work packages and their sectoral areas (e.g. Carter, 2007; Carter, Jylhä, Perrels, Fronzek, & Kankaapää, 2005; Peltonen et al., 2005).

The role of research into climate change impacts and adaptation options was recognised early on in the preparation of the NAS and after its introduction, and there were calls for research programmes that would support the implementation of adaptation. These calls were met in the form of the Climate Change Adaptation Research Programme (ISTO) that focuses on providing practical research knowledge and tools for the implementation of the national strategy. The programme was prepared in cooperation between government ministries in 2005 and initiated in 2006. Research needs identified during the preparation of the Strategy and relevant research programmes were taken into account in the planning of the research programme.

Within ISTO, 16 projects are funded under the Environment Cluster Program of the Ministry of the Environment, as well as by the Ministry of Agriculture and Forestry and the Ministry of Transport and Communications. The work is undertaken in cooperation with relevant research bodies and other parties with the aim to ensure science-policy interaction, which is the main function of public research organisations. ISTO further contains a steering group of representatives of relevant ministries and scientific and funding institutions that steers research work, monitors its progress and organises its evaluation. In 2009, the ISTO steering group was

[2] The work packages under FINADAPT included (1) co-ordination, (2) climate data and scenarios, (3) biodiversity, (4) forests, (5) agriculture, (6) water resources, (7) human health, (8) the built environment, (9) transport, (10) energy infrastructure, (11) tourism and recreation, (12) economic assessment, (13) urban planning, and (14) a stakeholder questionnaire.

transformed into a national adaptation network that consists of representatives of different ministries and other relevant stakeholders.

The role of the ISTO programme was highlighted as a positive factor in the implementation of the national strategy (Ministry of Agriculture and Forestry, 2009). The multidisciplinary nature of the whole programme was considered a strong point, although more dialogue across the research projects was considered necessary to improve the programme (Valli & Sierla, 2008).

Furthermore, multidisciplinary projects within the programme are considered useful though rare so far, as much emphasis is still placed on the natural sciences and the impacts of climate change. Secondly, the aim of the programme in providing applied research knowledge and practical information for adaptation has been to a certain extent met. So far, socio-economic research into adaptation measures and their impacts has been more limited, and can also been seen in the projects that have been part of the wider research programmes. This was also identified as a gap in the evaluation of the ISTO programme (Valli & Sierla, 2008).

More importantly, the programme suffered considerably from the cuts in funding that were made after its launch. This shortfall was mainly due to cuts in research funding in all ministries and thus not specifically an adaptation-related decision. Available funding is now roughly a third of what was considered necessary during the planning of the research programme (Valli & Sierla, 2008). This has led to downscaling research efforts in terms of scope of research, length of projects and the ability to hire new researchers or train younger researchers, which has meant that not all the ambitious targets of the research programme can be met in the long term.

Overall, the analysis of research efforts with regards to climate change and adaptation in Finland shows how adaptation research has evolved towards the need for more specific and practical information. The first research programmes on climate change focused on providing scenarios and reliable climate data for Finland, after which the focus has been on specific climate impacts for specific sectors or regions. In more recent years, especially since the publication of the NAS, efforts have been on funding and conducting more applied research within a specific governance sector in order to provide knowledge and tools for implementation of the NAS. This does not, by any means, imply that basic climate data research is no longer necessary. Rather, the aim is to better coordinate basic climate research (e.g., ACCLIM II) and make it into a common platform for all other research projects under the ISTO umbrella.

4.5 The Role of the European Union in Climate Policy

Finland joined the European Union in 1995 and this has had a significant effect on Finnish policy-making, including climate issues. Since 2001, the National Climate and Energy Strategies have been revised every few years in order to accommodate the commitments agreed to in term of emissions reductions, with the biggest changes arising from the introduction of the EU Emissions Trading System. The emphasis on renewable energy in the 2020 EU Climate Package has additionally increased the visibility of renewable energy in the latest Finnish national strategy. It is thought that the upcoming Government Foresight Report in 2009 will further contribute to the long term vision of Finnish climate policy in line with the EU Climate Package for 2020 (The Prime Minister's Office, interview).

With regards to adaptation, the impact of the EU has been less significant, mainly due to the fact that Finnish action has preceded that of the EU. Furthermore, Finland was consulted in the preparation processes of the EU Papers in relation to the NAS (Ministry of Agriculture and Forestry, interview 2). As Finland was the first country within the EU to produce a national strategy prior to EU guidelines or directives, the extent to which the Green and White Paper influence the implementation of the NAS is yet unclear (Ministry of Agriculture and Forestry, interview 1&2). In terms of reviewing the NAS in 2013, support for indicators for measuring adaptation are considered to be something that could be addressed at the EU level (Ministry of Agriculture and Forestry, interview 2).

Individual sectors within government have already acted in terms of EU directives that relate to adaptation, among which the Flood Directives has been the most prominent. Although land use is an issue under national jurisdiction, the Flood Directive influences land use indirectly through water management guidelines. The EU directives have also had the effect of replacing some national environmental regulation. The Ministry of Transport and Communications noted that environmental regulations that previously came from the Ministry of the Environment now come directly from the European Union (Ministry of Transport and Communications, interview).

In addition, it was noted that the EU is not the only international actor that influences national adaptation policies in specific sectors. For instance within the agricultural sector, participation of the Ministry of Agriculture and Forestry in the work of the United Nations Food and Agricultural Organisation (FAO) and Organisation for Economic Cooperation and Development (OECD) on adaptation has enabled the Ministry to build its capacity in terms of information and networks (Ministry of Agriculture and Forestry, interview 1).

At the regional and local level, the effects of the EU are less visible in terms of directives. Mostly, their effects are felt through directives and changes in national legislation that then is left for implementation by the municipalities or through regional development programmes funded by the EU (Uusimaa Regional Council, interview 2). However, as the following paragraphs will show, the role of the EU is considered to be very important in terms of enabling adaptation, fostering networks within the wider regions and building capacity in the absence of direct national support.

4.6 Other Actors in Climate Change Adaptation

Aside from the government at various levels of social organisation, there are only a few non-governmental actors that have engaged with climate change adaptation. The Association of Finnish Local and Regional Authorities (ALFRA) promotes and coordinates climate activities of municipalities, organised the first climate conference for municipalities in 1997 and an additional three conferences since (Kuntaliitto, 2001). ALFRA also provides the main avenue for municipalities to share information on climate change and energy issues, mainly through conferences, workshops and the internet (ALFRA, 2007). These include best practice cases in a variety of municipal functions, such as construction, energy production and waste management.

Since 1997, the focus on ALFRA activities has very much been on mitigation (Kuntaliitto, 2001). In a review conducted by ALFRA municipal experiences, it was concluded that most municipalities have succeeded in reducing their carbon emissions and have had positive experiences with the climate strategies that were drawn up (Hakanen, Luoma, & Mynttinen, 2008). Adaptation to climate change was included fully into ALFRA activities in 2006 and was discussed in the 4th Climate Conference for Municipalities in 2007.

In 2008, ALFRA published new guidelines for climate change mitigation and adaptation (ALFRA, 2008). These new guidelines outline recommendations to municipalities, to the central government and those to be adopted by ALFRA itself. For the municipalities, the main measures recommended include the formulation of a climate strategy and the incorporation of climate concerns in everyday planning. ALFRA encourages the central government to pay more attention to flood control activities and to increase the funding for research into climate change impacts. ALFRA itself has selected climate change mitigation and adaptation as one of the strategic main focuses of its activities and has pledged to assist municipalities in adapting to climate change and to assure that responsibilities and benefits are shared in an equal manner nationwide. The background memo for the new guidelines focuses more on the role that municipalities have to play in climate change adaptation (ALFRA, 2008). Despite the fact that adaptation is considered in here, the main emphasis is still on mitigation of climate change. Main emphasis regarding adaptation here is put on land use planning and building regulations, especially in areas that are prone to flooding. Other technical measures are also specified.

4.7 Sub-national Level Adaptation Work

Regional and local action is crucial in climate change adaptation, including both planning in terms of suitable local adaptations as well as implementation through acceptable solutions and practices. So far, some research has been conducted on the extent to which climate change has been taken into account in land use planning from both adaptation and mitigation points of view, although some examples do exist on land use planning in general (Harmaajärvi, 2005) and in specific localities in Finland (Wahlgren, Kuismanen, & Makkonen, 2008). As well as studying horizontal integration at the national level, Kivimaa and Mickwitz (2009) also studied the vertical integration of climate concerns through a case study in Finland. The study indicates that integration of climate issues into regions' general plans is rather limited. The preferred mode of integration is to include climate concerns into regular planning and operations rather than design specific climate budgets. Most of the sub-national adaptation work has been taking place within specific projects, often in collaboration with research institutes and universities.

Out of the 20 Regional Councils at the regional level, the majority mention climate change and climate change adaptation in their regional development plans and a few in their regional development programmes. Seven Regional Councils stand out as having taken climate change and adaptation into account not only in their regional development plans for the long term, but also in their shorter term implementation strategies, i.e. the Regional Development Programmes. These are the Regional Councils of Lapland, Pirkanmaa, Pohjanmaa, Pohjois-Karjala, Itä-Uusimaa and Uusimaa (Itä-Uudenmaan liitto, 2007; Lapin liitto, 2005; Pirkanmaan liitto, 2007; Pohjanmaan liitto, 2006; Pohjois-Karjalan liitto, 2005; Uudenmaan liitto, 2006). The majority of the regional environmental centres have also recognised climate change as a threat to the global, national and regional environment in the strategies. However, most of the regional actions in the implementation strategies focus on mitigation, emphasising energy efficiency, the role of land use planning in reducing energy use in industry and traffic. Only a few RECs have outlined direct measures for climate change adaptation in their regional plans, outside their legislative duties within the environment administration. The most advanced measures for adaptation are undertaken by the REC of Uusimaa, South Western Finland and Pirkanmaa, for more details of regional strategies (cf. Haanpää, Tuusa, & Peltonen, 2009).

At the beginning of 2009 there were 348 municipalities in Finland, with the number slowly decreasing as the trend of municipal mergers continues. With regards to municipalities, ALFRA launched the Cities for Climate Protection campaign in Finland in 1997 as part of the ICLEI – Local Governments for Sustainability network. By the year 2000, approximately 40 municipalities had joined; presently there are over 50 members (Hakanen & Luoma, 2008). In comparison to the municipalities that are taking part in the LA21 implementation (over 270), climate appears not to attract as much attention. However, in terms of percentage of population, some 50% of the total population of Finland live in the 50 municipalities taking part in the climate campaign, as the municipalities that have joined represent the most population rich in Finland.

In order to identify and analyse good practices of climate adaptation in Finland in this case study, the Uusimaa region was chosen for this case study (see Fig. 4.1). Uusimaa is located in Southern Finland on the Baltic Sea and the 6,366 km^2 region has approximately 1.4 million inhabitants, over a quarter of the country's population. The region consists of 24 different municipalities, ranging from the only metropolitan area of the country, including the capital city to smaller, more sparsely populated rural municipalities. The economic sector is service industry-oriented with over 80% of employment within that sector; agriculture accounts for less than one percent of the economic output of the region (Uudenmaan liitto, 2009).

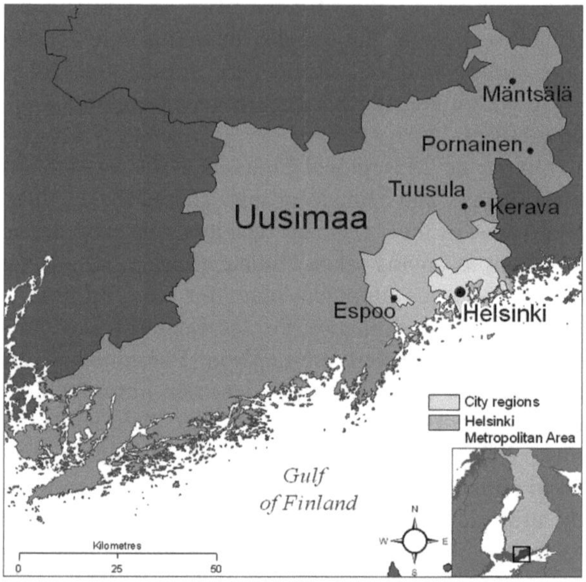

Fig. 4.1 The Finnish case study areas

The Uusimaa region represents an ideal opportunity to explore climate change adaptation and adaptive capacity in a nested case study. There are several climate strategies and adaptation processes that are happening in the case study region. Some of these focus on adaptation at the regional level, while others concentrate on a municipal cooperation or a single municipality (see Table 4.1 for more details on the case studies). All of these are voluntary project-funded initiatives that aim to produce a climate strategy for the stakeholders taking part. The case study region represents and highlights many opportunities and challenges in terms of cooperation across scales, varying capacities to respond to climate change and differing priorities in terms of adaptation measures.

4.7.1 Uusimaa Regional Council Climate Strategy

The Uusimaa Regional Council has been involved in climate issues since 2005 when it became involved in an EU-funded research project. Since then, the Council has been involved in calculation of emissions within the region and has done so for all municipalities (Huuska, 2006). With regards to adaptation, the council was a participant in the *Developing Policies & Adaptation Strategies to Climate Change in the Baltic Sea Region* (ASTRA) project, which was co-financed by the Baltic Sea Region's INTERREG III B Program of the European Union. The main objective of the ASTRA project was to assess regional impacts of the ongoing global change in

Table 4.1 Details of selected sub-national cases

Case studies	Climate change strategy development
Uusimaa Regional Council	The preparation of climate strategy that includes both mitigation and adaptation components for the region currently in progress.
Helsinki Metropolitan Area Council	The HMAC published a climate strategy in 2007 that focused on mitigation. The council is currently in preparing an adaptation strategy that includes the cities that form the council.
KUUMA Municipal Cooperation	The KUUMA cooperation includes six municipalities in the Uusimaa region and a climate strategy that includes both mitigation and adaptation is currently under way.
Espoo municipality	Espoo municipality was one of the first municipalities to put together a climate adaptation preparedness report that outlines measures that need to be undertaken across various municipal sectors.

climate and to develop strategies and policies for climate change adaptation (Hilbert, Hilbert, Mannke, & Schmitdt-Thome, 2007). A flood map of the Uusimaa region was prepared as a part of the project, which highlighted flood prone areas in the region. The project stressed that further mapping of flood risks is necessary in some areas, and that this work should begin as soon as possible (Hintsala, 2007).

The preparation of the regional council's climate strategy is currently underway, of which adaptation will be a component (Uusimaa Regional Council, interview 1). The preparation of the strategy is in early stages, with a main focus thus far on identifying mitigation measures and particularly the role of spatial planning in reducing carbon emissions (Uusimaa Regional Council, interview 1). The actual format and scope of the adaptation part of the strategy is yet to be decided but a workshop has been held in order to develop the adaptation component further. Although adaptation is seen important, the need for a separate strategy was questioned, as it was stated that climate issues including adaptation have become or will in the future be part of considerations and decision-making within the Council already (Uusimaa Regional Council, interview 2).

Nevertheless, the Council sees its role as one that activates other stakeholders and brings them together both in terms of adaptation and in other unrelated actions. This includes not only municipalities within the region, but also business and chambers of commerce. Main challenges were considered to be the ability to negotiate between various stakeholder interests, mainly in terms of economic and environmental priorities of land use. An underlying issue is the nature or mandate of Regional Council in terms of committing people to their work that they are doing, as they have no power to do that but rather rely on voluntary cooperation and willingness to participate on the side of the regional stakeholders. Further concerns were raised about short term funding in terms of the actual implementation of the strategy and whether it will remain a strategic document with no real measures to steer the region towards adaptation (Uusimaa Regional Council, interview 2).

4.7.2 Climate Strategy for Helsinki Metropolitan Area Council 2030

The Helsinki Metropolitan Area Council (HMAC), together with experts from the cities of Helsinki, Espoo, Kauniainen and Vantaa, prepared a *Climate Strategy for Helsinki Metropolitan Area 2030*, approved by the HMAC board in 2007. The aim of the Climate Strategy is to reduce greenhouse gas emissions and to produce a joint view of the cities on ways to reduce greenhouse gas emissions (YTV, 2007). In addition, the Strategy aims to find practical ways to meet the reduction commitment and to prepare an agreement of the HMAC cities and other parties on common ways to reduce greenhouse gas emissions. In order to implement the strategy, HMAC started a three-year Julia 2030 project, jointly funded by the EU Life + programme and the HMAC. The project budget is 2.1 million Euros, with half of the money covered by the HMAC and its partners.

The preparation of the adaptation component of the strategy was begun in 2009 as a part of two projects, both partially supported by the EU for a period of three years. Beyond the Julia 2030 project, the preparation of the strategy is also supported by a research project, *Climate Change: Impacts, Costs and Adaptation in the Baltic Sea Region* (BaltCICA) that is partially funded by the EU and expected to be finished in two years. Currently, the process has moved towards identifying stakeholders that are to be involved in the preparation of the strategy, as well as tools that can be used to integrate these various views into the planning of the strategy.

HMAC benefits from a long tradition of environmental and air quality work that has been going on in the council for a long time. In fact, it was highlighted by all sub-national actors as a good example that has influenced and helped others to begin work on adaptation within the region. The HMAC's focus has broadened from air quality to mitigation and now to adaptation, with a team of people assigned specifically to adaptation issues. Furthermore, the HMAC team has good connections to the adaptation research community in Finland and is able to access research networks and information.

Concerns in terms of executive power over HMAC municipalities were expressed (HMAC, interview) as the municipalities will not be bound by the recommendations of the strategy once it is published. Therefore, the aim of the adaptation team is to engage the stakeholders from the cities' administration in order to ingrain the importance of the strategy from the beginning and thus influence and improve the implementation process. Also, it was noted that improved cooperation between regional bodies in terms of climate strategies would benefit the uptake of adaptation within Uusimaa (HMAC, interview).

4.7.3 Climate Strategy for the KUUMA-partnership

Six municipalities of Järvenpää, Kerava, Mäntsälä, Nurmijärvi, Pornainen and Tuusula in the central Uusimaa region have formed a partnership to improve the quality and delivery of municipal services in the area. The municipalities have experienced a sharp rise in population as people have moved to the area over the last decade, and this trend is considered to continue (KUUMA, 2004). The aim of joint action is to further the competitiveness of the sub-region in terms of the economy but also to improve municipal services across different municipal sectors. Three of the municipalities, Kerava, Järvepää and Tuusula are more urbanised with distinguishable city centres while the remaining three, Pornainen, Mäntsälä and Nurmijärvi have a more dispersed and rural settlement structure.

As part of the municipalities' cooperation in matters concerning the environment, the partnership begun preparations for a climate strategy in 2007 (KUUMA, 2007). The initial work that has been carried out has so far focused on mitigation, including seminars on energy efficiency, calculations of greenhouse gas emissions and identification of best practice cases that can be used to engage and activate other stakeholders. The initial strategy mainly includes measures for mitigation, but a preliminary analysis for adaptation was carried out in the autumn of 2008 and will eventually lead to an adaptation strategy. The implementation plan of the strategy is due to be completed during 2009 and will be used in implementing the measures outlined in the strategy beginning 2010 (KUUMA, 2004).

As a tangible result of climate work through the cooperation, all KUUMA municipalities have joined or have made the commitment to join the Cities for Climate Protection Campaign in Finland. In addition, an environment programme for municipal technical services has been conducted and is due to be reviewed every two years. In terms of the climate strategy, the aim has been to engage the environment sector of the participating municipalities, after which the aim is to broaden sectoral cooperation within the municipalities to include technical services and schools in terms of energy efficiency measures (KUUMA Climate Strategy, interview). The strategic goal of the climate plans now focuses on the year 2050 with a mid way point measures for 2020. Adaptation priorities that have been identified so far concentrate on the risks posed by heavy rainfall and extreme storm events and their impact of urban structures and built environments. Similarly, issues of land use planning and building regulations are necessary to take into account in the planning of the strategy. One of the tools used in the preparation was the collection of best practices across the municipal and private sectors in order to demonstrate the benefits of including climate concerns into normal practices. There are also plans to start up a private sector forum that will discuss climate and energy issues that could also provide a venue for voicing adaptation issues (KUUMA Climate Strategy, interview).

Overall, interviewees within the KUUMA municipalities considered that preparing a joint climate strategy not only strengthened the cooperation between the municipalities, but that it was necessary in terms of the threat that climate change poses to individual municipalities. A regional strategy helps to deal with regional impacts of climate change (HMAC; Regional Council, interviews), supplemented by further linkages with other actors in the region. Furthermore, regional cooperation has the advantage of pooling resources to improve the adaptive capacity of the municipalities. As preparation of climate strategies within municipalities is currently a voluntary initiative and outside of the normal functions of municipal environmental work, there is no state funding available. Thus, only by pooling their resources are smaller municipalities such as the KUUMA municipalities able to consider adaptation (City of Kerava, Mäntsälä, Pornainen, Tuusula, interviews).

In general, it was felt that there is enough information of climate change and adaptation but that time and human resource constraints did not allow for the municipalities to utilise these to the full extent. However, as the municipalities within the KUUMA partnership are not on the coastline, it was considered that the most immediate impacts such as sea level rise, were not a direct threat and therefore the identification of adaptation options is not an urgent task. Moreover, as severe climate impacts had not yet been felt in these municipalities, the nature of adaptation and its implications were to some extent still under discussion. It is yet unclear to what extent adaptation is considered as more of a change in consumption and organisation of social life in general, or whether it consists of purely technical measures that need to be implemented by municipalities.

4.7.4 Espoo Preparedness Strategy

The city of Espoo is located on the South Coast in the Helsinki Metropolitan Area (see Fig. 4.1). The city has an extensive stretch of coastline and suffered from exceptional flooding in July 2004 and January 2005. However, the risk of flooding not only occurs on the coast, but within rivers and streams in the area as well. National and EU regulations suggest that no residential buildings should be built below 3m of the sea level. Currently, there are approximately 2,400 buildings and 800,000 km^2 below that recommended height and it is estimated that once in a 100 years flooding may cause significant economic damage in these areas (Espoon tulvatyöryhmä 2005).

Discussions of global environmental issues have long traditions in Espoo, with the earliest mentions of climate change in city environment plans in the early 1990s (Espoo Environment Centre, interview 2). Climate work in Espoo was begun in 2005 when the issue was more widely discussed across the city's administrative sectors and led by the Environment Centre. The city of Espoo appears to have the most advanced adaptation strategy when it comes to individual cities in Finland. A Climate Change Preparedness Strategy for Espoo city was conducted as a part of the wider HMAC Climate Strategy (Soini, 2007). The Strategy was based on a range of interviews with the city officials from several different sectors, including the city planning department, estate management, city technical department, city water authority, emergency rescue services of Western Uusimaa, and procurement department.

For the Preparedness Strategy, a risk management approach was taken in terms of current and future risks arising from climate change impacts and extreme weather events. This is also influenced by the fact that Espoo has an environment security working group that has focused on climate change as a risk. It is admitted that the preparedness plan still maintains a very technical approach to climate change adaptation (Espoo Environment Centre, interview 1), though there are discussions on widening the approach to include issues such as health.

The report sets out the main likely impacts of climate change and their consequences for infrastructure, environment and the built environment. It further sets out the main adaptation measures to be taken by the different city departments. Although the report deals with Espoo exclusively, the main conclusion is that the all cities in the Helsinki Metropolitan Area ought to work within the framework of the Climate Strategy for Helsinki Metropolitan Area Council 2030 as this is likely to yield the best results.

Although awareness of climate and the need to take adaptation into account is instructed in the preparedness strategy, it was nevertheless noted that the existence of a strategy does not in itself equal implementation and uptake of measures (Espoo Environment Centre 1 & 2; Espoo Environmental Council, interviews). Designating people with specific responsibilities is considered to be useful to ensure commitment and ensuring the degree of mainstreaming with responsibilities across city's administrative sectors. Further, although the report acknowledges the difficulties of estimating the impacts of climate change in Espoo due to inherent uncertainty, it nevertheless concludes that there is enough of a knowledge base to begin implementing adaptation measures (Soini, 2007). It should, however, be noted that despite the fact that adaptation has been a concern for Espoo in terms of impacts such as sea level rise and heavy rainfall, mitigation remains a priority alongside adaptation.

As with other municipalities in Finland, the lack of climate legislation was seen to hinder the identification of adaptation options and uptake of measures. As long as there is no legislation that municipalities need to take into account, they are to a large extent unable to do so. Espoo, on the other hand, has been able to access resources through its own networks. These include the regional networks with HMAC as well as EU projects in cooperation with research institutes within Finland. The profile of adaptation and other environmental issues is also supported by the large number of Green League council members in the City Council. In terms of importance, climate change and adaptation rank amongst the four top issues that the Council has dealt with within the last few years (Espoo Environmental Council, interview 2). Amongst other awareness raising activities, the Environmental Council of the Espoo city council organised a showing of the *Inconvenient Truth* to all its council members in order to bring home the message of climate change, and has to some extent contributed to this high ranking.

In terms of climate information, it was agreed that enough information exists on a national level but that more detailed information is needed in order to weigh up adaptation options. Constraints on human resources were indicated as constraints to the collection and utilisation of existing climate information. With access to external funding networks, Espoo has allowed for some of its staff to engage in 'free thinking' (Espoo Environment Centre, interview 1&2) that has in turn enabled the consideration of a wider range of environmental issues and the pursuit of strategies in these areas. A new member of staff was hired for a fixed term to work on media and information in terms of climate within the Environment Centre, demonstrating the ability of municipalities to increase their capacity to adapt building on external networks.

4.7.5 Integration of Adaptation into Policy-Making

Overall, the sub-national level in Finland exemplifies the multi-level nature of climate change governance, and this case study reaches similar conclusions to other recent studies (Haanpää et al., 2009; Swart et al., 2009). Within one region, four climate change strategies are in progress, of which some are interlinked but not all. This demonstrates the interest and level of awareness of adaptation at the sub-national level of the importance of climate change issues and the need for local responses (Uusimaa Regional Council, interview 2). Several of the respondents highlighted municipal decision-makers and council members that have been active in getting the issue into public discussion. Although climate change issues are considered very important within the Green League that has strong support in the region, mitigation and adaptation nevertheless receive support across party lines and are not confined to a strict environmental agenda.

As many of these strategies are still in their early stages, one of the common dilemmas faced by the actors was the scope of the adaptation strategy to be prepared and the tools to be used in engaging stakeholders and implementation adaptation measures. As the strategies are based on voluntary initiatives, it was also in some cases unclear to what extent other stakeholders could be tied to the strategies, whether the strategy would remain more an overview to guide action, or whether it could provide concrete measures to be implemented. This is especially true in terms of the Regional Council and the HMAC, both of which are essentially instruments of cooperation for and by the municipalities and lack the political power to force their members to abide to the recommendations, thus constraining action on adaptation.

In all these climate strategies, the organisations have approached adaptation differently to a certain extent than other environmental issues that they have dealt with in the past. Municipal environmental centres mainly focus on implementing measures that are required of them in environmental legislation. Strategies for dealing with climate change, or any new environmental issues for that matter, are not yet required by law and thus represent a challenge to the municipalities in terms of organisation of work and funding. Drawing up strategies that engage multiple stakeholders across sectors is challenging, though new tools and methods for doing so have been identified and utilised. In Espoo, the city's environmental centre utilised ways of cooperation in order to reach other sectors and stressed the importance of adaptation measures mainly related to flooding (Espoo Environment

Centre, interview 1). This has included bringing stakeholders together from different sectors of the city administration for cross-sectoral dialogue.

Similarly, the HMAC (in cooperation with the city of Helsinki) has begun a process through which they are identifying the adaptive capacities of their own sectors within their organisations, drawing on the UKCIP's LCLIP to that enables participants to build a local climate impacts profile within the BaltCICA project. Unusual weather events have been identified during the last decade within the region, and sectors have been asked to outline measures that were undertaken during those events and the likelihood of those reoccurring or strengthening in the context of a changing climate. This enables the not only the collection of data for possible measures, but also engages the sectors in a meaningful dialogue in terms of their own expertise, as well as the overall direction of adaptation measures within the organisations.

The KUUMA municipalities are also looking to the private sector in the preparation of their climate strategy. For the mitigation part of the strategy, good practice cases have been identified within the sub-region in terms of energy efficiency and reduction in emissions. It is acknowledged that businesses are often faster to act on environmental issues that result in economic gains, and that the business sector could be used as an example for the municipal sector (KUUMA Climate Strategy, interview). Perhaps this has been easier since the focus has so far centred on mitigation, but this form of approach can be considered for adaptation as well, once the strategy progresses to that point.

4.8 Challenges and Opportunities of Multi-Scale Adaptation

The early action on adaptation on all levels of government in Finland can partially be explained by good horizontal co-operation at the national level, high awareness of the climate change threat and as well as by the existence of the highly efficient environment administration. Many of the sub-national actors were aware of the early action on adaptation at the national level and considered it as a good example for lower levels to engage in climate change activities. However, issues specifically related to adaptive capacity need to be looked at more carefully. In all, adaptive capacity can be considered fairly high all across the levels of governance, although a more careful analysis reveals differences between sectors and levels that help to explain the slow emergence of implementation of adaptation. Though adaptive capacity has been defined in various ways, the following discussion roughly uses those categories generally acknowledged to be important in the literature, such as the political support, social capital and institutions, availability and access to financial resources, technology and information and human capital (Smit & Wandel, 2006; Yohe & Tol, 2002).

It is evident that there is political interest and support for climate change issues, both mitigation and adaptation, on all levels of social organisation in Finland. Nearly all of those interviewed considered that the political climate was extremely favourable to issues related to adaptation, although many brought up the possible negative effects of the global financial crisis that was unfolding during the fieldwork period. Although there perhaps is still more focus on adaptation in relation to mitigation, it was thought that the importance of adaptation had also been recognised. At both the national and sub-national levels, adaptation has been supported by political actors, as well as in city councils where there is a good representation of Green Party councillors within the region.

Furthermore, the early publication of the NAS can also be attributed to the political culture of Finland, as exemplified by one of the respondents. Adaptation is not seen as a rival or an alternative to mitigation, but a necessary issue that needs to be dealt within the government.

> The Finns are pragmatic- the types that have already thought of [adaptation]. In some European countries they say that if you talk about adaptation it means... that you have given up on mitigation. It is a fact that temperatures are rising. Adaptation should be thought of... and studied more. So the Finns take things as they are and leave the politeness to others. We do not have those burdens here, maybe so action adaptation can be more advanced. (Prime Minister's Office, interview)

Also, adaptation is considered less political than mitigation (Ministry of the Environment; Ministry of Employment and Economy, interviews). This can be partially explained by the fact that so far, no severe climate change impacts have been felt and adaptation measures have been targeted towards governmental structures with little impact on business or the economic sector so far.

Traditions in forward-looking environmental policy and management have also contributed to the early action on adaptation, particularly at the sub-national level. The existence of innovative actors and the importance of the region at the national scale were regarded as important in the early emergence of adaptation within Finland. The Helsinki Metropolitan Council has been involved in environmental issues in the metropolitan region for a long time, and its climate strategy work published in 2007 was seen to be inspiring by the other climate actors in the area (Uusimaa Regional Council; City of Kerava; City of Espoo, interviews). This has also allowed the HMAC to build its adaptation work on a solid foundation of climate knowledge and understanding of strategic projects. Learning from those processes is seen to be crucial in preparing the adaptation strategy (HMAC, interview). Similarly, the Espoo environment centre has a long tradition in taking global environmental change into account, with mentions of climate change as a threat in the environment plans as early as 20 years ago (Espoo Environment Centre, interview 2).

Another important factor that was stressed by many interviewees was the importance of horizontal networks at the national level. Many participants in the adaptation network have also participated in the preparation of the NAS and the NCES, in global climate negotiations, in the work of the Intergovernmental Panel on Climate Change and in the implementation of adaptation measures within their own administration (Valli & Sierla, 2008). Cooperation between representatives of Ministries was considered one of the reasons why the NAS had been a successful process, one which builds on the traditional way of preparing strategies in Finland. Multi-sectoral cooperation and involvement of different sectors in preparing national strategies is a long tradition in Finland, taking advantage of each sector's speciality and knowledge.

At the sub-national level, cooperation with partners was considered crucial in motivating and making it possible to pursue new issues, such as climate change adaptation. Without the existence of cooperation, it would have been unlikely that especially the smaller municipalities would have been able to participate in developing climate strategies. This was especially true for the KUUMA cooperation, where the network formed between six municipal environment centres was considered crucial in getting the strategy going and garnering support. Networks with the research institutes and researchers were also considered important (HMAC, interview), although this was very much dependent on the actor in question. Examples of adaptation in Finland or abroad and results of adaptation research were followed by all actors, since adaptation presents a new challenge for local administration.

A close linkage between science and policy has been a contributing factor in advancing the adaptation agenda. At the national level, the public research organisations under each Ministry have provided research into impacts of climate change and possible adaptation options directly relevant to their respective sector. The active involvement of the ministries in the ISTO steering group has enabled the different administrative sectors to utilise research knowledge quickly and the ministries to engage in a dialogue with the research community with regards to the direction and focus of research (Valli & Sierla, 2008). Furthermore, research programmes have helped to bring together multiple disciplines in efforts to foster multidisciplinary work. Although these have been somewhat slow in emerging, they nevertheless provide a good starting point for further collaboration. In addition, two of the four sub-national cases have directly benefited from research collaboration through the participation in research programmes, mainly through the EU INTERREG programmes that help European regions to form partnerships and work together on common issues.

This EU regional collaboration has not only been beneficial in sharing and exchanging knowledge but also crucial in supplementing financial resources for strategy work at the sub-national level. Overall, majority of those interviewed did not consider financial resources to be constraining. At the national level, funding for research was considered crucial and often times had been pooled in order to fund research projects (Ministry of the Environment, interview 1). There has been no additional funding available for the implementation of the NAS, but it is also argued that this is not a crucial since the aim is to mainstream adaptation into existing practices. Available money for adaptation has been channelled into adaptation research to public research organisations under each ministerial sector. However, concerns were raised in terms of the availability of financial resources for the implementation of the strategy, which could result in a less than satisfactory outcome.

Although adaptation is considered important in Finland, mitigation remains a priority for many in the country as much can be still be done to reduce emissions. Furthermore, adaptation is seen as an environmental issue to a large extent, rather than something that needs to be considered by all sectors of society. This may, in the long run, have an impact on the implementation of the NAS within sectors that do not explicitly deal with the environment. Horizontal integration of adaptation into different sectors is still emerging on all levels in this case study. At the national level, this is reflected in the fact that those Ministries who have made most progress on adaptation are those that directly deal with the environment or natural resources, i.e. the Ministry of Environment, Ministry of Agriculture and Forestry and certain sub-sectors of the Ministry of Transport and Communications. As the evaluation of the implementation of the NAS highlighted, cross-sectoral integration and adoption of measures can be difficult to evaluate within the different administrations (Ministry of Agriculture and Forestry, 2009). The environment administration has attempted circumvent this problem by assigning responsibilities for every measure in their action plan in order to make sure that mainstreaming actually leads to concrete actions and results (Ministry of the Environment, interview 2).

One of the constraining factors of the NAS in terms of implementation of adaptation across Finland is considered to be the lack of adequate consideration given to sub-national actors during the preparation of the strategy (Ministry of Agriculture and Forestry, interview 2). During the preparation, horizontal integration of sectors was successful but vertical linkages received less attention, and this is to certain extent hindering the uptake of adaptation in areas where adaptive capacity is lower than the Uusimaa region (Peltonen, 2007). This is also something that is likely to be addressed when the NAS is reviewed in the next couple of years (Ministry of Agriculture and Forestry, interview 2).

Similarly, there seems to be a lack of certainty over roles that different organisations can play as the definition of adaptation and its implications keep evolving and shaping at the sub-national level. This is also demonstrated at the local level where integration across sectors can prove challenging (Espoo Environment Centre, interview 1 & 2). This is especially true in cases where climate change impacts are not already felt and there is no perceived urgency in implementing the measures. In some regards, the Finnish municipalities are faced with the same dilemma as with 'sustainable development' where the message was clear but the actual content and implementation proved demanding (Niemi-Iilahti, 2001).

One of the concerns raised by the sub-national actors is the voluntary nature of climate strategies below the national level. Currently, all initiatives are funded either by the organisations themselves or the funding is secured from external sources, reflecting the general mode of working in local administration under the new public administration paradigm (Rantala & Sulkunen, 2007). Governance increasingly happens through short-term project planning in order to achieve efficiency and cost-effectiveness. In this case, there are concerns that as the short term project funding for the preparation of the strategy runs out, the implementation of the measures may fall short of the outlined targets due to lack of resources that were previously provided for the project. Although this is a valid concern, some actors have been able to acquire further funding for the implementation of the strategy (e.g., the Julia 2030 project within the HMAC).

4.9 Conclusions

Adaptation to climate change is an issue that requires the engagement of actors across multiple scales and sectors. In Finland, the national level responded early to the need to design and implement a national adaptation strategy. Since its completion in 2005, adaptation measures are currently taking place across various sectors of government ministries with varying degrees of speed. The measures undertaken so far have focused on change in legislation and the need for more research knowledge on not only impacts but also on different adaptation options that are available. At the sub-national level, there are several voluntary climate initiatives ongoing, with the case study region of Uusimaa particularly active in this regard. The biggest short-comings are lack of vertical integration of policies and levels of governance in general, as also identified by Kivimaa and Mickwitz (2009). Similarly, the lack of additional funding is a constraint for implementation of adaptation measures, although the mainstreaming approach conveniently does not rely on additional funds to be successful.

A recent study suggests that degree of integration of adaptation into sectoral policies is important in terms of successful adaptation (Puppim de Oliveira, 2009). Naturally, the evaluation of this is difficult and this is also admitted by the first evaluation of the NAS in 2009. However, so far it appears that the need to adapt has been recognised across the ministries in Finland and some sectors are already implementing measures. Although adaptation is still a recent concept for many municipalities, it is nevertheless taken seriously at least within the environmental sector. As highlighted in the literature, autonomy to make decisions, institutional capacities and competing demands hinder the progress of adaptation at the regional and local levels (Puppim de Oliveira, 2009). The review of the NAS in the coming years is likely to focus more on implementation across sectors and vertically, which is currently posing some problems. Regional and local actors have shown their capacity to develop their own strategies through different networks and with multiple partners. Although the capacity of Finnish municipalities to deal with environmental issues is growing (Niemi-Iilahti, 2001), this also needs to be strengthened in the future for successful adaptation.

References

Interviewees

Association for Local and Regional Authorities. Helsinki, January 16, 2009.
City of Espoo Environment Centre (1). Espoo, January 21, 2009.
City of Espoo Environment Centre (2). Espoo, January 21, 2009.
City of Espoo Planning Office. Espoo, February 2, 2009.
City of Espoo, Environment Committee. Helsinki, January 22, 2009.
City of Kerava & KUUMA Climate Committee. Kerava, January 27, 2009.
Helsinki Metropolitan Area Council. Helsinki, February 14, 2009.

KUUMA Climate Committee. Kerava, January 27, 2009.
Ministry of Employment and the Economy. Helsinki, January 9, 2009.
Ministry of Food and Agriculture (1). Helsinki, January 12, 2009.
Ministry of Food and Agriculture (2). Helsinki, February 16, 2009.
Ministry of the Environment (1). Helsinki, January 9, 2009.
Ministry of the Environment (2). Helsinki, January 29, 2009.
Ministry of Transport and Communication. Helsinki, January 29, 2009.
Municipality of Tuusula. Tuusula, January 28, 2009.
Municipality of Mäntsälä. Mäntsälä, January 29, 2009.
Municipality of Pornainen. Pornainen, January 26, 2009.
Prime Minister's Office, Finnish Government. Helsinki, January 12, 2009.
Regional Council of Uusimaa (1). Helsinki, January 20, 2009.
Regional Council of Uusimaa (2). Helsinki, January 28, 2009.
Regional Council of Uusimaa (3). Helsinki, February 16, 2009.
Uusimaa Regional Environment Centre. Helsinki, February 16, 2009.

Secondary References

ALFRA 2007. Local Authorities and Climate Change. Urban Planning, Infrastructure and the Environment. Helsinki, ALFRA.

ALFRA 2008. Ilmastonmuutoksen hidastaminen ja sopeutuminen- Kuntaliiton linjaukset. Helsinki, ALFRA.

Carbon Committee 1991, *Committee memorandum* 21. The Government of Finland, Helsinki.

Carter, T. (2004). Special Issue: FINSKEN: Global change scenarios for Finland in the 21st century. *Boreal Environment Research, 9*(2), 89–198.

Carter, T. (2007). *Assessing the adaptive capacity of the Finnish environment and society under a changing climate. A summary for policy-makers*. Helsinki: Finnish Environment Institute Mimeographs.

Carter, T., Fronzek, S., Mitikka, V., Suomalainen, M., Leikola, N., Heikkinen, R., et al. (2004). *Integrated assessment modelling of global change impacts and adaptation (FINESSI): Progress report*. Helsinki: Finnish Environment Institute (SYKE).

Carter, T., Jylhä, K., Perrels, A., Fronzek, S., & Kankaapää, S. (2005). *Alternative futures for considering adaptation to climate change in Finland*. Helsinki: Finnish Environment Institute Mimeographs.

Carter, T., & Kankaapää, S. (2003). *A preliminary examination of adaptation to climate change in Finland*. Helsinki: The Finnish Environment.

Espoon tulvatyöryhmä. (2005). Tulvaongelma Espoossa: Teknisen ja ympäristötoimen johtoryhmä.

Esty, D., Levy, M., Srebtnjak, T., & de Sherbinin, A. (2005). *2005 Environmental Sustainability Index: Benchmarking National Environmental Stewardship*. New Haven: Yale Center for Environmental Law & Policy.

Finnish Environment Institute. (2008a). Climate change scenarios for Finland. Retrieved July 4, 2009, from http://www.environment.fi/default.asp?contentid=178197&lan=EN

Finnish Environment Institute. (2008b). *Environmentally related energy taxation in Finland*. Helsinki: Finnish Environment Institute.

Forsström, J., & Lehtilä, A. (2005). Skenaarioita ilmastopolitiikan vaikutuksista energiatalouteen. VTT Working papers, Helsinki.

Haanpää, S., Tuusa, R., & Peltonen, L. (2009). Ilmastonmuutoksen alueelliset sopeutumisstrategiat. READNET- hankkeen loppuraportti. Yhdyskuntasuunnittelun tutkimus- ja koulutuskeskuksen julkaisuja. No. C75. Espoo: Helsinki University of Technology.

Hakanen, M., & Luoma, K. (2008). *Kuntien ilmastokampanja. Raportti kyselystä 2008*. Helsinki: Kuntaliitto.

Hakanen, M., Luoma, K., & Mynttinen, M. (2008). Kuntien ilmastokampanja. Raportti kyselystä 2008 (pp. 1–14). Helsinki: Kuntaliitto.

Harmaajärvi, I. (2005). *Ilmastonmuutoksen haasteet yhdyskuntasuunnittelulle*. Paper presented at the Maanmittaustieteiden päivät, November 17–18, 2005, Espoo.

Hilbert, K., Mannke, F., & Schmitdt-Thome, P. (2007). *Towards climate change adaptation in the Baltic Sea region* (pp. 1–57). Espoo: Geological Survey of Finland.

Hintsala, J. (2007). *Uudenmaan rannikkoalueiden yleispiirteinen tulvakartta*. Helsinki: Uudenmaan liitto.

Hirvelä, J. (2008). *Tienpitäjä ja ilmastonmuutos*. Paper presented at the Ilmastopolitiikan toimikunnan avausseminaari, Liikenne- ja viestintäministeriö, May 14, 2008, Helsinki.

Honkatukia, J., Kemppi, H., & Kerkelä, L. (2005). Energia-strategian kansataloudellisista vaikutuksista. Helsinki: VATT-keskustelunaloitteita.

Huuska, P. (2006). *Uudenmaan kasvihuonekaasupäästöt 1990 ja 2003*. Helsinki: Uudenmaan liitto ja YTV.

Ilmatieteenlaitos. (2009). Suomen ilmaston tulevat muutokset mallitulosten perusteella. Muutos Suomessa. Retrieved June 4, 2009, from http://fmi.fi/ilmastonmuutos/suomessa_18.html

Itä-Uudenmaan liitto. (2007). Itä-Uudenmaan maakuntakaava: Itä-Uudenmaan liitto.

Jalasto, P., Linkama, E., Lampinen, S., Saarlo, A., Tuominen, A., & Järvi, T. (2007). Liikenne 2030: Liikenne- ja viestintäministeriö.

Järvi, T., & Laurikko, J. (2008). *Liikennesektorin vaikutukset ilmastonmuutokseen ja mahdollisuudet hiilidioksipäästöjen vähentämiseksi*. Paper presented at the Ilmastontoimikunnan avausseminaari, Liikenne- ja viestintäministeriö, May 14, 2008, Helsinki.

Jauhiainen, J. S., & Niemenmaa, V. (2006). Alueellinen suunnittelu. Tampere: Vastapaino.

Kivimaa, P., & Mickwitz, P. (2009). *Making the climate count. Climate policy integration and coherence in Finland*. The Finnish Environment. No. 3. Helsinki: Finnish Environment Institute.

Kok, M. T. J., & de Conick, H. C. (2007). Widening the scope of policies to address climate change: directions for mainstreaming. *Environmental Science & Policy, 1*(7–8), 587–599.

Kuntaliitto. (2001). Kunnat ilmaston puolesta 28.-29. 11. 2001. Tampere: Kuntien ilmastonsuojelukampanjan julkaisu.

Kuntaliitto. (2009). Maakuntien liitot. Retrieved July 6, 2009, from http://www.reg.fi/

KUUMA. (2004). KUUMA-seutu. Retrieved June 3, 2009, from http://www.kuuma.fi/alue.asp

KUUMA. (2007). Keski-Uudenmaan Ilmastostrategia ja toimintaohjelma: Keski-Uudenmaan kumppaanuus.

Länsi-Suomen ympäristökeskus. (2007). *Länsi-Suomen ympäristöstrategia 2007–2013. Visiona eurooppalainen kestävän kehityksen esimerkkialue*. Unpublished manuscript, Vaasa.

Lapin liitto. (2005). Eläköön Lappi. Lapin maakuntasuunnitelma 2022: Lapin liitto.

Lounais-Suomen ympäristökeskus. (2007). Lounais-Suomen ympäristöstrategia 2020: Lounais-Suomen ympäristökeskus.

Maa- ja metsätalousministeriö. (2001). *Maatalouden kehitysarvio kansallista ilmasto-ohjelmaa varten* (Vol. 2001:2). Helsinki: Maa- ja metsätalousministeriö.

Maa- ja metsätalousministeriö. (2006). *Tulevaisuuskatsaus vuoteen 2015, Maa- ja metsätalousministeriön toimiala* (Vol. 2916). Helsinki: Maa- ja metsätalousministeriö.

Maa- ja metsätalousministeriö. (2008a). *Kansallinen metsäohjelma 2015. Lisää hyvinvointia monimuotoisista metsistä*. Helsinki: Maa- ja metsätalousministeriö.

Maa- ja metsätalousministeriö. (2008b). *Pitkän aikavälin skenaariot- maatalous, metsätalous ja maankäyttö* (Vol. 32626). Helsinki: Maa- ja metsätalousministeriö.

Marttila, V., Granholm, H., Laanikari, J., Yrjölä, T., Aalto, A., Heikinheimo, P., et al. (2005). *Finland's national strategy for adaptation to climate change*. Helsinki: Ministry of Agriculture and Forestry.

Metsäntutkimuslaitos. (2004). *Selvitys Suomen metsiä koskevista nieluarvioista* (Vol. 10/11). Helsinki: Maa- ja metsätalousministeriö.

Ministry of Agriculture and Forestry. (2009). *Ilmastonmuutoksen kansallisen sopeutumisstrategian toimeenpanon arviointi*. Helsinki: Ministry of Agriculture and Forestry.
Ministry of Employment and Economy. (2008). *The long term climate and energy strategy*. Helsinki: The Government of Finland.
Ministry of Finance. (2009). *Regional administration reform 2010*. Helsinki: Ministry of Finance.
Mustamäki, E. (2008). *Ilmastopolitiikan toimikunnan seminaari: FINSTASHIP* Paper presented at the Ilmastopolitiikan toimikunnan avausseminaari, Liikenne- ja viestintäministeriö, May 14, 2008, Helsinki.
Niemi-Iilahti, A. (2001). Finland. In search of New Implementation patterns. In W. M. Lafferty (Ed.), *Sustainable Communities in Europe*. London: Earthscan Publications Ltd.
Niemimuukko, O. (2008). *Ratahallintokeskuksen mahdollisuudet ilmastonmuutoksen hillinnässä*. Paper presented at the Ilmastopolitiikan toiminkunnan avausseminaari, Liikenne- ja viestintäministeriö, May 14, 2008, Helsinki.
Panula-Ontto-Suuronen, A. (2005). *Etelä-Savon ympäristöohjelma 2005–2010*. Helsinki: Etelä-Savon ympäristökeskus.
Peltonen, L. (2007). *Adapting planning for climate change adaptation: Policy development and implications for the planning system in Finland*. Paper presented at the Association of European Schools of Planning. AESOP Conference.
Peltonen, L., Haanpää, S., & Lehtonen, S. (2005). *The challenge of climate change adaptation in urban planning*. Helsinki: Finnish Environment Institute Mimeographs.
Pirkanmaan liitto. (2007). Pirkanmaan maakuntaohjelman toteuttamissuunnitelma 2008–2009. Tampere: Pirkanmaan liitto.
Pirkanmaan ympäristökeskus. (2001). *Pirkanmaan ympäristöohjelma*. Tampere: Alueelliset ympäristöjulkaisut.
Pohjanmaan liitto. (2006). Pohjanmaan maakuntaohjelma 2007–2010: Pohjanmaan liitto.
Pohjois-Karjalan liitto. (2005). Pohjois-Karjalan maakuntasuunnitelma 2025: Pohjois-Karjalan liitto.
Pohjois-Pohjanmaan ympäristökeskus. (2005). *Pohjois-Pohjanmaan ympäristöstrategia 2005–2015*. Unpublished manuscript, Oulu.
Puppim de Oliveira, J. A. (2009). The implementation of climate change related policies at the subnational level: An analysis of three countries. *Habitat International, 33*(3), 253–259.
Rantala, K., & Sulkunen, P. (Eds.). (2007). *Projektiyhteiskunnan kääntöpuolia*. Helsinki: Gaudeamus.
Rantanen, J. (2008). *Sektoritutkimuksen selvitysraportti*. Helsinki.
Raunio, T. (2004). The changing Finnish democracy: Stronger parliamentary accountability, coalescing political parties and weaker external constraints. *Scandinavian Political Studies, 27*(2), 133–152.
Regina, K., Lehtonen, H., & Esala, M. (2008). Arvio perusuran ja politiikkaskenaarion mukaisesta maatalouden kasvihuonekaasupäästöjen kehityksestä vuoteen 2050 menessä, Vol. 13/3/2008. Helsinki: Maa- ja elintarviketalouden tutkimuskeskus.
Sairinen, R., & Teittinen, O. (1999). Voluntary agreements as an environmental policy instrument in Finland. *European Environment, 9*(2), 67–74.
Salonen, K. (2008). *Ilmailuhallinnon näkökulma ilmastopolitiikkaan*. Paper presented at the Ilmastopolitiikan toimikunnan avausseminaari, Ilmastopolitiikan toimikunnan, May 14, 2008, Helsinki.
Smit, B., & Wandel, J. (2006). Adaptation, adaptive capacity and vulnerability. *Global Environmental Change, 16*(3), 282–292.
Soini, S. (2007). *Ilmastonmuutos ja siihen varautuminen Espoossa*. Espoo: Espoon ympäristökeskus Monistesarja.
Swart, R., Biesbroek, G. R., Binnerup, S., Carter, T., Cowan, C., Henrichs, T., et al. (2009). *Europe adapts to climate change. Comparing National Adaptation Strategies*. Sastamala: Vammalan Kirjapaino Oy.

The Government of Finland. (2001). *National climate strategy Finland. Government report to the Parliament*. Helsinki: The Government of Finland.

The Government of Finland. (2006). *Kohti kestäviä valintoja. Kansallisesti ja globaalisti kestävä Suomi. Kansallinen kestävän kehityksen strategia. Suomen kestävän kehityksen toimikunnan asettama strategiaryhmä*. No. 5/2006. Helsinki: The Government of Finland.

Tirkkonen, J. (2000). *Ilmastopolitiikka ja ekologinen modernisaatio. Diskursiivinen tarkastelu suomalaisesta ilmastopolitiikasta ja sen yhteydestä metsäsektorin muutokseen. Acta Universitatis Tamperensis 781*. Tampere: Tampere University Press.

Tuomenvirta, H. (2004). *Reliable estimation of climatic variations in Finland*. Helsinki: Finnish Meteorological Institute.

Tynkkynen, O. (2008). *Pitkän aikavälin ilmastopolitiikka ja liikenne*. Paper presented at the Ilmastopolitiikan toimikunnan avausseminaari, Liikenne- ja viestintäministeriö, May 14, 2008, Helsinki.

Urwin, K., & Jordan, A. (2008). Does public policy support or undermine climate change adaptation? Exploring policy interplay across different scales of governance. *Global Environmental Change, 18*(1), 180–191.

Uudenmaan liitto. (2006). Uudenmaan maakuntaohjelma 2007–2010. Tavoitteena Uudenmaan hyvinvointi ja kilpailukyky. Vol. A 19- 2007. Helsinki: Uudenmaan liiton julkaisuja.

Uudenmaan liitto. (2009). Uusimaa facts. Retrieved June 3, 2009, from http://www.uudenmaanliitto.fi/index.phtml?l=en&s=1087

Uusimaa Regional Environment Centre. (2007). *Our shared environment 2020. Environmental programme for Uusimaa Region*. Helsinki: The Finnish Environment

Uusivuori, J., Kareinen, T., Penttilä, T., Pohjola, J., Salminen, O., Sievänen, R., et al. (2008). Arviot metsien kasvihuonekaasutaseista baseline- ja politiikkaskenaarioissa sekä ilmastonmuutosta hillitsevistä toimista metsätaloudessa Helsinki: Metsäntutkimuslaitos (METLA).

Valli, R., & Sierla, L. (2008). *lmastonmuutoksen sopeutumistutkimusohjelman väliarviointi*. Helsinki: Maa- ja metsätalousministeriö, Sito Oy.

Valtioneuvosto. (2005). *Lähiajan energia- ja ilmastopolitiikan linjauksia- kansallinen strategia Kioton pöytäkirjan toimeenpanemiseksi* (pp. 1–122). Helsinki: Valtioneuvoston selonteko eduskunnalle.

Valtioneuvosto. (2008). *Valtioneuvoston päätös valtakunnalisten alueidenkäyttötavoitteiden tarkistamisesta*. Helsinki: Valtioneuvosto.

von Bergmann-Winberg, M.-L. (2000). Finnish Regions and Municipalities in a Nordic Setting: Processes of Restructuring, Internationalisation and Integration. In J. Gidlund & M. Jerneck (Eds.), *Local and regional governance in Europe: Evidence from the Nordic regions*. Cheltenham: Edward Elgar.

Wahlgren, I., Kuismanen, K., & Makkonen, L. (2008). *Ilmastonmuutoksen huomioiminen kaavoituksessa- tapauskohtaisia tarkasteluja*. Helsinki: VTT Tutkimusraportti

Wilenius, M., & Tirkkonen, J. (1997). Climate in the making: Using Delphi for Finnish climate policy. *Futures, 29*(9), 845–862.

WWF. (2004). *Living Planet Report*. Gland World Wildlife Fund.

Ympäristöministeriö. (2009). Alueelliset ympäristökeskukset. Retrieved July 7, 2009, from www.ymparisto.fi/ayk

Ympäristöministeriön työryhmä. (2008). *Ilmastonmuutokseen sopeutuminen ympäristöhallinnon toimiallalla*. Helsinki: Ympäristöministeriö.

Yohe, G., & Tol, R. S. J. (2002). Indicators for social and economic coping capacity–moving toward a working definition of adaptive capacity. *Global Environmental Change, 12*(1), 25–40.

YTV. (2007). *Pääkaupunkiseudun ilmastostrategia 2030. Ilmastonmuutoksen hillintä keskeiseksi osaksi kaupunkien suunnittelua ja päätöksentekoa*. Helsinki: Pääkaupunkiseudun yhteistyövaltuuskunta.

Chapter 5
Adapting to Climate Change in Sweden: National Policy Development and Adaptation Measures in Västra Götaland

E. Carina H. Keskitalo

Abstract Located in south-western Sweden, the Västra Götaland region is often seen as one of the areas of the country most vulnerable to flooding and erosion, and will be highly impacted by flooding and sea level rise as a result of climate change. Drawing upon a literature study and semi-structured interviews with actors in climate policy, this chapter reviews the development of adaptation policy in Sweden. The chapter focuses particularly on the Commission on Climate and Vulnerability (2007) and a government bill *An Integrated Climate and Energy Policy*: *Climate* (2009) in which suggestions by the Commission were included. The chapter describes the development of adaptation policy and measures on the regional and local levels in Västra Götaland and within select municipalities. The study illustrates the national distribution of responsibility through which municipalities are given a large role in integrating adaptation measures, and describes some of the differentiated responses such responsibilities may elicit on the local level.

Keywords Adaptation · Climate change · Sweden · Västra Götaland · Gothenburg

5.1 Introduction

Sweden has traditionally been characterised by both strong local governments and a central government of corporatist orientation, with strong ties to (among other actors) trade unions and employers' organisations. Some have noted that as a reaction to severe budget deficits from the early 1990s, Sweden has begun to 'reveal a general shift towards the Regulatory State' (Christensen & Lagreid, 2005, p. 27), and an increasing focus on privatisation and regulation within the welfare state system. However, new public management methods such as performance assessments have not been implemented in the same manner as in states such as the

---

E.C.H. Keskitalo (✉)
Political Science/Department of Social and Economic Geography, Umeå University, Umeå 90187, Sweden
e-mail: Carina.Keskitalo@geography.umu.se

E.C.H. Keskitalo (ed.), *Developing Adaptation Policy and Practice in Europe: Multi-level Governance of Climate Change*, DOI 10.1007/978-90-481-9325-7_5,

United Kingdom (UK), but rather have been merged within existing structures (cf. Temnes, 1998).

Although an unitary state, Peters and Pierre (2005) note, however, that Sweden may in some ways be 'more multilevel than a classical federal arrangement, in that the lowest level[s] of government . . . [are] rather . . . autonomous political systems in their own right' (p. 46). Local governments also hold the right to local taxation and have a so-called 'general competence', or the ability to select issues upon which to act for the benefit of its citizens within the existing legal framework. This is demonstrated principally through the so-called planning monopoly held by Sweden's 290 municipalities, through which all binding local planning is undertaken at the municipal level. In contrast to such extensive decentralisation to the local level, the regional level is characterised by the county administrative boards, which were historically instituted as the regional arm of the state and focus on the implementation of state regulations (cf. Östhol and Svensson, 2002).[1]

In the environmental policy arena, Sweden has regularly been characterised as a leader by example and as an early actor in the implementation of a number of pollution control measures and mitigation policies (cf. Börzel, 2002; Lundqvist & Biel, 2007; Vail 2008). However, while Sweden has tended to adopt environmental policy early on, Swedish intiatives on domestic climate change adaptation have developed relatively recently in response to local and international pressures, in contrast especially to early adopters such as the UK. This chapter describes adaptation to climate change in Sweden with a focus on the Västra Götaland region[2] and a selection of its municipalities, including the second largest city of Sweden, Gothenburg (see Fig. 5.1). The Västra Götaland region is a relatively low-lying, landslide-prone area with a high proportion of clay soils, and has been the site for the largest landslides in Sweden since the 1950s, as well as extensive floods in 2001. The area is situated downstream from Lake Vänern, Sweden's largest lake, which drains mainly through the Göta River and its tributaries flowing into the sea near Gothenburg.

In addition to Gothenburg, the study assesses other local governments, among these Mölndal, a smaller municipality within the greater Gothenburg region, and the town of Trollhättan, both of which achieved a high rank in a Swedish Society for Nature Conservation (SSNC) survey on municipal climate change action (principally focused on mitigation but with some consideration of adaptation) (SSNC,

[1]The division of responsibility in Sweden across the local, regional and national levels has recently been assessed through the regional investigation (*ansvarskommitténs utredning*) on the division of responsibility, which have suggested the creation of larger and more unified regional structures (Ansvarskommittén 2007).

[2]In the two regions of Västra Götaland (comprising 49 municipalities) and Skåne, directly elected regional councils are taking over responsibilities for regional development from the County Administrative Boards (SALAR 2009) as well as the coordination of health care, which is generally divided from the County Administrative Boards and addressed through a specific body dedicated to health care (VG, 2009a). While the Västra Götaland region has developed a climate strategy, this is targeted at mitigation (VG, 2009b).

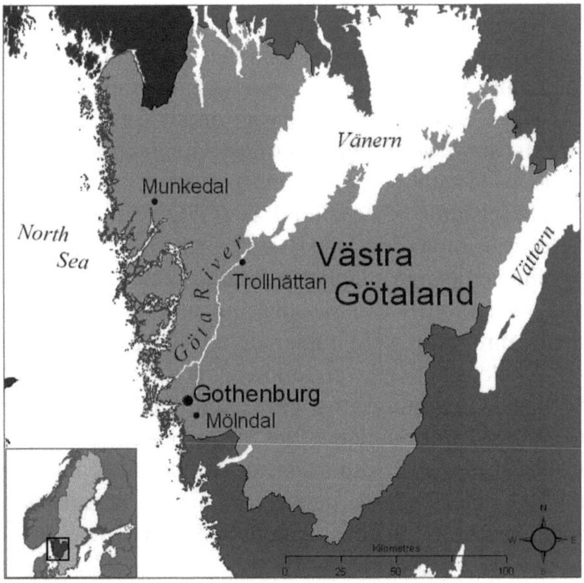

Fig. 5.1 Map of the Västra Götaland case study region and the selected case study municipalities in the region

2007). The study further includes the small municipality of Munkedal, in which one of the largest landslides in Sweden took place in 2006 (SGU, 2009). (For more information on methodology and theoretical framework of the study, see Chapter 1).

Projections for the Västra Götaland area and the Gothenburg region in particular foresee an increase in local average temperatures by 3–4°C, an increase in average precipitation during the winter, spring and autumn, and a decrease in average precipitation during the summer. An increase in extreme precipitation during all seasons is also anticipated, as well as an increase in sea level of 0.1–0.9 m above the current average to 2100. These changes are expected to cause, among other things, changes in flood risk and dam security, an increased risk of erosion, and an increased risk of storm felling of forest, yet also to improve preconditions for hydroelectric power generation, forest growth and agriculture (Gothenburg City Office, 2006; Commission on Climate and Vulnerability, 2007). The likely positive effects of climate change include a lengthened summer season with the possibility of several harvests, as well as more limited heating needs in winter (Commission on Climate and Vulnerability, 2007).

With regard to water, a warmer future climate is also projected to result in more unstable winters and higher flows of water from Lake Vänern through the Göta River; at the same time, summer flows may decrease, 'possibly leading to the maximum tapping into the Göta River being utilised to a higher extent than today' (Gothenburg City Office, 2006, p. 12, author's translation) and to increased landslides. In addition, the increased development foreseen for the Gothenburg region (and largely in low-lying areas) makes the issue of development and flood risk in low-lying areas particularly relevant. The area has also recently been affected by storms, including major storms that have impacted large parts of southern Sweden. While the Västra Götaland area was not among those worst hit by the major storms Per and Gudrun, 'the January storm 2005 ("Gudrun") [was] one of the three worst storms in western and innermost Götaland since 1900' (Gothenburg City Office 2006, p. 11, author's translation).

The chapter is structured as follows: firstly, actions on the national level are described, including legislation, regulation, and state commissions with relevance to climate change adaptation, with a specific focus on the Commission on Climate and Vulnerability and the subsequent 2009 Climate Bill in which suggestions from the Commission were implemented. The regional level is then discussed, including state actions at that level, followed by a description of the municipal level, including both state regulations that play out at the local level and the independent actions of the select local municipalities in the region. Finally, the integration between these levels in terms of responsibility is discussed, including a treatment of the role of the EU, actors outside public administration, and the potential of policy transfer to support adaptation development in Sweden.

5.2 National Level

5.2.1 Swedish Work on Climate Change of Particular Relevance for Adaptation

Sweden has generally treated climate change as an issue alongside other environmental issues, focusing on pollution reduction (or mitigation) through high emission reduction targets over the last two decades. Stabilisation of carbon dioxide emissions has been a Swedish policy aim since 1988 (Swedish Energy Agency and Swedish EPA, 2007), but gained momentum in the 1990s. In 1997, a climate strategy for the energy sector that focused on the support of renewable energy-based electricity production was developed (Ministry of Sustainable Development, 2005). The following year, in 1998, the Swedish Parliament set a target of stabilising emissions from the transport sector to 1990 levels by 2010. Mitigation was also the focus in the 2002 Swedish Climate Strategy Bill (Swedish Government, 2002). Sweden has further aimed to mainstream environmental concerns through non-binding aims in the form of environmental quality objectives, established in 1999. To date, Sweden has adopted 16 environmental quality objectives to be achieved by 2020, including objectives for the reduction of climate impacts through mitigation and water safety; however, adaptation is not explicitly identified (Swedish Energy Agency and Swedish EPA, 2007).[3]

[3] Among these, the reduced climate impact goal aims to stabilise greenhouse gas emissions in such a way as to avoid a harmful impact on the climate system. For the period 2008–2012, Swedish emissions are projected to achieve levels 4% below those of 1990, without the use of carbon sinks or flexible carbon-reduction mechanisms provided through international climate change agreements.

Current policy measures with regard to climate change in Sweden include a number of environmental tax and support measures, such as the carbon dioxide tax (1991) and the 'green tax switch' (2000) (Commission on Climate and Vulnerability, 2007; Ministry of the Environment, 2008). In 1997 and 2002, energy policies were implemented to provide grants for electricity-saving measures and to support the creation of municipal energy advisers (Ministry of Sustainable Development, 2005). As part of the 2002 climate strategy, a general information campaign on climate change was organised (Swedish EPA, 2004). Follow-up and evaluation of the Swedish mitigation targets in so-called 'control stations' that monitor implementation is also ongoing (cf. Swedish EPA, 2009).

Since 2005, a number of administrative structures with relevance for climate change have been developed. These have principally targeted mitigation, and include the creation of the Commission on Oil Independence to decrease oil dependency and use by 2020 (Commission on Oil Independence, 2006). After 2006, several bodies were set up by the newly elected government, including the Commission for Sustainable Development, the Scientific Council on Climate Issues, the Commission on Climate Change and Development and a parliamentary climate committee on climate change and the preparation of the Climate Bill (Swedish EPA, 2008). A Commission on Climate and Vulnerability was also continued from the previous government, dedicated to vulnerability, risk and adaptation issues. These bodies were generally organised according to the traditional form of governmental investigation, where reporting and consultation are undertaken in the form of governmental commissions (*kommittéväsendet*), the reports of which are passed along to a large number of relevant government agencies and organised interests (so called *remiss*) (Hall and Montin, 2007). Interviews indicated a broad consensus that changes in the political majority did not as such affect the treatment of climate change as an issue, and that the commissions established at this time were a result of the progress of climate change as an issue on the international stage. Events such as the Gudrun storm and floods may also have played a role for the development of the Commission on Climate and Vulnerability (Prime Minister's Office, interview).

The first of these commissions, the Commission on Sustainable Development, is an advisory commission set up 'to promote efforts across sectors, adopting an international perspective and taking into account ecological, social and economic aspects. Climate change will be the focus of the Commission's work' (Government Offices of Sweden, 2007, para. 4). The Commission aimed to provide guidance on support to national and EU strategies for sustainable development, international cooperation on sustainable development, and the 2008 review of climate policy in preparation for the period after the Kyoto protocol, as well as preparations for Sweden's EU presidency during the second half of 2009 (Government Offices of Sweden, 2008). The output of the Commission consists of discussions and advice to government, both through open and internal meetings (Commission on Sustainable Development, interview; cf. Government Offices of Sweden, 2008). The Scientific Council on Climate Issues (*Vetenskapliga rådet för klimatfrågor*) was created specifically to support climate change work and to provide the scientific basis for a Bill on climate. The report of the Scientific Council, focused on mitigation, was released in 2007 (Scientific Council on Climate Issues, 2007). The Parliamentary Climate Drafting Committee (*Klimatberedningen*) was further commissioned to assess the country's ability to reach national mitigation targets for 2008–2012, and to develop suggestions for objectives and measures for climate policy.

However, despite a partial focus on climate change, adaptation has not been a consideration for any of these commissions. An interviewee at the Commission on Sustainable Development noted the limited importance ascribed to adaptation as a possible rationale: 'I don't feel that there is a pressure and a large problem in the case of adaptation. And in that regard, those types of issues have ... not reached the higher levels of the system' (Commission on Sustainable Development, interview). In 2008, the Parliamentary Climate Drafting Committee presented its *Swedish Climate Policy* report (Klimatberedningen, 2008), noting that the report had not included issues of vulnerability and adaptation since this was the task of the Commission on Climate and Vulnerability.

Beyond these Committees, Sweden has addressed various facets of climate risk through means not directly related to climate change. These include: a commission and a bill on security and awareness (both in 2001); a 2005 bill on the coordination of crisis situations; a 2006 report on security; and a commission appointed in 2007 on the planned development of a coordinated agency on crisis management. With some exception of the last, domestic climate change adaptation is not explicitly a target in the reports (Commission on Climate and Vulnerability, 2007). With respect to water issues, preparedness with regard to future climate is partly addressed through the Swedish National Audit Office report on Swedish water supplies, which among other things shows the need for increased preparedness given projected increases in extreme weather events (Commission on Climate and Vulnerability, interview).[4]

One area in which Sweden has included a notable focus on adaptation is in the provision of overseas development assistance. The Commission on Climate Change and Development was set up in 2007 to investigate the linkages between climate change impacts (including adaptation) and development in developing countries. Led by the Swedish Minister for International Development Cooperation and supported by EU and international expertise, the Commission was launched by the Swedish government in conjunction with the UN General Assembly Special Session on Climate Change. The final report was delivered in the spring of 2009, with the aim to 'climate-proof' official development assistance (Commission on Climate Change and Development, interview). The Commission has no domestic focus on adaptation, which, as one interviewee notes, is a result of, among other things, a focus on addressing development issues in the context of the European Union during Sweden's EU chairmanship over the second half of 2009 (Commission on Climate Change and Development, interview). At the Prime Minister's Office, one

[4]In addition, the National Food Administration has been active in developing water quality regulation and crisis preparedness for national water supplies, including the provision of workshops and advice to municipalities (Commission on Climate and Vulnerability, interview).

interviewee explained that '[t]he major focus we have right now with regard to adaptation is how to finance adaptation in developing countries... [and] how the EU should find that funding' (Prime Minister's Office, interview).

The Swedish focus on climate change has thus largely rested on mitigation and on adaptation in developing states rather than on integrating adaptation into different policy areas or within measures for sustainable development. The exception to this is the Commission on Climate and Vulnerability, whose work is described below. The perception that there is little need for domestic adaptation may have several origins. One interviewee noted that 'Sweden has to be said to be one of the countries that will be relatively spared, even from the most significant climate change impacts' (Swedish Insurance Federation, interview). Another interviewee noted that the likely benefits of climate change for Sweden may explain why a limited focus or integration has been placed on adaptation so far (Prime Minister's Office, interview).

5.2.2 The Commission on Climate and Vulnerability

The absence of policy initiatives on domestic adaptation at the national level in Sweden was first addressed with the appointment of the Commission on Climate and Vulnerability (Climatools, interview). Appointed in June 2005 following the government's receipt of the *National Climate Policy in Global Cooperation* bill (Proposition 2005/06:172), the Commission finalised its main report, *Sweden Facing Climate Change–Threats and Opportunities*, in October of 2007 (Commission on Climate and Vulnerability, 2007).

The creation of the Commission on Climate and Vulnerability may be seen as a shift in Swedish policy orientation: 'Earlier ... one felt that adaptation was a bit like giving up [on emissions reduction] ... but as the years pass one has realised more and more that whatever we do, there will be climate change' (Commission on Climate and Vulnerability, interview). Many interviewees at the national scale noted that this increased focus on adaptation may have also been prompted to a large extent by international events. An interviewee at the Commission on Sustainable Development noted that the launching of a commission dedicated specifically to domestic climate impacts and adaptation could be seen as a way to manage

> the media attention that the climate change issue had received when the Stern report came, when the first reports on IPCC work came in November 2006 ... If you look ... at the alliance government's election platform very little was said about climate change. (Commission on Sustainable Development, interview)

Beyond the international focus on climate, political pressure from select regional actors with regard to flooding and the increased risk of extreme events as a result of climate change may have been important catalysts in the development of national initiatives on adaptation. In particular, extensive floods in 2001 prompted county administrations around Lake Vänern (including Västra Götaland) to send a written request to the national government demanding action on the issue. An interviewee at the Commission on Climate and Vulnerability described the situation:

> There had been very large floods in 2001 in both Mälaren and Vänern in Sweden ... there were only ten centimetres left at the power plant by the Vänern outlet before it would have flooded ... and we were really concerned whether there would be landslides along the Göta River valley .... As a result, the political leadership of the counties around Vänern wrote to the government saying that this was a national issue: "[despite that] we are several counties, we cannot solve this [on our own]" (Commission on Climate and Vulnerability, interview).

One interviewee indicated that this written request to some extent both forced political action on adaptation and supported a re-framing of flood risk in the context of climate change: 'those questions went to the Department of Defense and then one saw that this had to do with climate change' (Commission on Climate and Vulnerability, interview). A second focusing event was the Gudrun storm in 2005, which caused major storm felling and electricity outages and resulted in a situation where

> demands were posed on climate change ... Göran Persson, who was the Swedish Prime Minister, went out and said that we would start a new Commission to see how vulnerable Sweden is and what adaptations we need to make to climate change (Commission on Climate and Vulnerability, interview).

The resultant Commission on Climate and Vulnerability consisted of three main working groups and 25 sub-groups, with the participation of a total of some 150 experts (Commission on Climate and Vulnerability, interview; Commission on Climate and Vulnerability, 2007). An advisory committee additionally included civil servants from the Departments of Defense, Agriculture, the Environment, Finances and Industry (Commission on Climate and Vulnerability, 2007). The scientific basis was supplied by a new climate scenario developed by the Swedish Meteorological and Hydrological Institute (SMHI). This was used alongside the existing regional climate scenarios that have long been developed in Sweden through projects such as the Swedish regional climate modelling programme *Sweclim* (1997–2003). Given that it was partly motivated by recent flooding events, the Commission was also assigned the task of supplying an interim report on the consequences of flooding in Lakes Vänern, Mälaren and Hjälmaren. The interim report was finalised in 2006 (Commission on Climate and Vulnerability, 2006), prior to the 2007 final report. An interviewee noted that the focus of the interim report was a result of the fact that

> the counties had written to the Government, to the Department of Defense. They had to reply and do something about this ... They had a letter, a memorandum [*skrivelse*] lying there, and if nothing was done from the Government side and there was new flooding, that would not be good, so they were anxious for us to take that issue up. (Commission on Climate and Vulnerability, interview)

The interim report (Commission on Climate and Vulnerability, 2006) states that the 2001 flooding of Lake Vänern was so extensive that the local county administrative board took over the responsibility for regulation of the lake in accordance with the law on rescue services [*Räddningstjänstlagen*], and ordered an increase in the amount of water to be withdrawn. Despite these measures, however, the flooding resulted in damage of several hundred million Swedish kronor (SEK), or tens of millions of euros. The 2006 report also takes account of several proposals that were later to be included in the 2009 Bill on climate and energy, as well as in the

Commission's full 2007 report, and centred on the need for measures to increase the drawing of water to limit flood and landslide risk in the future.[5]

The final report of the Commission on Climate and Vulnerability in 2007 is an extensive document of some 670 pages (excluding basic data reports included in the appendices) that provides an overview of vulnerability to climate change in Sweden in different areas and sectors. Major impacts of climate change described in the report include an increase in forest production, improved conditions for agricultural production, and increased risks for flooding, landslides and erosion (Commission on Climate and Vulnerability, 2007). Impacts on water quality, mountain environments including reindeer husbandry and tourism, risks of pests and disease, and the potential decrease in domestic energy use for heating are also described. The report suggests potential measures on a sectoral basis, often including a review of existing legislation and regulation as well as suggestions for the development of support systems, monitoring, research, and improved dissemination of information and support.

The Commission further notes the need to clarify responsibilities for preventative measures and targets the issue of distributing responsibility between individual property owners, the municipality and the state. The report concludes that the municipalities' jurisdiction over planning and construction indicates that the municipality should be primarily responsible for preventative measures for existing property. To support adaptation, the Commission suggests that county administrative boards should be assigned regional coordinating roles and the responsibility for the development of regional analyses of long term water supply (Commission on Climate and Vulnerability, 2007).[6]

The Commission also suggests strengthening municipal responsibility for climate change adaptation and using state funding only for larger-scale adaptation investments to decrease vulnerability to extreme weather events and long-term change (Commission on Climate and Vulnerability, 2007). The Commission proposes that the Swedish Environmental Protection Agency (EPA) be made the responsible body for national monitoring and reporting of adaptation (in addition to its similar existing responsibilities for mitigation). Other roles are also assigned: the SMHI is designated the responsible body for knowledge development on climate change, while national sectoral agencies are assigned responsibility for adaptation to climate change in their own issue area.[7]

---

[5]These proposals included the need for creating voluntary agreements with Vattenfall, the holder of the water court's judgement, to be able to increase the drawing of water from Vänern through the regulated Göta River; measures to avoid landslides that would require greater water withdrawals; and the possible construction of a tunnel from Vänern to the sea to draw water should the risk of flood occur again (Commission on Climate and Vulnerability, 2006).

[6]The Commission also suggests the development of specific climate adaptation delegations at the county administrative board level (a suggestion that is ultimately not taken forward in the 2009 Climate Bill).

[7]A specific research centre on adaptation was also proposed (a recommendation that is ultimately not included in the Bill). In a supporting report to the Commission on Climate and Vulnerability, additional recommendations include the creation of a central climate coordinator (ultimately assigned in 2008) and the development of boundary organisations and arenas for learning through

With regard to issues of flooding, high water, erosion and landslides (particularly for the Lake Vänern and Göta River valley areas), the Commission concludes that 'among the most serious consequences we have been able to identify are risks for landslides and flooding' (Commission on Climate and Vulnerability, 2007, p. 642, author's translation). In this regard, the Commission suggests a number of measures, many of which are ultimately forwarded in the 2009 Climate Bill in different forms. These include directing the Swedish Rescue Services Agency (*Räddningsverket*) to map areas at risk for flooding, landslides and erosion and to continue work on an environmental hazards database. The Swedish Mapping, Cadastral and Land Registration Authority (*Lantmäteriet*) is instructed to create an improved national topographic data base, while the Swedish Geotechnical Institute (SGI) is assigned the task of investigating landslide risks.[8]

The Commission reveals that a large number of Sweden's major surface water sources are vulnerable to risks of microbiological and other forms of pollution during periods of flooding or heavy rain. As maintaining good-quality ground water is one of the environmental objectives issued by the Swedish State, the Commission highlights the need to 'control pollution risks in catchment areas' (Commission on Climate and Vulnerability water supply issues, interview). This issue was addressed through, among other things, reformulation of the Swedish EPA guidelines for fresh water protection areas [*vattenskyddsområde*] (Commission on Climate and Vulnerability, water supply issues, interview).

5.2.3 *National Legislation, Regulation and Measures Taken on Adaptation*

5.2.3.1 Implementation of the Commission on Climate and Vulnerability: The 2009 Bill, Regulation Letter Appointments and Budget Allowances

In March 2009, the Bill on *An Integrated Climate and Energy Policy Climate* was published (Government Offices of Sweden, 2009; hereafter 'the Bill'), following consultation on the suggestions in the Commission with relevant agencies and other organisations. While the majority of the Bill deals with mitigation, the *Adaptation*

---

the development of a specific research centre. Given the risk of fragmenting existing expertise, the Bill does not ultimately take this suggestion forward.

[8]The Commission also suggests the reassessment of the need for the review of water court decisions with regard to climate change, and the inclusion of landslide risks in the legislative framework to which municipalities must adjust their planning. According to the Commission, municipalities should also be able to undertake measures on others' land in order to protect surrounding building infrastructure (Commission on Climate and Vulnerability 2007).

to a Changed Climate chapter addresses adaptation and the suggestions made by the Commission on Climate and Vulnerability. The Bill as a whole was reviewed by a parliamentary committee issuing its statement in May 2009 following a review process (MJU, 2009). The parliamentary committee agreed with the main suggestions regarding adaptation in the Bill, many of which were also implemented already in the annual governmental regulation letters to state bodies, and the committee's suggestion was subsequently accepted by parliament in June 2009 (Swedish Parliament, 2009).

As suggested by the Commission, the Bill allocates regional coordination responsibility to the county administrative boards and attributes sectoral adaptation to sectoral bodies (such as the Swedish EPA and the Swedish Forest Agency) for their respective areas of responsibility. However, the Bill further calls for county administrative board work to be coordinated with existing crisis management and water authority tasks, rather than in separate climate adaptation delegations (as suggested by the Commission but not favoured by county administrative boards when circulated for comments). The Bill further suggests that national monitoring of regional and sectoral work is to be developed in parallel and coordinated with existing measures for mitigation follow-up coordinated by the Swedish EPA. Authorities responsible for the Swedish environmental quality objectives are also required to assess how and whether priorities are relevant in a changing climate (Government Offices of Sweden, 2009).

The Bill additionally includes a section on the need for legislative change in the existing planning and building law (PBL, Law 1987:10, cf. Planning and Building Act, 2009). This is based partly on a Bill 2006/07:122, which among other things addressed the increasing environmental risks described in the interim report of the Commission on Climate and Vulnerability in 2006. Subsequent changes in planning and building legislation in force from January 1, 2008, concern the requirement for consideration not only of health and security (as previously required in the law), but also of 'flooding and erosion' in planning and local infrastructural development. Any municipality that does not fulfil the demands of this legislation may be hindered in providing planning permission for such development by the relevant county administrative board under its supervisory role. However, an indication that the county administrative board will deny planning permission is generally signalled during consultations between a municipality and the supervisory board, which then results in changes in the proposed development and often avoids overt conflict (cf. Nyström, 2003).

The Bill further appoints the National Board of Housing, Building and Planning (*Boverket*) to assess whether language regarding landslides should also be added to legislation, or whether these can be considered adequately covered by existing language on hazards. As they are defined in the Civil Protection Act (*Lagen om skydd mot olyckor*, Law 2003:778), the responsibility for the prevention of hazards is deemed the responsibility of the property or facility owner (Swedish Civil Contingencies Agency, interview). However, given projected changes in risk, the Bill requires an assessment as to 'whether it is suitable to, mainly through changes in planning and building legislation, provide municipalities with the right to undertake actions on other's land in order to protect surrounding buildings' (Government Offices of Sweden, 2009, p. 191, author's translation).

This issue of the distribution of responsibilities introduced in the Commission is also implemented in the Bill, with consequences for such issues as the funding of preventative measures at the municipal level. Following the suggestion of the Commission on Climate and Vulnerability, the Bill reviews the total annual grant of SEK 40 million currently available to municipalities upon competitive application for particularly urgent preventative measures with regard to landslides and other environmental hazards. The Bill lowers the percentage that the grant may cover from 80 to 60% of the cost of measures from 2010 onward. In the referral process preceding the Bill, however, many municipalities indicated a preference for increasing state funding for prevention on the basis that the grant has only covered the cost of physical investment and that all applicant municipalities have not had their applications granted. The result has been the availability of sufficient funding for as few as 10–15 municipalities each year: 'If we are to prevent hazards, much more than 40 million per year is needed' (Swedish Civil Contingencies Agency, interview).[9] In the Bill, it is stated that local responsibility is the primary incentive for the retained level and decrease in percentage of funding: 'The motivation is the principle of own responsibility. The municipality has the primary responsibility for planning and building permission and has a decisive influence over building localisation' (Government Offices of Sweden, 2009, p. 196, author's translation).

The Bill thus considers prevention of environmental hazards within the scope of municipalities' existing responsibilities and as a part of their regular responsibility for citizen protection. For individual properties, however, property owners are ultimately responsible unless it can be shown that building permission has been granted by the municipality on faulty grounds, for instance, without paying sufficient attention to flood risk. For such a situation, extending municipal responsibility over a longer period of time was initially proposed by the Commission, but rejected in the Bill.[10] The Commission also suggests that while certain flaws in insurance protection currently exist, these are not significant enough to warrant specific state support for measures, but could be handled by individual insurance companies. It is stated that in the future, private insurance protection might need to be developed with regard to environmental hazards (Commission on Climate and Vulnerability, 2007).

---

[9]Crisis management administration has been revised during the period of study. In January 2009, the Swedish Civil Contingencies Agency (*Myndigheten för samhällsskydd och beredskap, MSB*) was created from the Swedish Emergency Management Agency (*Krisberedskapsmyndigheten*), the Swedish Rescue Services Agency (*Räddningsverket*) and the Swedish National Board of Psychological Defence (*Styrelsen för psykologiskt försvar*). The new authority is responsible for all work related to public safety, emergency management and civil defence, and has two positions focused on adaptation to climate change as well as additional experts on environmental hazards (Civil Contingencies Agency, interview).

[10]The Commission suggested increasing the period for municipal responsibility for detail plans and building permission to 20 years, a suggestion that was not passed into the Bill (Commission on Climate and Vulnerability, 2007).

With regard to responsibility, however, the Bill does accept the Commission's suggestion that the state should take a part in financing large-scale investments beyond the scope of municipalities, including those that concern the Lake Vänern area:

> We have some municipalities that have been affected, such as Kristianstad and Arvika, and they have received some funding, but there are several such needs that we take notice of, especially in Vänern and the Göta River Valley; something would need to be done there. (Commission on Climate and Vulnerability, interview)

This focus is reflected in several of the Bill's larger appropriations. The largest novel funding arrangements in the Bill give three authorities in particular special allocations. Following the traditional means of allocating responsibility, these roles were assigned through the issuance of regulation letters (*regleringsbrev*), which essentially constitute governmental decisions that are made annually to steer the work of authorities in the coming year (the underlying government instruction for the authority, on the other hand, is valid until changed). Regulation letters are based on the parliamentary decision on the national budget and government ambitions, and are discussed with the relevant department. The letters contain demands regarding implementation and reporting, as well as the financial framework for the appropriation in question.

The first of these larger appropriations concerned the county administrative boards, which will collectively receive an annual sum of SEK 25 million from 2009 to 2011 to work with adaptation issues:

> [t]he suggestion was ... that the county administrative boards should be a motor in adaptation work at the regional level. So the county administrative board role is to support municipalities, drive municipalities on and perhaps distribute some funding to common projects. (Commission on Climate and Vulnerability, interview)

This follows from the Commission's suggestions that county administrative boards should provide support to municipalities, undertake regional analyses of climate change impacts and summarise climate impact information, follow up sectoral and private adaptation work, and initiate the development of catchment level groups (*älvgrupper*) (Commission on Climate and Vulnerability, 2007).

Second, the Swedish Mapping, Cadastral and Land Registration Authority (*Lantmäteriet*) will receive an annual sum of SEK 40 million for three years in order to improve and develop a detailed topographic data base [*höjddatabas*] for the whole country. This appropriation was emphasised in the Bill as an issue on which a majority of relevant agencies and organisations agreed. The Bill further notes that while a majority of these agencies also proposed cost-free access to the database, the Swedish Mapping, Cadastral and Land Registration Authority has been instructed to assess potential models for database accessibility (Government Offices of Sweden, 2009). The need for such information is reflected in concerns expressed by interviewees in this study for absolving municipalities of the responsibility for topographical studies (Gothenburg City Building Office, interview).

> Mapping is very much dependent on good topographical data ... and we have been pressing for topographical data in Sweden to be improved ... we have [so far] had to make rather

> coarse limitations and assessments with regard to flooding and threatened settlement and risks connected to that. (SGI, interview)

For the Västra Götaland County Administrative Board, the development of an improved database will further 'make it possible to compare levels with one another, since around Vänern a number of local topographical data bases exist, and so it is very difficult to communicate where the bar should be placed for new building infrastructure' (Västra Götaland County Administrative Board, interview).

Third, the Swedish Geotechnical Institute (SGI) is allocated an annual SEK 35 million for three years in order to map landslides in the Göta river valley, taking into consideration the risks resulting from a changed climate (SGI, interview). This commission stems from the SGI's prior responsibility for geotechnical issues and specific responsibility for handling landslide and erosion issues, as well as its coordinating role with regard to coastal erosion since 2002. The coordinating responsibility for assessment of coastal erosion has also included specific responsibility for stability issues and reviews of all development proposals in the Göta River Valley, given its high susceptibility to landslides and its high levels of development (SGI, interview). Prior to the proposal of the Commission on Climate and Vulnerability, in 2004, the SGI had also been instructed by the Swedish government to develop an action plan for preventative work on landslide risks, within which the SGI included flooding and related environmental consequences. The SGI submitted its report to government in February of 2006 (Gothenburg City Office, 2006) and it became a foundation for both work in the Commission on Climate and Vulnerability and the recent government commission (SGI, interview).[11]

With relevance to the SGI assignment, studies for the Commission on Climate and Vulnerability had shown that increasing water flows could require reinforcements in the Göta River Valley at a cost between one to six billion Swedish kronor. This is the single largest assessment of requirements for infrastructural development in the Bill: while flooding in Stockholm may to some extent be managed by the less costly re-building of existing infrastructure, flooding in Lake Vänern and the Göta River requires larger infrastructural measures. Suggestions for the management of flood risks in the Göta River have included the construction of a tunnel from Vänern through which high water flows could be directed, estimated at a cost of some five billion SEK. The present government commission to the SGI is intended to clarify such risks, measures and costs (SGI, interview):

> More detailed geotechnical analyses are needed to more exactly determine the location of any tunnel and provide more exact measurements of building costs ... Before more detailed technical analyses are initiated regarding the prerequisites for a tunnel, the possibilities for increased withdrawal of water through the Göta River [must] be assessed. (Government Offices of Sweden, 2009, p. 188, author's translation)

---

[11] In February 2006, *Svenska Kraftnät*, the state utility that administers the national electrical grid, Vattenfall AB, municipalities and county administrative boards in the Göta River Valley, the SGI and the Swedish Maritime Administration also started a joint effort to develop basic information for dam risks and high flows, with the aim of finishing the project by 2008 (Gothenburg City Office, 2006).

One existing measure to avoid large-scale flooding has been to change the water withdrawal strategy in Vänern. To this end, the Västra Götaland County Administrative Board was assigned the task of negotiating with Vattenfall AB, an electricity company partly owned by the Swedish state and possessing the relevant right to water electricity generation, to lower high water levels. The so-called water court rulings [*vattendom*] constitute legislatively binding frameworks that describe the right and extent to which water can be drawn at different times, including minimum and maximum levels. The water court judgements are generally based on the environmental situation at a historical point in time, thus assuming a given flow and usage rather than changes in these that may occur for instance with climate change (cf. Commission on Climate and Vulnerability, 2007). To protect public safety in high risk situations, the County Administrative Board may supersede existing water court decisions to draw a higher amount of water. Negotiations to change the flow can also be conducted with the party having the right to withdraw water from the lake, in this case for electricity generation.

With regard to the situation in Västra Götaland, an agreement between the County Administrative Board and Vattenfall AB to lower water levels by up to 40 cm was reached in October 2008 (County Administrative Board Västra Götaland, 2008; Government Offices of Sweden, 2009). Discussions have also been held on the possibility of changing the relevant water court decision from 1937; however, this has since been deemed too difficult given the need to contact all relevant stakeholders for any re-negotiation of water and withdrawal rights (cf. Government Offices of Sweden, 2009).

As a result, the Swedish Rescue Services Agency has been given coordinating responsibility for the continued investigation of possibilities for water withdrawals at Vänern, while the SGI has been given the assignment to assess possible preventative measures for erosion (Government Offices of Sweden, 2009). Following these assessments and a decision on long-term solutions, the state will then appoint a negotiator (*förhandlingsman*) to develop recommendations for financing selected measures. Financing is to be shared between the state and relevant actors, including municipalities and holders of the water court judgements, possibly in the form of public-private cooperation (Government Offices of Sweden, 2009).[12] The state thus takes on some responsibility for measures deemed beyond the scope of any single municipality; however, responsibility for financing is to be distributed across a larger array of actors.

Other, smaller assignments in relation to climate change adaptation include the 2008 issuance of a regulation letter to the National Board of Housing, Building and Planning to assess how planning and building could be adapted to minimise negative impacts of climate change (National Board of Housing, Building and Planning, interview). The Board's participation in the Commission on Climate

[12]In addition, for 2010–2011, an annual SEK 20 million is also provided to increase society's ability to withstand flooding and landslides (Government Offices of Sweden, 2009).

and Vulnerability has also resulted in a report on building in a changed climate now used for adjusting building regulations (National Board of Housing, Building and Planning, interview). The Swedish Rescue Services Agency also received a governmental assignment in 2006 to develop a national platform for work with environmental hazards, in which 15 organisations are currently represented (Government Offices of Sweden, 2009). Further appropriations were also made to authorities in 2006 and 2008 (summarised in the Bill), including assignments to the Swedish Road Administration, the Swedish Rail Administration, the Swedish Maritime Administration, the Swedish Energy Agency and *Svenska Kraftnät* (the state authority that administers the national electrical grid) regarding the assessment of risks due to increases in extreme events, flooding and water levels. The Bill further provides funding to the Swedish Forest Agency for, among other things, information dissemination with regard to climate change risks to land and water use (Government Offices of Sweden, 2009).

Finally, the Bill also assigns national coordinating responsibility for drinking water to the National Food Administration, noting that relevant authorities must develop guidance on including sustainable water provision concerns in planning. This is a particular response to concerns such as the following raised by interviewees:

> Water supply has a fragmented responsibility. The Swedish EPA technically protects the catchment area . . . the water quality that comes out of the water plant is the supervisory responsibility of the National Food Administration . . . But there is no authority that is concerned with how we build water supply systems or what sort of pipe material or security or leakage we can have . . . there it has fallen between the chairs (Commission on Climate and Vulnerability water supply issues, interview)

The problem of dispersed authority with regard to water was also noted by another interviewee in terms of need for integration:

> Departments and agencies . . . may not consider that their issues have a bearing on ground water, a case in point being the Swedish Board of Agriculture, whose decisions and guidelines also have significance for ground water (Geological Survey of Sweden, interview).

5.2.3.2 Independent Initiatives at the National Level

So far, the national level in Sweden has largely focused on the distribution of responsibility for adaptation to climate change. However, awareness-raising and supporting work with regard to adaptation has also developed voluntarily among state agencies, including the development of tools for adaptation to climate change and an information web portal. For example, the Swedish Defense Research Agency developed the Climatools project in response to a call for research projects by the Swedish EPA. Projected to run 2006–2011, Climatools is aimed at developing tools for adaptation to climate change that can be used at the local level. The tools are intended to support the engagement of scenario methodology, the management of conflicts between different aims, and the development of economic or cost-effectiveness tools for calculating the impact of extreme weather events on health

(Climatools, interview). Tools are also to be integrated into existing processes such as risk and vulnerability analyses, planning processes, and health and environmental impact assessment. In particular, inspiration for this project was drawn from UK work on tools for local community adaptation (including the LCLIP local climate impacts profile developed by the UKCIP; cf. Chapter 3) and the use of expertise from Hampshire County Council, and future plans exist to facilitate connections between progressive municipalities in Sweden and the UK (Climatools, interview).

Current phases of the project include the beginning of pilot tests; however, some doubts as to the project's continuation exist. While the Climatools interviewee suggested that tools may be made available on the website of the Swedish EPA or on a climate adaptation portal website, 'it is not clear who will manage or market this tool box [after the end of the project]' (Climatools, interview).

Awareness-raising also exists in the form of a voluntary network established to develop the *Anpassningsportalen* web portal on adaptation to climate change. The portal was developed by the Swedish EPA (the funding agency for Climatools) in cooperation with the National Board of Housing Building and Planning, the SGI, the new Swedish Civil Contingencies Agency (MSB) (which replaced the Swedish Rescue Services Agency, the Swedish Emergency Management Agency, and the Swedish National Board of Psychological Defence on January 1, 2009), the SMHI, and, to a lesser extent the Swedish Energy Agency. Initiated by the Climate Policy Unit at the Swedish EPA and launched in 2007, the project has had no explicit assignment from the government to work specifically with adaptation, but has done so voluntarily within authorities' existing general instructions. The group has also held seminars with actors such as the county administrative boards to determine the forms of information required from authorities (Swedish EPA, interview).

The Swedish EPA additionally holds the secretariat under the Environmental Quality Objectives Council [*miljömålsrådet*] and over the course of 2009 assessed the impact of climate change on the environmental quality objectives. A revision of the environmental monitoring system to improve the identification of changes in the environment that may be the result of climate change is also under way. However, 'so far, we have not placed a very large focus on [adaptation] work ... the role so far ... has been to spread information and coordinate' (Swedish EPA, interview).

5.3 Regional Level

5.3.1 Impacts and Authority at the Regional Level

Västra Götaland is one of the two regions in the country in which a separate regional organisation is gradually taking over responsibilities both for health care (from a special body) and responsibilities for regional development (from the County Administrative Board) (cf. SALAR, 2009). During the time of this study, this body, called Västra Götalandsregionen, did not hold particular responsibilities with regard to adaptation to climate change, although a climate change strategy targeted at mitigation was published (VG, 2009a; VG, 2009b). However, it is possible that in the future the regional body will also develop priorities for adaptation for the region, especially if adaptation comes to be seen as an issue with impacts on regional development. Given this background and the defined role of the county administrative board, this section will mainly focus on the county administrative board as a coordinating body for adaptation.

5.3.1.1 General Responsibilities of the County Administrative Board

The county administrative board constitutes the regional implementing body of the Swedish state and has limited intrinsic policy-making capacities. Within the planning system, county administrative boards supervise municipal planning and have the right to hinder municipal development plans if they do not sufficiently take health, security and, since 2008, erosion and flood risk into account. In addition, the county administrative board also ensures that the interests of national priority areas [*riksintressen*] are respected, including the provision of drinking water, interests requiring coordination between municipalities [*mellankommunala intressen*], and the adherence to environmental quality norms. On the basis of any of these, the Board may hinder planned municipal development (cf. Nyström, 2003). Coordination between the Board's supervisory role and local municipal development is intended to be undertaken during the municipality's comprehensive planning, prior to the detailed planning regarding specific developments. However, updated comprehensive plans do not exist in all municipalities, despite existing legislation to that effect (the legislation does not specify any sanctions on municipalities that do not fulfil these aims) (cf. Planning and Building Act, 2009; Nyström, 2003). This absence of updated comprehensive planning may sometimes result in conflicts between the solutions advocated by municipalities within the framework of the municipal planning monopoly and the board's supervisory role (cf. Pettersson, 2008).

Other tasks of relevance for adaptation for the county administrative board include assessments of risk and vulnerability, the implementation of environmental quality objectives, and its role as water authority. Risk and vulnerability assessments must be conducted according to the regulation on crisis preparedness and increased preparedness (2006:942), and include aspects on general risks, needs for improvement, and cooperation (County Administrative Board Dalarna, 2008). The assessments are intended to be used as basic data for planning and measures for the county administrative boards, municipalities, industry and other actors in the county. Since 2005, the boards have also been obliged to develop regional action plans to implement environmental quality objectives (Swedish EPA, 2008b).

On a more general level, the county administrative boards also work with regional development programmes (overarching planning tools in each county, although not always practically integrated with comprehensive planning) and function as umbrellas for more specific development through regional growth programmes, structural fund programmes, and/or county programmes for regional transport infrastructure. In the region, some of these functions are being transferred to the Västra

Götaland regional body. In general, regional development programmes are developed through partnerships in order to set targets (in line with parliamentary goals) for regional development policy and coordination across counties, and include plans for follow-up and evaluation of work (SALAR, 2009).

The implementing role of the county administrative boards also requires close cooperation with state agencies. For example, the National Board of Housing, Building and Planning cooperates with the county administrative boards in determining important issues to the counties, using both inquiries into planning and building as well as annual planning discussions. County administrative boards may also call for cooperation or consultation with national level agencies or participation in a reference group for a particular issue (National Board of Housing, Building and Planning, interview).

5.3.1.2 Coordination of Water Issues

With regard to water issues, responsibility is given to select county administrative boards under the designation of five Water Authorities, created in 2004 for the implementation of the EU Water Framework Directive. Water authorities coordinate work in their respective districts to set environmental quality norms and develop administration and action programmes with regard to water (Swedish Water Authorities, 2009). The water authority for the *Västerhavet* water district is the Västra Götaland County Administrative Board. An additional 55 associations for water conservation (*vattenvårdsförbund*) also work in the country's major coastal areas, watercourses, and lakes. Today, these associations may elect to be transformed into water councils, bodies with an increased coordinating role for water protection and related issues in cooperation with the water authorities (County Administrative Board Västra Götaland, 2009b; Härryda Municipality, 2009). The associations are typically stakeholder bodies such as the Lake Vänern Society for Water Conservation, a voluntary, non-profit organisation of 59 members comprising local authorities, county councils, government bodies, companies and NGOs (County Administrative Board Västra Götaland, 2009a).

In addition to these, river valley coordination groups (*älvgrupper*) exist as cooperation fora for stakeholders in river valleys where communities upstream may have an impact downstream, these including such actors as municipalities, power companies, and the Swedish road, rail and maritime administrations. The coordination group along Lake Vänern and the Göta River includes stakeholders both from Västra Götaland and Värmland counties, including the power companies Fortum and Vattenfall as well as the types of stakeholders listed above (Västra Götaland County Administrative Board, interview).[13] The regulatory and implementation framework with regard to water issues is therefore relatively dense and includes

[13]The different tributaries to Lake Vänern also have separate river valley coordination groups that usually meet twice a year and focus on information sharing and emergency prevention (Västra Götaland County Administrative Board, interview).

some participation by stakeholder bodies despite the central role of the County Administrative Board.

With regard to the focus on the particular vulnerability of Lake Vänern and the Göta River area within the national context, it can be noted that the Göta River Valley constitutes the main outflow area, rendering the municipalities along the river valley immediately impacted by high flows. The entire Göta River Valley is additionally characterised by unstable ground conditions and frequent landslides (Västra Götaland County Administrative Board, interview). The need for better knowledge on conditions of flooding and soil instability in the Göta River Valley was highlighted following two large landslides that occurred in 1950 and 1957. The events additionally prompted the formation of the SGI, as well as the regulation of building permits issued without prior geotechnical investigation (Västra Götaland County Administrative Board, interview). Flooding in the area also carries the risk of contamination of drinking water supplies along the river valley.

As described earlier in the chapter, negotiations have been held between the Västra Götaland County Administrative Board and Vattenfall AB on preventative water withdrawals during high water events. This has been suggested as a way to potentially avoid the expensive procedures required for changing the water court decisions (Västra Götaland County Administrative Board, interview) whereby drawing rights are allocated. However, some actors consider the agreement between the County Administrative Board and Vattenfall AB to be insufficient (Commission on Climate and Vulnerability, interview). Such protests are based on the idea that significant changes in environmental conditions have occurred since the establishment of original water judgements in the early to mid-1900s; for instance, precipitation has increased by 10% in Västra Götaland over the last fifteen years (Västra Götaland County Administrative Board, interview). Further, the previous focus on producing hydroelectricity for national development is now considered to require increased consideration of cultural and environmental values in the areas (Västra Götaland County Administrative Board, interview).

While designated as largely coordinating bodies and implementers of state legislation, the overarching nature of such tasks may present a challenge especially to smaller county administrative boards. As an interviewee at the County Administrative Board in Västra Götaland noted:

> This coordinating role is very much characteristic of the county administrative board ... we are both to persuade the holder of a water court's ruling to draw water where the least damage is caused and at the same time [to act] as a supervisory agency for water use and see to that the holder follows the court's ruling ... We follow this up in river valley coordination groups where we follow the process so that we know what the consequences are if we intervene or not.

Situated in one of the large Swedish regions, however, the County Administrative Board of Västra Götaland noted that 'with a large ... organisation of 700 employees, we can afford to have both a specialist in water protection, in dam safety, and perhaps also in crisis management' (Västra Götaland County Administrative Board, interview).

With regard to adaptation, the Boards' coordinating role foreseen through the Commission on Climate and Vulnerability and the changes in regulation letters have yet to fully develop. However, the Västra Götaland County Administrative Board has started preparatory work and identified 11 different units of the Board that are affected by and need to take into account climate change issues. The Board concluded that Västra Götaland is 'perhaps the county ··· most affected by climate impacts' (Västra Götaland County Administrative Board, interview). Together with municipalities, the Västra Götaland County Administrative Board has formed coordination groups in order to discuss common prerequisites for planning in areas such as the coastal and waterfront zone to avoid creating competition between interests wanting to locate close to the water (Västra Götaland County Administrative Board, interview). In response to the Commission on Climate and Vulnerability, the Västra Götaland County Administrative Board has also started to discuss coordination solutions for roads and built infrastructure (Västra Götaland County Administrative Board, interview). However, in both the Västra Götaland region and its municipalities, work explicitly related to climate remains focused on mitigation, including that set out in the proposal for a regional climate strategy (VG, 2009a; VG, 2009b).

5.4 Local Level

5.4.1 Municipal Climate Change Networks and Responsibilities

Sweden's 290 local authorities have a high degree of autonomy, which is reflected in discussions of the allocation of responsibility for climate change adaptation, as well as in the differences between municipalities in terms of how far local climate work has proceeded. National requirements for initiatives relevant to adaptation at the municipal scale have come in the form of legislation created especially following the creation of the Commission of Climate and Vulnerability. With regard to crisis preparedness, municipalities are also required to identify how risks and vulnerability will be decreased according to the law on extraordinary events (*Lagen om extraordinära händelser* 2002:833) and the Civil Protection Act (*Lagen om skydd mot olyckor* 2003:778), and to develop and report risk and vulnerability analysis to the Country Administrative Board following Law 2006:544 (Town of Trollhättan, 2005). In an investigation reported in the Commission on Climate and Vulnerability, municipalities suggested that adaptation at municipal level would require, among other things, central government clarification of responsibilities, regulation on the treatment of flood risks, clearer climate scenarios, increased coordination between municipalities and agencies, and resources for charting risks and performing detailed measurements in areas of high risk to climate change (Commission on Climate and Vulnerability, 2007).

While specific state adaptation support measures targeted at municipalities are yet limited, instruments for the support of mitigation and environmental aims have generally taken the form of grant programmes. Programmes with relevance for climate mitigation include the Local Investment Programme (LIP) and the Climate Investment Programmes (Klimp). The first was introduced in 1997 by the government and was intended to encourage municipalities to cooperate with local companies and organisations in the competition for LIP grants to cover up to a third of local environmental investments. During the period 1998–2002, nearly 270 municipalities applied for LIP grants, of which 161 received funding (some more than once). Approximately half of the total LIP grants of a total sum of SEK 6.2 billion were used for energy-related measures (Swedish EPA, 2005).[14] In 2003, the LIP programme was replaced by Klimp, a programme set up in a similar fashion but focused exclusively on climate and energy issues (Swedish EPA, 2008a). As a result of the Klimp programme, several municipalities have adopted local climate objectives strategies with local emission reduction targets (Swedish EPA, 2008a). In several cases, these strategies refer to the local implementation of Swedish environmental quality objectives on mitigation. Some interviewees suggested that state steering through economic incentives such as the LIP and Klimp programmes could also be implemented more broadly to include climate adaptation goals (GR, interview).

More recently, the Swedish Government has allocated a smaller grant of a total of SEK one billion for the period 2008–2010 for climate-related activities under the heading of *Sustainable Municipalities* (Ministry of the Environment, 2008). The programme follows up on activities initiated by the Swedish Energy Agency in 2003 that focused principally on sustainable energy use in five test communities in the first five-year phase, followed by an additional 62 municipalities in the second period. Today, the programme is managed through the National Energy Authority as well as the Regional Energy Offices. The programme briefly notes that adaptation to climate change is included within the framework of physical planning (Sustainable Municipality Programme, 2009).

Beyond national legislation and grant programmes, municipal initiatives on climate change in Sweden have been developed to a large extent through municipal and non-governmental networks that link and share information between municipalities. Organisations of relevance include the Swedish Association of Local Authorities and Regions (SALAR), an interest organisation for local and regional government. The adaptation issue was made especially relevant to SALAR following the publication of the EU Green Paper on adaptation in 2007, after which the newly appointed chairperson announced a focus on adaptation and the creation of a position for the management of adaptation to climate change in planning. Since 2008, SALAR has identified the reduction of climate impacts as a priority issue and is currently developing a leaflet on water management in physical planning (SALAR, interview).

---

[14]In 1999, the Swedish Institute for Ecological Sustainability (IEH) was established as a national authority to bridge the gap between research and stakeholders (e.g., local authorities), and provide information on best practices and support on LIP and LIP applications (IEH, 2003). In 2005, IEH was transformed into the Swedish Council for Sustainable Development, which, in turn, ceased operations in March 2007.

Another municipal network relevant to climate change is the Climate Municipalities (*Klimatkommunerna*), a network of 20 municipalities focused mainly on mitigation. The network works principally with information dissemination, arranging seminars in 2007 and 2008 on adaptation in conjunction with Climatools, and a regional seminar on adaptation in Skåne (southernmost Sweden) in 2007 (Climate Municipalities, interview). However, adaptation has yet to be fully integrated within the programme:

> There exists no natural platform for adaptation work in the municipalities ... emissions reduction and such may be addressed in environmental management or ... in the environmental strategy unit ... but the adaptation question does not have such a natural [placement] ... We have not built that competence [on adaptation]. (Climate Municipalities, interview)

In addition, there exists the National Association of Swedish Eco-municipalities (SEkom), a network established in 1995 and currently consisting of 70 municipalities who have agreed to adopt a strategic plan and programme for achieving local sustainability. The network is based in Agenda 21 and the sustainability principles of the 'Natural Step' organisation developed in Sweden in the 1990s. The Association has a small secretariat that organises events such as workshops, seminars and a course (and associated handbook) in climate communication, as well as initiatives on sustainable procurement. Despite some interest following the release of Al Gore's *An Inconvenient Truth*, however, the organisation has not noted that adaptation has yet become 'anything that our member communities request ... most municipalities are right now in an initiation phase [with regard to adaptation]' (SEkom, interview). Similarly, both SALAR and the Climate Municipalities also emphasised that issues must be member-driven: they noted that as adaptation has yet to reach a high profile among municipalities, they had limited possibilities to allocate resources for the issue.

Other networks that have focused on climate change mitigation include the BLICC corporate network and the Klimax grass-roots organisation, as well as international networks such as the Climate Alliance, Energie-Cités, and Cities for Climate Protection. Local Agenda 21 organisations, which were extensively developed in Sweden during the 1990s (Eckerberg et al. 1997), have also played a role in municipalities where Agenda 21 coordinators are still present. However, according to the Swedish Society for Nature Conservation, local work specifically on adaptation to climate change has only recently begun (SSNC, 2007).

Organisation relevant to adaptation also exist in the greater Gothenburg subregion (indicating a level of functional regional organisation separate from that of the county). The Gothenburg Region Association of Local Authorities (GR) is a cooperative organisation financed by member fees from 13 participating municipalities. The GR is one of four regional associations of local authorities in the Västra Götaland County working together under an organisation called Västkom. The GR also works closely with the Business Region Göteborg organisation for trade and commerce in the Gothenburg region (GR, 2009). Using a regional perspective on municipal planning, the GR has developed a regional water supply plan, and has participated in EU sewage treatment and water projects with the support of the West Sweden EU and Representation Office in Brussels. The GR also sits on the Water Council for the Göta River as well as the reference group for the 2009 SGI special governmental commission for the Göta River Valley. Following the report of the Commission on Climate and Vulnerability, the GR's steering group for environmental and societal issues (*miljö och samhällsbyggnad*) invited the SGI to describe the Göta River Valley situation, resulting in the development of a GR general action programme for infrastructure and water supply in the Göta River that included some mention of adaptation to climate change (GR, interview). Adaptation policy development is also here, however, at an introductory stage.

5.4.2 Municipal Case Studies

In addition to examining municipal requirements for adaptation developed by the Bill and the potential for initiatives through existing grant programmes and networks, this section provides a closer look at the capacity for municipal adaptation. The cases below illustrate varying vulnerabilities, priorities and adaptive capacities with regard to climate change, ranging from the relatively high vulnerability and extensive development of adaptation in Gothenburg, to vulnerable municipalities with fewer resources such as Mölndal and Munkedal. Trollhättan, on the other hand, may fall at the other end of the spectrum: while the municipality has not identified particular vulnerabilities, it could potentially exhibit a relatively high adaptive capacity with regard to climate change.

5.4.2.1 The Development of Adaptation Measures in Gothenburg

The second largest city of Sweden and located at the mouth of the Göta River, Gothenburg is considered to be 'one of the municipalities that work with climate adaptation issues in a more structured way' (Commission on Climate and Vulnerability, water supply issues, interview). In particular, Gothenburg has focused on the risks of flooding and sea level rise. Gothenburg's involvement with adaptation is spurred not only by its vulnerability to such impacts of climate change, but by increasing building pressures. Municipal growth in central parts of the city over the next 20 years is expected to occur principally in low-lying areas (Gothenburg City Environmental Administration, interview).

However, while Gothenburg is identified as a municipality with existing issues with regard to flooding and erosion, interviewees also noted the city's higher adaptive capacity in comparison to the typical municipality given the relatively large pool of human resources dedicated to climate change issues (SALAR, interview). The current climate team in the City of Gothenburg consists of twelve people from different administrations and companies, including the traffic office, park administration, environment administration, building administration, the state railway administration, rescue services, and the municipal Gothenburg Water (*Göteborgsvatten)* and Gothenburg Energy (*Göteborgsenergi)* companies. Monthly meetings are held, from which each representative brings relevant issues back to his or her own administration (Gothenburg City Building Office, interview). The existence of such a municipal stakeholder group on climate change adaptation issues is noted as a relatively unique venture: '[t]here was no one else in Sweden who was working in this way when we started' (Gothenburg City Building Office, interview).

The development of adaptation in Gothenburg began principally as a capacity-building issue that drew upon both local extreme events and existing indicators. In 2000, the Gothenburg City Building Office developed a temporary comprehensive plan for water as a thematic area, undertaken within an EU project. The project was undertaken at the civil servant level and based on national environmental quality objectives. Over the course of the development of the temporary comprehensive plan, however, it was noted that water levels projected in IPCC reports would cause a sea level rise that threatened the security margins for high water levels (Gothenburg City Building Office, interview). Following the Gudrun storm, these concerns were acted upon: building limits in local regulations were raised, and a crisis and hazards coordination group *(kris och katastrofsamordningsgrupp)* was established to act in the case of similar events (Gothenburg City Office, 2006).

The focus on adaptation in Gothenburg was also institutionalised at the local level in project form. Concern over increasing vulnerability in a private member's bill in the Municipal Council (*kommunfullmäktige*) eventually prompted the Gothenburg Municipal Council to give the Municipal Executive Committee (*kommunstyrelsen*) the assignment to investigate Gothenburg's preparedness for extreme weather events and required measures (Gothenburg City Office, 2006). The resultant *Extreme Weather Phase 1* project was coordinated by the Gothenburg City Office, which issued its report in 2006. The report found that according to the city topographical assessment system, the city could cope with an extreme high water level of 11.80 m, above which a security margin of 50 cm should be instituted. In the final decision, an additional 0.5 metres above the recommended security margin was implemented (Gothenburg City Office, 2006). In 2008, the municipal executive committee initiated an assessment of the need to further raise the lowest building elevation in response to the expected sea level rise, in which the Traffic Office recommended an additional increase of one meter above previous assessments (Gothenburg Traffic Office, 2008).

An important issue raised by these revisions is the distribution of responsibility between the municipality and citizens, and the need to follow the elevations stipulated in the comprehensive plan within detailed plans and building permits. For instance, an interviewee noted that 'if an individual property owner is affected and we haven't made her or him aware of these elevations, we may face liability for damages' (Gothenburg City Building Office, interview). Such issues were addressed in 2007 under the coordination of the Building Office in the *Extreme Weather Phase 2* project. In this phase, an assessment of an area of the city as a case study of sea level rise impacts was completed, as well as an overview of measures necessary to increase preparedness at the municipal, national, and individual levels. The report from this phase of the project discusses the need for early warning and monitoring measures, the short-term development of computer flood modelling, additional water level measuring devices, the publication of results on the web, and the development of an organisation for risk and crisis management. Over the longer term, the report suggests that laws and regulations need to be clarified, to allow for changes to water court judgements and to increase state funding for municipality preparedness measures (Gothenburg Building Office, 2009).[15]

Interviewees in Gothenburg also noted substantial vulnerabilities beyond those directly relevant to flooding and sea level rise, including the threat of loss of municipal electricity during storm events and contamination of water supplies through bacterial growth during periods of high temperatures. However, appropriate adaptations have yet to be identified, especially with regard to water supply issues which one interviewee described as highly vulnerable: 'The Göta river is our water supply ... we have a reserve supply ... which lasts for three weeks, and then we are out of water ... and in Gothenburg you have some 500,000 people' (Gothenburg City Building Office, interview). A potential site for additional supply is owned by a gravel mining company, but processes to investigate this option have been halted by limited coordination between the company, the county administrative board and the municipality (Gothenburg City Building Office, interview).

Interviewees also note that though the costs of more extensive flooding adaptation measures would be larger, they would be limited in relation to the value of property at risk of flooding in the city centre and require realisation of some of the proposed measures in the Commission and Bill (Gothenburg City Building Office, interview). The calls for an investigation into possibilities for water diversion from Vänern in both the Commission and Bill are therefore echoed in relation to required measures at the local level:

> We have calculated the costs for measures ... but ... the issues of responsibility have to be clarified first ... because the state also has facilities in this area, if the sea [level] rises then for instance the railway station will become acutely threatened ... the Road Administration has tunnels in the city ... so in some ways measures have to be distributed ... and the responsibility for individual property owners has to be defined. (Gothenburg City Building Office, interview)

The fact that Gothenburg Municipality has been able to independently begin developing adaptation responses (in contrast to many other Swedish municipalities) is attributed by interviewees both to Gothenburg's vulnerable situation and to a number of capacities. Interviewees in Gothenburg emphasised traditions and social capacities in Gothenburg, as well as informal leaders who acted as catalysts for the city's relatively early response. Gothenburg has had a number of climate awareness-raising campaigns, including campaigns in connection with Al Gore's receipt of the international Gothenburg Award for Sustainable Development. Gothenburg was also one of the first municipalities in the country to measure air pollution levels, demonstrating its tradition in environmental policy development (Gothenburg

[15]The report also notes that the city building office and Gothenburg City may be liable for damages if building permission is given with a elevation lower than the now set levels, which means that detail plans may not be set below this elevation to fulfil the requirements of the revised PBL legislation concerning risks for flooding and erosion (Gothenburg Building Office 2009).

City Environmental Administration, interview). Several interviewees further noted that Gothenburg is an internationally oriented city with a 'Gothenburg spirit' of well-developed cooperation between different municipalities, regional bodies and industries in the Gothenburg region on environmental issues (GR, interview). An interviewee noted that 'since we are rather far ahead in our work, we have also received an informational role ... presenting for the Swedish EPA and for other municipalities' (Gothenburg City Building Office, interview).

As the same interviewee noted, however, 'policies are rather easy to develop ... but bringing them into implementation is more difficult ... it requires a lot of information, education and [taking] personal contact' (Gothenburg City Building Office, interview). As a result, few practical measures have yet been taken beyond raising the lowest building elevation, potentially as a result of the ongoing discussion of responsibilities. One person described in particular that practical measures may have been delayed since actors may not have wanted to take on specific measures before knowing whether these would fall under their area of responsibility (Gothenburg City Building Office, interview). Planned developments include the installation of further water level measuring devices, the development of a 3D model for simulating flooding events, and a website that collects data on water level measurement. Despite the relatively low cost of such measures, implementation has been delayed to 2010:

> It is a limited cost in the context of Gothenburg ... you could simply gather the administrations and municipal companies we have in the [climate] group and say let's do this. ... But it cannot be done because it is not within their missions. So then you have to go centrally to the municipality. (Gothenburg City Building Office, interview)

It was thus noted that adaptation might have been able to progress further, but has been impacted by the difficulty of fitting a cross-sectoral issue within the budget and system:

> Municipal organisation is an establishment, and [adaptation] is a new issue for many. This means that it doesn't fit into any organisation, any slot ... and you are rejected in budget considerations ... in competition with other issues. (Gothenburg City Building Office, interview)

In sum, however, Gothenburg can be characterised by a number of capacities and features that support its development of adaptation. It has been possible to address its high vulnerability especially with regard to infrastructure and building development through support from early response to events and indicators; relevant policy development; relatively large financial and human resources; and local traditions with regard to environmental awareness and cooperation. On the other hand, adaptation remains limited through its nature as a cross-sectoral issue and that it has so far not been integrated into budget considerations. The development of adaptation actions was also limited at the time of the study by the absence of national decisions on responsibility for measures.

5.4.2.2 The Development of Adaptation Measures in Mölndal, Trollhättan and Munkedal

The three smaller municipalities in this study illustrate widely differing capacities and priorities with regard to adaptation. As smaller municipalities, they each illustrate issues of limited capacity in terms of financial, human and informational resources. However, while the relatively vulnerable municipality of Mölndal receives considerable support through its proximity to Gothenburg, Munkedal illustrates both the potential for high sensitivity to climate change and severely limited resources to deal with the challenges arising from it. The municipality of Trollhättan illustrates a more limited perception of vulnerability and thus development of the issue, but also a well-developed environmental policy organisation that could potentially support the development of adaptation if it were to be defined as a priority.

Mölndal Municipality

Mölndal Municipality has been engaged in climate change mitigation through its climate strategy (in coordination with the state environmental quality objective on limited climate impact), and as a part of the Climate Municipalities network. Given its proximity to and impact on watercourses it shares with Gothenburg, Mölndal often works in coordination with Gothenburg, but has not yet integrated the issues of sea level rise risk to the same extent as Gothenburg has (Mölndal Municipality, interview).

As in Gothenburg, actions in Mölndal have largely been undertaken in response to extreme events: in 2006, the Mölndal River (*Mölndalsån*) flooded the centre of the municipality, including the basement of the municipal office. This event resulted in implementation of measures including increased dredging of the river in cooperation with Gothenburg municipality, and the reinforcement of river banks and building of embankments. The event also prompted Mölndal to add 30 cm to the lowest building elevation as a security precaution in the event of a similar flood event, as well as develop greater cooperation on flooding issues with municipalities upstream. Water court rulings have also been reviewed with an eye to the potential to limit risks for flooding (Gothenburg City Environmental Administration, interview). Water authorities in Mölndal are currently developing action plans to address the need to assess vulnerable areas and make empirical measurements in order to be able to plan for and respond to flooding events (Mölndal Municipality, interview). Ongoing extreme events also continue to highlight the need for risk avoidance: 'there was a [smaller] landslide in Gothenburg some months ago, and the Swedish Rescue Services Agency called and asked how our municipality was doing and were there any risks . . . so that creates an additional pointer on that issue' (Mölndal Municipality, interview).

An interviewee noted that most of this work has been driven by highly committed civil servants (Gothenburg City Environmental Administration, interview). However, limitations in terms of knowledge and resources to work independently in Mölndal are also present. As one interviewee noted: '[w]e don't have sufficient funding to hire people who have knowledge on these issues [and] we also don't have investment funding to undertake the measures that we would need to' (Mölndal Municipality, interview). From the perspective of a smaller local municipality, such issues may also be rather daunting given the limited ability to influence state decisions on Lake Vänern, the Göta River and the adjacent watercourses currently under investigation: 'perhaps there are no solutions for where this water should go, why we have built basements in central Mölndal – it's not possible to do anything about it today' (Mölndal Municipality, interview).

Trollhättan Municipality

Despite (or perhaps as a result of) its strong focus on industry, Trollhättan Municipality has been relatively advanced when it comes to mitigation and issues of integrating the environmental quality objectives. In relation to emissions reduction, the municipality portrays itself as a 'good practice' case (Trollhättan Municipality, interview). In assessments of climate change work (mitigation in particular), Trollhättan has been ranked among the top municipalities by the Swedish Society for Nature Conservation (SSNC, 2007; Trollhättan Municipality, interview). The focus in Trollhättan has largely centred on developing municipal aims and resource and energy plans, and this work has been supported in part by large LIP and Klimp grants. Trollhättan has also had an environmental policy since the early 1990s, and has two Agenda 21 coordinators, which is relatively unusual (Trollhättan Municipality, interview). With regard to implementation of the environmental quality objectives, one interviewee noted that 'this is probably where Trollhättan actually aims to follow directions; if one is to have an energy plan, we have it. There are some municipalities that still don't' (Trollhättan Municipality, interview).

Issues of adaptation to climate change, however, have received less strategic attention (Trollhättan Municipality, interview). Though one interviewee noted that risks were particularly significant in 2000–2001 and again in 2006, other upstream municipalities had experienced greater problems during that time. Landslide risks are not notable in central Trollhättan, but rather downstream; given its location, Trollhättan will not be threatened by rising sea level (Trollhättan Municipality, interview). Municipal planning for 2020–2030 indicates some waterfront development is to be included in the municipality's risk and vulnerability analysis, but does not reveal a particular focus on adaptation (Trollhättan Municipality, interview). Trollhättan thus presently has no particular organisation for adaptation, although a monitoring group may be developed with representation from different municipal sectors. Given that the municipality has not identified adaptation as a priority, an interviewee indicated that the municipality will look to the county level for guidance (Trollhättan Municipality, interview).

Munkedal Municipality

The final local case, Munkedal Municipality is a small municipality and the site of the most recent and extensive clay land slide in Sweden. The slide took place on

December 20th, 2006, in Småröd, five kilometres south of the central town, with 'devastating consequences for road and railway traffic ... over several months' (SGU, 2009, para. 31, author's translation). The major impact of the landslide prompted the Swedish Road Administration to appoint a special commission, which reported that the cause of the landslide had not necessarily been the recent heavy rainfall, but rather ongoing construction work that caused extra stress on the clay-based ground (Swedish Road Administration, 2006). The Munkedal area itself is rich in clay soils and at risk of landslides, as well as the formation of rifts, as have occurred in other similar areas. The municipality is also characterised by low-lying housing that is frequently flooded (Munkedal Municipality, interview).

Work on climate change in Munkedal has taken place mainly as a result of demands made by the county administrative board for mitigation issues to be included in planning. The municipality has developed both a protection plan and climate strategy, which centre mainly on changes in the energy system (Munkedal Municipality, interview; cf. Munkedal Municipality, 2008). An established cooperation network between municipalities in this part of the country further focuses on defining environmental quality objectives in local action plans 'in order to get about the same requirements, so that no one [municipality] gains or loses advantages' (Munkedal Municipality, interview). Flooding and sea level rise risks are mainly included in comprehensive planning, revised according to law every fourth year. The interviewee also noted that the extensive developments requiring detailed plans often concern coastal areas (Munkedal Municipality, interview).

Limits on adaptive capacity are noted in terms of ongoing development and funding concerns. For instance, 'funding is a limiting factor ... [and] there is the challenge of finding residential ... locations in attractive areas ... that are not especially negatively impacted by climate changes' (Munkedal Municipality, interview). Though the municipality has applied for funding from the Swedish Rescue Services Agency for emergency prevention, it has not received funding (Munkedal Municipality, interview). The small size of the municipality has additionally constrained the ability to dedicate human resources to the issue: 'almost all of those of us who work in municipalities handle our particular jobs alone ... we're generalists ... to go deeper into something means that one often needs help' (Munkedal Municipality, interview).

5.5 Actors Beyond Government and Administration

5.5.1 The Role of the Private Sector

While the different levels of national public administration are highlighted in both the Commission and Bill, as well as by the interviewees, adaptation necessarily also involves other levels and actors. As compared to the attention paid to issues of responsibility between national and local levels, the role of industry and enterprises has received relatively little attention in either the Commission on Climate and

Vulnerability or the subsequent Bill. An interviewee at the Commission expressed the rationale behind the state-focused scope:

> We felt that the responsibility must be placed on ... the large industries and sectors ... [to] themselves undertake assessments ... It is on the municipal level and on the county administrative board level that one has to do something to provide the background. [For the private sector] ... there, the state's role is primarily to provide information on what climate changes will occur, scenarios and such. (Commission on Climate and Vulnerability, interview)

One exception to the limited discussion of the private sector has been the insurance industry. The insurance industry has been made especially relevant given that property owners are responsible for damage to their individual properties unless it can be shown that building permission has been issued by the municipality without sufficient regard to risk. In Sweden, residential housing is insured according to a template and without assessing specific site characteristics and, unlike other Nordic countries, also covering general protection from environmental hazards (Swedish Insurance Federation, interview).[16] Interviewees discussed whether the insurance system may be used in a positive sense as a self-regulating measure apart from state or municipal measures to limit development in high-risk areas through differentiated premiums. Caveats include whether any such development would have negative effects on existing housing in areas at risk (i.e. exclusion from insurance), or on the municipality (i.e. the issuance of building permits in areas at risk in response to citizen preferences) (Prime Minister's Office and National Board of Housing, Building and Planning, interviews).

During the Commission on Climate and Vulnerability process, insurance industry representatives were seen as somewhat unwilling to establish differentiated premiums with regard to climate change, as establishing such a system would require costly individual assessments (Commission on Climate and Vulnerability, interview). An interviewee at the Swedish Insurance Federation (the trade association for insurance companies in Sweden) instead noted that planning and building regulation should take these risks into account. In addition, the interviewee noted that information and monitoring of risks of increased precipitation and high water levels required further development, including an improved topographical data base, extended funding for county administrative boards, and improved mapping for landslides (Swedish Insurance Federation, interview). These measures were to some extent addressed in the Bill; however, an interviewee at the Commission on Climate and Vulnerability suggested that:

> there are some things that could be done but that weren't suggested ... simple things, such as the idea that real estate agents could be required to describe whether the building is an

[16]The interviewee from the Swedish Insurance Federation further noted that flooding in other Nordic countries is not included in normal insurance, and referenced the Norwegian environmental hazards insurance system (where a given price is applied e.g. for fire insurance). Finland was noted as a system that does not insure flooding damage, 'perhaps as they have not had so much flood damage before' (Swedish Insurance Federation, interview).

> area recognised by the municipality as being flood-prone. (Commission on Climate and Vulnerability, interview)

Given the potential role of the insurance industry, some interviewees further suggested an increase in cooperation between county administrative boards and insurance companies. Such cooperation could be used to determine whether the insurance sector could support municipalities that did not grant building permits in specific areas, using the motivation of higher premiums or, alternatively, no insurance responsibility. As noted by one interviewee,

> If the municipality knows this, it means that no conflict needs to arise regarding using this particular vulnerable land area for exploitation. So in that case, two parties – where the insurance industry is actually acting on the behalf of the individual – can jointly find a solution that is beneficial for the whole. (Västra Götaland County Administrative Board, interview)

International developments, such as the 2008 Nordic climate conference for insurance companies, have additionally been used to 'try and develop proposals for intentions within the sector' (Swedish Insurance Federation, interview). The Swedish Insurance Federation has also begun to describe best practices on climate change-related issues among Swedish insurance companies, following such examples as the UK Association of British Insurers' 'climate wise' list (Swedish Insurance Federation, interview).

Some cooperation with regard to the issue of infrastructure also currently exists on a voluntary basis, for example, through the dialogue project 'Building-Living and Property Management for the Future' (*Bygga-Bo*), active since 1998. The dialogue is a cooperation between companies, municipalities, and national authorities, with the aim of improving sustainability in the building and property sector by 2025 in the areas of indoor environment, energy use and natural resources (National Board of Housing, Building and Planning, 2009). The National Board of Housing, Building and Planning has also recently applied for funding under the environmental objectives to work with adaptation to climate change in existing buildings (SALAR, interview).

5.5.2 Impacts of the EU Level

Discussions of the EU level with regard to adaptation are relatively limited in both the Commission on Climate and Vulnerability and the Bill. Interviewees discussed the EU level to a very limited extent, which may reflect the hitherto limited development of adaptation approaches in which specific impacts from the EU level might be identified. The Commission on Climate and Vulnerability does note, however, that Sweden should support recommendations in the EU Green Paper on adaptation to conduct a survey of EU regulation to ensure support for adaptation. The report also notes the absence of concrete EU measures for the protection of biological diversity in a changing climate, and proposes review and improved implementation of the EU Natura 2000 network, the prioritisation of nutrient leakage to the Baltic Sea, as well as a review of the Habitats Directive. In addition, the report advocates the inclusion of adaptation into EU finance mechanisms, including the structural funds (Commission on Climate and Vulnerability, 2007).

Interviewees described the EU's impact on adaptation in terms of the Swedish implementation of EU directives and relevant reorganisation of Swedish structures, particularly with regard to the Water Framework Directive and water protection. Interviewees also noted that EU projects on water and flood risk management have inspired some focus on climate change adaptation (Västra Götaland County Administrative Board, interview). However, potential concerns included the added complexity within the framework of water court decisions that could result from the implementation of the Water Framework Directive, including potential re-assessments of water court rulings (Gothenburg City Building Office, interview). Interviewees also noted that the implementation of the Water Framework and Flooding Directives would require improved oversight of the water system which, together with the EU Green Paper on adaptation, could bring up issues that would support the development of adaptation (Geological Survey of Sweden, interview). An interviewee from SALAR highlighted that EU regulation requires different implementation in different countries:

> In Sweden, it may be that we need changes in legislation for what can be included in detailed plans at the municipal level, whereas in another country [such issues] may be [items that] the regions can implement on the lower level (SALAR, interview).

EU regulations, including those without direct relevance to adaptation, may therefore need to be blended with Swedish approaches on adaptation, and are impacted by the distribution of responsibility in the Swedish system.

5.6 The Distribution of Responsibility for Adaptation to Climate Change

5.6.1 Responsibility for Adaptation Across Governmental Levels

The development of adaptation policy and measures in Sweden highlights a rapidly developing framework on adaptation. However, it also highlights the major issue of distribution of responsibility for adaptation, which was discussed to a considerable extent by interviewees. Table 5.1 illustrates the types of adaptation policy and organisation developed at different levels (cf. Chapter 1). As the table shows, many of the adaptation policies and measures that have been developed lie either on national or local levels. The Commission on Climate and Vulnerability and the 2009 Bill both placed significant emphasis on the state's distribution of responsibility for adaptation to climate change. This concern is echoed in comments to the Commission on Climate and Vulnerability, where municipalities indicated their preference for an increase in the state grant for preparedness measures on the municipal level. In addition, the extent to which the case study municipalities have acted

Table 5.1 *Main adaptation policy development and initiatives on different levels.* Overarching policy priorities and binding targets are those developed in the Commission on Climate and Vulnerability and the related Bill. For the Swedish case, policy priorities are reflected in the general principles of municipal vs. state responsibilities

	Type		
Level	National	Regional	Local (examples)
Policy priority	State responsibility for large scale measures beyond municipal scope Municipal responsibility for preparedness according to Bill	Nationally mandated coordination responsibility and responsibility for risk and vulnerability assessment	Voluntary temporary comprehensive plan with regard to water Identification of possible adaptation measures Inclusion of adaptation in risk and vulnerability assessments
Binding measures	Changes in Planning and Building Act and hazards legislation; Development of topographical data base; Landslide mapping; Agency investigation of withdrawal strategies in Vänern	Negotiation of water withdrawal strategies	Raising of minimum building elevation
Adaptive capacity-building measures	Commission; Web portal (voluntary); Tools development (voluntary)	Coordination groups	Municipal and EU projects Municipal cooperation organisation seminars and workshop Flood prevention measures
Dedicated issue-specific organisation	–	–	–
Main-streaming in existing organisations	Coordination responsibilities to county administrative boards; Sectoral responsibility to sectoral agencies	Coordination according to national requirement	Climate teams

on adaptation is largely in relation to their identified vulnerabilities and the prerequisites that form adaptive capacity (such as size and funding of local government). The issue of responsibility – and possibilities – for adaptation has thus been a major discussion point among interviewees: 'A very big . . . question is . . . who should pay for what' (SALAR, interview). This section aims to discuss more closely how responsibilities have been reviewed and discussed in the Bill and by interviewees, potentially impacting the development of policy and measures at different levels.

The 2009 Bill settled some aspects of the allocation of responsibility. Municipal authority over comprehensive and detailed plans for local development and infrastructure is reflected in the Bill through the allocation of responsibility for adaptation to the municipal budget. It is also reflected in the larger proportion of measures to be paid by localities, even if the total available state grant level is held constant. However, the state bears general responsibility for major infrastructure such as railway stations and tunnels. This division of responsibility is reflected in the Bill in that large-scale actions deemed beyond the capacity of several municipalities or even counties (as in the example of the possible tunnel from Lake Vänern) are to be researched, with subsequent proposals for financial options to be developed by an appointed negotiator. Municipal self-sufficiency through municipal taxation and the local planning monopoly may thus render the question of costing at different levels more difficult than it may be in a more centralised state.

Proposals in the Bill do not, however, settle all issues regarding the division of responsibilities and the associated development of measures. The responsibility for prevention attributed to the municipal level may provide difficulties for smaller municipalities with fewer resources. In some cases, municipalities may also be required to take a longer-term planning perspective than has been the norm. Issues such as these were discussed by nearly all interviewees. For example, one interviewee noted that a long-term perspective would be needed for water maintenance in order to develop reserve supplies: 'the challenge there is to gain political acceptance ... because it is connected with costs' (Geological Survey of Sweden, interview). Interviewees also noted issues of limited staffing capacities at both the local and county level, as well as the fact that conflicting messages may be received from the national level regarding municipal planning:

> Very often we need to manage questions that are in the planning system treated by the National Board of Housing, Building and Planning and the Swedish EPA ... their views can be entirely contrary ... The Swedish EPA says that it should be done in one way, and they talk with their target group which can be people in environmental administration. ... The National Board of Housing, Building and Planning ... talks with planners ... it doesn't make it easier on the local level. (SALAR, interview)

Such issues are not necessarily resolved by the Bill, but are instead deemed to be issues for municipal planning. In this regard, the municipal planning monopoly is often viewed as both an empowering and a limiting feature of the Swedish system, from the point of view of central government and municipality actors alike. Decision-making power at the local level 'means that the municipalities ... must make an effort to coordinate their work with others. Then it is a challenge for the county administrative boards ... to take a coordinating role' (Climate Municipalities, interview). Another interviewee suggested that the decentralised planning responsibility structure may have particular consequences for a complex issue such as adaptation:

> [It] means partly that there are very many local politicians that have to understand the issue. And that [is something] you can identify as a weakness unless you ... have the resources to develop the competences of many (SALAR, interview).

The limited ability of the Swedish government to steer implementation was also noted, both through examples from the implementation of the environmental quality objectives on the local level provided by interviewees, as well as through the results of a consultancy study in preparation for the national environmental objectives evaluation in 2007–2008. One interviewee noted that the implementation of state environmental quality objectives relies mainly on 'naming and shaming' rather than the ability to force municipalities to implement targets. One interviewee noted: 'When the state sets up environmental quality objectives, for instance ... the state can steer that they are implemented on county level ... but they cannot steer further [down]... it is largely voluntary' (Gothenburg City Environmental Administration, interview). The current national framework for steering adaptation is comprised of changes in legislation (such as in amendments to planning and building law, described above) and through the county administrative boards' rights to prevent municipal planning decisions with regard to a restricted number of parameters. However, the regional level is relatively limited in what it can enforce:

> In other [countries], you can have a ... regional level that mandates that no shopping malls may be established more than ten kilometres outside the city centre ... If you would try to develop anything like that in Sweden, there would be an uproar. (SALAR, interview)

As a result, while each municipality can choose to include planning principles with regard to climate change in their comprehensive plan, 'many municipalities ... have no current comprehensive plan ... even though you have to have it according to law ... because they have not seen it as necessary or not had the resources' (SALAR, interview).[17] While considerations could be written into detailed plans at each development, these plans are often not established until a development is taking place, potentially limiting the extent to which each development is considered in the context of other ongoing or future development. Against this background, a governmental commission (*Miljöprocessutredningen*) has investigated processes of how broader interests are to be taken into account in detailed planning in municipalities (cf. Andersson, 2009). Interviewees indicated, however, that any modification to the system will need to start from the Swedish institutionalised local planning monopoly. While critical, one interviewee noted: 'I still want to believe in our system in the long run, because it is possible to start from local preconditions, which of course are different at different places' (SALAR, interview).

Given its current structure, the planning system thus has particular implications for adaptation. While some municipalities have started to discuss the use of stricter limits in the allocation of building permission, this has largely been in response to

[17]The same interviewee noted that funding for wind power from the state had prompted many municipalities who did not have a current comprehensive plan to start working on one, demonstrating a clear connection between municipal action on comprehensive plans and state incentives (SALAR, interview).

identified vulnerabilities or to ensure that the municipality will not be held liable in cases of, for instance, flooding (SALAR, interview), resulting in significant differences in how municipalities deal with adaptation. Issues have also been raised for the county administrative boards, including the concern that future adaptation needs in the region could pose difficult choices between addressing risks of flooding for roads, infrastructure, and buildings and those for nature reserves, Natura 2000 and cultural heritage areas, outdoor recreation, fauna and birdlife. The development of practical measures could thus become very complex as these would entail managing collective risk and placing preservation interests against security interests (Västra Götaland County Administrative Board, interview). Impacts on planning may extend beyond what is currently foreseen where, for instance, cities may be too densely planned and built to allow for sufficient green space for filtration to avoid or lessen flooding (National Board of Housing, Building and Planning, interview). Adaptation may thereby result in more extensive cross-municipal cooperation in the future, especially in areas close to municipal borders.

5.6.2 Policy Transfer and 'Lesson-Drawing'

Given the nature of adaptation as an issue under development and the attribution of responsibilities for adaptation in Sweden, the potential for policy transfer may exist both at the domestic level via networks (e.g., between municipalities), and in the transfer of lessons from existing international examples. The general orientation in Sweden on climate change can be seen in the prevalence of mitigation on the national level, as well as in 'good practice' municipalities that focus particularly on mitigation (such as Trollhättan). Nationally developed tools and networks for the direct transfer of adaptation practices between municipalities are so far limited in their development, and lack an established long-term institutional context. The use of such tools for the development of adaptation actions that can be transferred between different municipalities may also be limited by the capacity of municipalities to access and apply them, especially given a limited focus on adaptation. An interviewee at Climatools noted that: 'If tools are really going to be used by . . . small municipalities . . . they cannot be too complex. So we have considered whether to develop a tool . . . and then make a light version' (Climatools, interview).

With regard to the broader transferability of approaches to adaptation, most interviewees noted that examples could be taken from elsewhere. The Climatools project has drawn extensively on examples from the UK in several ways, including an invited presentation from Hampshire County Council, the employment of students to test UK methods for local adaptation, and attempts to apply the UKCIP Local Climate Impacts Profile (see Chapter 3). The expertise of the UK was heralded by interviewees for instance: for instance, as 'Great Britain stands out, as they have done something on the municipal level, they have worked together' and that 'there are municipalities that have themselves taken action, [it is] not only that there has been research developed' (Climatools, interview). Cooperation or lesson-drawing has also occurred as a result of common features or impacts in different areas that have led to collaboration across national borders; for example, joint work on steep terrain environments by the Swedish Geotechnical Institute and Austrian authorities (Swedish Geotechnical Institute, interview).

However, interviewees noted differences in domestic contexts, and the special circumstances presented by Sweden's local planning monopoly in particular: 'you cannot directly transfer what is being done in England or Germany ... but have to adapt it to the system we have here' (National Board of Housing, Building and Planning, interview). Differences also exist with respect to the kinds of lessons that can be taken from particular contexts, in the sense that the development of such areas as rescue services, physical planning, environmental insurance and hazard relief differ considerably between countries: 'Italy for instance does perhaps not act so much pre-emptively, but they have perhaps the foremost organisation for acting when [hazards] take place' (Commission on Climate and Vulnerability, interview). Differences were also noted between types of states (e.g., federal and unitary states), where the threshold for influence across different regions of a country may be higher or lower. One interviewee noted, for instance, that while processes of regionalisation may be leading Sweden in a direction with stronger political power at regional level, strong national states may be more capable of acting decisively on issues such as adaptation (Prime Minister's Office, interview).

5.7 Conclusion

The development of adaptation policy in Sweden began to emerge most markedly in 2005 as a more limited priority in the shadow of mitigation. Among national-level interviewees, adaptation to climate change was not generally considered a priority, although interviewees also noted that the issue was not a partisan or conflicted topic. The limited priority given to adaptation was instead largely the result of the relatively positive effects of climate change anticipated on national level for Sweden, a perception that has drawn on climate impacts research and existing Swedish scenarios. However, adaptation policy is under development and has now even been mainstreamed through the 2009 Climate Bill and its associated legislative and regulative changes.

In this context, the development of a Swedish agenda on adaptation has to some extent been the result of international policy developments that have increased the focus on climate change (including adaptation), as well as large-scale events that have been perceived as related to climate change. In some instances, specific events have served as major 'focusing events' for policy-making and agenda-setting at the national scale, as in the case of the floods in 2001 which prompted strong reactions from local policy-makers and to some extent spurred the development of the Commission on Climate and Vulnerability, including its interim report in 2006. Actions developed in the Commission and Bill have drawn on knowledge and established actors in the areas of landslide and flood risk mapping, actors that were established as a result of hazard events in the first place. For example, the formation of the SGI in the late 1950s occurred as a result of extensive landslides in the Gothenburg area, while a national platform for work with environmental hazards in the Swedish Rescue Services Agency (now the Swedish Civil Contingencies Agency) was developed in reaction to landslides in 2006.

Events have also created important momentum at the local and regional levels, for example, bringing the risk of higher water levels than previously thought possible to light. While adaptation has been a more limited priority at national level, it has gained increasing focus in areas where development pressures have rendered local low-lying areas even more vulnerable. Such municipalities and counties have acted to alert the state to the challenges that lie beyond the scope of single municipalities and that require concerted effort and support. In response to EU projects that have included a focus on adaptation (as in Gothenburg) or in relation to broader international awareness-building on adaptation, concerns over flooding impacts have come to be connected to climate change.

Following its establishment as an issue on both national and local political agendas, the focus of discussions on adaptation has turned to the division of responsibility for adaptation within the decentralised Swedish unitary state system. While the state is taking responsibility for select changes in the regulative framework and for major infrastructural measures, municipalities are obliged to act preventatively, utilising, among other things, tools of local planning (an issue emphasised by the reduction in allocations for local preparedness development covered by the state). The distribution of responsibility in this manner has raised further issues of capacity in that areas with greater adaptive capacity (in terms of such resources as available funding, personnel and an established focus on flood risk and/or environmental policy) may be able to develop adaptation measures, while smaller municipalities may be hindered even in the face of identified vulnerabilities. State policy on adaptation has therefore largely developed within the system of political responsibilities and existing assumptions of a decentralised system, and to some extent even reinforced the distribution of roles in the existing system.

At the regional level, the county administrative boards have been given the responsibility for coordinating adaptation, a task in line with their existing roles as the coordinators and implementing agencies for state policy. Since 2006, county administrative boards have also been responsible for the annual development of risk and vulnerability assessments on county level and the supervision of local planning, and in cases such as Västra Götaland hold significant responsibilities as water authorities. With respect to issues of water management, some interviewees noted the need for greater coordination (a need that will likely be addressed during the implementation of the EU Water Framework Directive) and any revision of existing water legislation. Here, the potential for further stakeholder involvement may lie in existing organisations such as conservation societies and in their potential reshaping into water councils under the Water Framework Directive.

At the local level, interviewees particularly noted a considerable difference between municipalities in terms of the extent of policy development on adaptation. Gothenburg has developed a relatively large strategic stakeholder group on adaptation in response to high development pressures and identified vulnerabilities with regard to flooding and sea level rise. The participation of the municipality in an EU project also helped the city to begin early action on adaptation measures, including an increase in minimum building elevation and the *Extreme Weather* projects. Interviewees also noted a number of underlying capacities in Gothenburg: general climate change awareness, a tradition in environmental work, and a 'cooperation spirit'. However, as a result of the undecided distribution of responsibilities for adaptation that would partly come though the Bill, many of the identified practical measures had yet to be taken at the time of the interviews. That authorities elected to wait until decisions were taken indicates the importance of the division of responsibilities and a clear framework for local action determined at the national level.

In contrast to the relatively well-developed understanding of vulnerabilities and adaptation needs in Gothenburg, the smaller case study municipalities exhibited more limited adaptive capacity, particularly with reference to issues of funding, staff and political leadership. In Mölndal, work on adaptation has so far been driven by civil servants and supported by work in Gothenburg, but has been constrained by limited funding to hire issue-specific staff or to undertake identified measures. By contrast, the municipality of Trollhättan exhibited a more limited focus on vulnerability and development on adaptation; however, given its established environmental policy development and vision of leadership in certain environmental areas, it is possible that its capacity to take on adaptation issues if prioritised may be relatively high (cf. Chapter 1). Finally, Munkedal has exhibited a high vulnerability to landslide events but reported limited financial and human resources to develop adaptation. So far, limited cooperation has occurred between municipalities on adaptation, as networks such as the Climate Municipalities have focused almost entirely on mitigation. The allocation of responsibilities between actors has also been almost exclusively focused on the public sector, with relatively little focus on the private sector (with some exception for the insurance industry).

With regard to the impact of the EU level, the Commission on Climate and Vulnerability makes reference to the need to review such elements as the Habitats Directive and the Natura 2000 network to include considerations of a changing climate. While EU projects in general are not emphasised by the interviewees, Gothenburg's participation in an EU project that led to the initial identification of climate change vulnerabilities may indicate possibilities for awareness-raising through the EU framework. A number of interviewees also noted the potential to draw experience from countries with more severe climate-related issues such as flooding, tempered by the need to adjust any lessons learned to the Swedish structure of local planning. The benefit of drawing on the experience of the UK (in particular the UKCIP LCLIP process) was highlighted by the Climatools programme.

Adaptation therefore constitutes an issue under development in Sweden, where it has so far been somewhat integrated into existing measures but has not prompted a significant re-consideration of tools, approaches or systems. In the future, adaptation may pose difficult choices as mechanisms for prioritising actions and informing those choices are currently limited, raising the worrisome prospect of severe impacts in low-lying areas in particular.

References

Interviews

Climatools (projekt vid Totalförsvarets Forskningsinstitut FOI), Swedish Defence Research Agency. Stockholm, March 10, 2009.
Climate Municipalities (Klimatkommunerna), Lund, February 4, 2009.
Commission on Climate Change and Development (Kommissionen för klimat och utveckling), Stockholm, February 6, 2009.
Commission on Climate and Vulnerability (Klimat- och sårbarhetsutredningen), Stockholm, February 12, 2009.
Commission on Climate and Vulnerability, Water supply issues (Klimat- och sårbarhetsutredningen, deltagare), Sundsvall, March 11, 2009.
Commission on Sustainable Development (Kommissionen för hållbar utveckling), Stockholm, February 6, 2009.
Geological Survey of Sweden (SGU: Sveriges geologiska undersökningar), Stockholm, January 30, 2009.
Gothenburg City Building Office (Göteborgs stads stadsbyggnadskontor), Gothenburg, January 28, 2009.
Gothenburg City Environmental Administration (Miljöförvaltningen i Göteborg), Gothenburg, January 24, 2009.
Gothenburg Region Association of Local Authorities (GR: Göteborgsregionens kommunalförbund), Gothenburg, February 11, 2009.
Munkedal Municipality, Munkedal, January 20, 2009.
Mölndal Municipality, Mölndal, January 22, 2009.
National Association of Swedish Eco-municipalities (Sveriges Ekokommuner, SEkom), Lund, January 27, 2009.
National Board of Housing, Building and Planning (Boverket), Karlskrona, February 3, 2009.
Prime Minister's Office (Statsrådsberedningen), Stockholm, February 10, 2009.
Regional Energy Agency of West Sweden (Hållbar Utveckling Väst), Gothenburg, January 23, 2009.
Swedish Association of Local Authorities and Regions (SALAR: Sveriges Kommuner och Landsting), Stockholm, February 2, 2009.
Swedish Civil Contingencies Agency. Myndigheten för samhällsskydd och beredskap, (MSB), Karlstad, January 29, 2009.
Swedish Environmental Protection Agency (EPA), Stockholm, February 9, 2009.
Swedish Geotechnical Institute (SGI), Linköping, January 30, 2009.
Swedish Insurance Federation (Försäkringsförbundet), Stockholm, March 10, 2009
Swedish Meterological and Hydrological Institute (SMHI), Norrköping, February 5, 2009.
Sustainable Municipality programme, Swedish Energy Agency (Uthållig kommun, Energimyndigheten), Eskilstuna, February 26, 2009.
Trollhättan Municipality, Trollhättan, January 21, 2009
Västra Götaland County Administrative Board, Gothenburg, January 22, 2009

Secondary References

Andersson, U. (2009). Minskad byråkrati med förslag från miljöprocessutredningen. *Plan: Tidskrift för samhällsplanering, 3–4*, 23–27.
Ansvarskommittén (2007). *Hållbar samhällsorganisation med utvecklingskraft. Ansvarskommitténs slutbetänkande.* Swedish Government Official Report SOU 2007: 10. Stockholm: Regeringskansliet.
Börzel, T.A. (2002). Pace-setting, foot-dragging, and fence-sitting: Member State responses to Europeanization. *JCMS, 40*(2), 193–214.

Christensen, T., & Laegreid, P. (2005). *Regulatory reforms and agencification* (Working Paper 6). Oslo: Stein Rokkan Centre for Social Studies.

Commission on Climate and Vulnerability (2006). *Översvämningshot. Risker och åtgärder för Mälaren, Hjälmaren och Vänern. Delbetänkande av Klimat- och sårbarhetsutredningen.* Stockholm: Swedish Government Official Report SOU 2006:94.

Commission on Climate and Vulnerability (2007). *Sweden facing climate change – threats and opportunities* [in Swedish]. Stockholm: Swedish Government Official Report SOU 2007:60.

Commission on Oil Independence (2006). På väg mot ett oljefritt Sverige (In Swedish). June 2006. Commission on Oil Independence, Prime Minister's Office.

County Administrative Board Dalarna (2008). *Regional risk och sårbarhetsanalys.* Rapport 2008: 25. Plan och beredskapsenheten. Dalarna: Länsstyrelsen.

County Administrative Board Västra Götaland (2008). Risk och sårbarhetsanalys för verksamhetsåret 2008. Rapport 2008:81. Länsstyrelsen Västra Götalands län, Enheten för skydd och säkerhet, Göteborg.

County Administrative Board Västra Götaland (2009a). Lake Vänern Society for Water Conservation: Activities. Retrieved June 6, 2009, from http://www.lansstyrelsen.se/vastragotaland/Projektwebbar/Vanern/English/activities.htm.

County Administrative Board Västra Götaland (2009b). About Sweden's Societies for Water Conservation. Retrieved June 6, 2009, from http://www.lansstyrelsen.se/vastragotaland/Projektwebbar/Vanern/English/swedens_societies.htm.

Eckerberg, K., Forsberg, B., & Wickenberg, P. (1997). Sweden: setting the pace with pioneer municipalities and schools. In W. M. Lafferty & K. Eckerberg (Eds.), *From Earth Summit to Local Forum. Studies of Local Agenda 21 in Europe* (pp. 51–84). Oslo: ProSus.

Gothenburg Building Office (2009). *Extrema väderhändelser fas 2.* Fallstudie Gullbergsvass, januari 2009. Gothenburg: City Building Office.

Gothenburg City Office (2006). *Extrema vädersituationer.* Hur väl rustat är Göteborg? Utredning maj 2006. Gothenburg: Stadskansliet.

Gothenburg Traffic Office (2008). *Havsnivåhöjning och samhällsviktiga anläggningar.* Report dnr 1117/08. Gothenburg: Gothenburg Traffic Office.

Government Offices of Sweden (2007). Members of the Government Commission on Sustainable Development. Retrieved June 2, 2009, from http://www.regeringen.se/sb/d/8177/a/78441.

Government Offices of Sweden (2008). Commission on Sustainable Development: Tasks. Retrieved June 2, 2009, from http://www.regeringen.se/sb/d/11018/a/111519.

Government Offices of Sweden (2009). *En sammanhållen klimat- och energipolitik. Klimat.* Regeringens proposition 2008/09:162. Stockholm: Government Offices of Sweden.

GR [Gothenburg Region Association of Local Authorities] (2009). Gothenburg Region Association of Local Authorities. Retrieved June 6, 2009, from http://www.gr.to/2.5f30b95110fd8ec51a80000.html.

Hall, P., & Montin, S. (2007). *Governance networks and democracy at regional and local level in Sweden* (Working Paper 2007: 9). Denmark: Centre for Democratic Network Governance, Roskilde University.

Härryda Municipality (2009). Vattenråd. Retrieved June 6, 2009, from http://www.harryda.se/miljo/hurmarmiljoniharrydakommun/vattenrad.4.75ce29a911dbe931e608000322.html.

IEH [Institutet för ekologisk hållbarhet] (2003). *Lokala investeringsprogram – vägen till ett hållbart samhälle.* Stockholm: Naturvårdsverket.

Klimatberedningen (2008). *Svensk klimatpolitik.* Betänkande av Klimatberedningen. SOU 2008:24. Stockholm: Miljövårdsberedningen.

Lundqvist, L., & Biel, A. (Eds.). (2007). *From Kyoto to the Town Hall: Making international and national climate policy work at the local level.* London: Earthscan.

Ministry of the Environment (2008). Ministry of the Environment. Retrieved June 6, 2009, from http://www.regeringen.se/sb/d/1471/m/header.

Ministry of Sustainable Development (2005). *Sweden's fourth national communication on climate change under the United Nations Framework Convention on Climate Change.* Ds 2005:55. Stockholm: Ministry of Sustainable Development.

MJU (Miljö- och jordbruksutskottet) (2009) Betänkande 2008/09: MJU28. Riktlinjer för klimatpolitiken m.m. Stockholm: Miljö- och jordbruksutskottet.

Munkedal Municipality (2008). Energiplan och klimatstrategi 2008–2012. Munkedals kommun.

National Board of Housing, Building and Planning (2009). Welcome to the Dialogue Project. Retrieved May 30, 2009, from http://www.byggabodialogen.se/templates/Page3477.aspx.

Nyström, J. (2003). *Planeringens grunder. En översikt* (2nd ed.). Lund: Studentlitteratur.

Östhol, A., & Svensson, B. (2002). The Swedish cases. Partnerships by State-led design. In A. Östhol & B. Svensson (Eds.), *Partnership responses – Regional governance in the Nordic states. Future challenges and institutional preconditions for regional development policy, Volume 4* (pp. 85–134). Stockholm: Nordregio.

Peters, B.G., & Pierre, J. (2005). Swings and roundabouts? Multilevel governance as a source of and constraint on policy capacity. In M. Painter & J. Pierre (Eds.), *Challenges to state policy capacity. Global trends and comparative perspectives* (pp. 38–51). Houndmills Basinstoke: Palgrave Macmillan.

Pettersson, M. (2008). Renewable energy and the function of law: a comparative study of legal rules related to the planning, location and installation of windmills. Doctoral Thesis 2008:65, Luleå University of Technology, Luleå, Sweden.

Planning and Building Act (2009). Planning and Building Act SFS 1987:10. Retrieved October 5, 2009, from http://www.notisum.se/Rnp/sls/lag/19870010.htm.

SALAR (2009). Regionala utvecklingsprogram (RUP). Retrieved June 6, 2009, from http://www.skl.se/artikel.asp?C=5940&A=45410.

Scientific Council on Climate Issues (2007). Vetenskapligt underlag för klimatpolitiken. Rapport från Vetenskapliga rådet för klimatfrågor (In Swedish). Rapport 2007: 03. Stockholm: The Environmental Advisory Council, Ministry of the Environment.

SGU [Geological Survey of Sweden] (2009). Skred och ras. Retrieved December 3, 2009, from http://www.sgu.se/sgu/sv/samhalle/risker/skred_s.htm.

SSNC (2007). Rapport. Klimatindex för kommuner 2007. Retrieved October 15, 2009, from http://www.naturskyddsforeningen.se/upload/Foreningsdokument/Rapporter/rapport_klimatindex.pdf.

Sustainable Municipality Programme (2009). Uthållig kommun – med energin som språngbräda för ett hållbart samhälle. Retrieved October 5, 2009, from http://www.energimyndigheten.se/sv/Om-oss/Var-verksamhet/uthallig-kommun/.

Swedish EPA (2004). *Den svenska klimatkampanjen – en del av Sveriges klimatstrategi. Slutrapport.* Stockholm: Swedish EPA.

Swedish EPA (2005). *Sju år med LIP och Klimp – en lägesrapport i september 2005.* Stockholm: Swedish EPA.

Swedish Energy Agency and Swedish EPA. (2007). Den svenska klimatstrategins untveckling. En sammanfattning av Energimyndighetens och Naturvårdsverkets underlag till kontrollstation 2008. Swedish Energy Agency, Eskilstuna and Swedish EPA, Stockholm.

Swedish EPA (2008a). Swedish EPA. Retrieved June 12, 2009, from http://www.naturvardsverket.se/sv/.

Swedish EPA (2008b). Effekter av investeringsprogrammen LIP och Klimp. Stockholm: Naturvårdsverket.

Swedish Government (2002). *Sveriges klimatstrategi.* Proposition 2001/02:55. Stockholm: Sweden.

Swedish Parliament (2009) *Riksdagens protokoll 2008/90*: 133, tisdag 16 juni. Stockholm: Sweden.

Swedish Road Administration (2006). Skredet i Småröd. Sammanfattning av rapporterna utarbetade av Vägverkets oberoende utredare docent Jan Hartlén. Retrieved June 18, 2009, from http://www22.vv.se/templates/NewsPage22985.aspx.

Swedish Water Authorities [Vattenmyndigheterna] (2009). Välkommen till Sveriges fem vattenmyndigheter! Retrieved July 6, 2009, from http://www.lansstyrelsen.se/vattenmyndigheten/Pressrum/Nyheter/2005/2005-10-03.htm.

Temnes, M. (1998). Finland and New Public Management. *International Review of Administrative Sciences 64*, 441–456.

Town of Trollhättan (2005). Plan för hantering av extraordinära händelser och handlingsprogram enligt lagen om skydd mot olyckor. Trollhättans stad.

Vail, B. (2008). Ecological modernization at work? Environmental policy reform in Sweden at the turn of the century. *Scandinavian Studies, 80*(1), 85–108.

VG [Västra Götalandsregionen] (2009a). Remiss: Smart Energi. Vägen bort från fossil energi. Förslag till klimatstrategi för Västra Götaland. Västra Götalandsregionen, Miljösekretariatet.

VG [Västra Götalandsregionen] (2009b). Klimatstrategi för Västra Götaland. Retrieved 15 December 15, 2009, from http://www.vgregion.se/sv/Vastra-Gotalandsregionen/startsida/Miljo/Prioriterade-miljoomraden/Klimat/.

Chapter 6
'Planning for Today': The Nature and Emergence of Adaptation Measures in Italy

Lisa Westerhoff

Abstract This chapter examines the Italian approach to climate change adaptation at the national scale, where the design of a formal adaptation strategy has yet to emerge out of preliminary stages. Adaptation discourses at the national level were initiated in 2007 at the time of the National Climate Change Conference, after which efforts to pull together a national adaptation strategy were considerably slowed by low prioritisation of climate change adaptation, changes in national administration and poor coordination. The ways in which adaptation is beginning to emerge at the regional and local scales is assessed in the Emilia-Romagna region and its province and municipality of Ferrara. Policy reviews and interviews with decision makers in the case study areas indicate that despite such slow progress, both national and regional actors have furthered adaptation discourses and activities at different scales. Issues that have typically hindered environmental policy development are partially overcome as strong political leadership, stakeholder involvement and strengthening vertical and horizontal networks are coupled within governments with long-standing interest in environmental issues and positive science-policy linkages. At all scales, adaptation has occurred in response to current risks and vulnerabilities with little consideration of and future projections and long-term planning.

Keywords Adaptation · Environmental policy · Governance · Italy · Planning

6.1 Introduction

Adaptation to climate change in the Italian Republic provides an interesting contrast to the ways in which several other European countries have approached a changing climate. Complex political and legal frameworks, a history of clientelism and corruption, and regional disparities despite strong central control have created an

L. Westerhoff (✉)
Department of Social and Economic Geography, Umeå University, Umeå 90187, Sweden
e-mail: lisa.westerhoff@gmail.com

E.C.H. Keskitalo (ed.), *Developing Adaptation Policy and Practice in Europe: Multi-level Governance of Climate Change*, DOI 10.1007/978-90-481-9325-7_6,

intricate context within which planned adaptations have only begun to develop. As with environmental policy formation, action on adaptation is emerging on several levels of governance, from State-led strategies to regional policies and networks linking climate-conscious local authorities. While the national level has had limited and intermittent success in pulling together a coherent adaptation strategy, opportunities for developing planned adaptation at sub-national scales of governance have begun to emerge as particular regions and local authorities independently take on adaptation projects and programmes. In cooperation with various EU and NGO networks, those local administrations with access to adequate financial and other forms of capital have been begun to address the need to adapt, albeit not without some difficulty.

This chapter will examine the ways in which the unique traits of Italy and its approach to both environmental and climate change issues are shaping the extent and direction of adaptation activities at multiple scales of governance. Global climate models predict that under the IPCC A1B scenario, Italy will undergo an increase in average annual temperature of 2–5°C and a reduction in total annual precipitation, the effects of which will include the increased magnitude of extreme precipitation events, heat waves and drought (MATTM, 2007a). Coupled with projected increases in sea level rise, impacts are expected in various sectors, including agriculture, energy production, water management, biodiversity and conservation, health, tourism and the coastal zone (MATTM, 2007a).

Sections 2 and 3 of the chapter provide a review of the development of environmental policy and legislation in Italy, highlighting capacity issues in terms of resource allocation, technical support and institutional fragmentation as they have manifest in Italy's entry into climate change politics. In Section 4, the challenges faced by Italian national administrations in its progress towards a coherent climate adaptation policy are then developed by following the evolution of the Italian National Communications and the culmination of national efforts with the advent of the National Climate Change Conference in 2007. Ongoing activities at the national and sub-national scales are then assessed, focusing on the region of Emilia-Romagna and its Province and Municipality of Ferrara. Section 5 provides a review of the factors that have enabled or constrained the development of adaptation measures at local scales, drawing out the different forms of capacity as well as the relevant interrelationships between scales of governance. The chapter concludes with an overview of Italy's progress and highlights areas of possible improvement for the successful development of adaptation strategies.

6.2 Italian Environmental Policy: Evolution and Current Issues

In order to explore the past and present issues that have been faced in the design and implementation of environmental policies and measures, and to highlight those that continue to affect the success of climate change adaptation measures, a brief review the course that environmental policy has taken in Italy is warranted. While concern for the environment has historically taken a backseat in the Italian political realm, recent administrations have begun to make concerted efforts to address the absence of environmental measures, developing an array of policies and tools over the last few decades. Despite frequent governmental instability and pervasive administrative ineffectiveness, the Ministry of the Environment and Territory has made slow but gradual progress towards addressing long-standing environmental problems and the implementation EU environmental directives. However, the complex and fragmented nature of Italian environmental research and policy design, coupled with the Ministry's limited power of implementation, have proved to be significant barriers to effective environmental policies.

Italy's first forays into the environmental policy arena occurred somewhat later than several of the other Western European countries, hindered by issues of low public awareness, minimal political interest and ineffective environmental institutions. Environmental legislation and public awareness grew throughout the 1980s, in large part the result of growing public opposition to nuclear power (spurred by the 1986 Chernobyl accident) and the resultant 1987 referendum that saw a decisive win by the green social movement (Rosenbaum, 1987). Political support for the newly-formed Green Party also grew significantly during this period, eventually leading to its entry into government following the election of Romano Prodi's centre-left coalition *The Olive Tree* in 1996 (Biorcio, 2002). However, environmental policy maintained a marginal role in Italian politics due to a lack of political and civil environmental education and the absence of a singular body under which environmental policies could be determined (Lewanski, 1998). Coordinated under the Ministry of Health and administered by Local Sanitation Agencies at the municipal scale, initial environmental concerns were framed and addressed as matters of public health and so received very little political visibility or importance at the national level. The creation of the Ministry of the Environment in 1986 heralded a new era in which environmental policies began to assume legitimacy through the establishment of a visible entity that dealt strictly in matters of the environment (Lewanski, 1998).

Throughout the 1990s, public awareness of environmental issues increased further via a growing environmental movement and a series of environmental crises that drove select administrations to take a more serious and direct approach to environmental policy (Biorcio, 2002). In 1993, the Interministerial Committee for Economic Planning (CIPE) approved a national plan for sustainable development, initiating a new series of actions at various scales, including the provision of support for the Local Agenda 21 bodies that were forming across the country. Under Romano Prodi's centre-left coalition government, the capacity of the national environmental policy arena enjoyed a series of upgrades through an increase in power and resources allocated to the Ministry of Environment, the creation of an environmental technical support body in 1994 (the National Agency for Environmental Protection, or ANPA), and the emergence of the first regional environmental protection bodies (the ARPA) (Ramieri, Wallace-Jones, & Lewanski, 2001). The decision to ban nuclear power from Italy's roster of energy production provided for the transfer of additional scientific capacity over to the Ministry, as the Italian National Agency for New Technologies, Energy and the Environment (ENEA) was restructured to address matters of both energy and the environment (ENEA, interview). The creation of these two supporting bodies provided much-needed scientific research capacity to the Ministry's activities; however, Italy's growing public demand for environmental action and initial obligations under the EU directives forced the government to quickly adopt policies and legislation in such a way that much initial environmental policy was fragmented and incomplete (Marchetti, 1996). Much of Italian environmental activities have thus taken a largely reactive approach, coping with and compensating for environmental crises as they arise through a lack of preventative measures.

In the present era of Italian environmental politics, matters are generally approached in the same reactive manner, though preventative approaches have been introduced in areas such as soil conservation and water management. Despite a series of funding cuts to environmental programmes in the early-mid 1990s, the Ministry of Environment and Territory (MATTM) has undergone an expansion in institutional and administrative capacity, adding human and financial resources throughout the late 1990s and early 2000s over several directorates created to address questions of sustainable development, flood risk management and natural disasters. These were accompanied by the development of a number of more effective policy tools, including the introduction of Environmental Impact Assessments, pollution tariffs and a carbon tax (OECD, 2003).

Notwithstanding these increases in the capacity and effectiveness of the Ministry, decisions regarding environmental policy continue to be shared among several ministries. The CIPE represents the primary body through which matters of economic importance, including environmental issues, are jointly determined by Italian ministries and various technical agencies. Strong private lobbies such as Confindustria (an organisation of private Italian companies) constitute a strong influence over the policy process and frequently present a considerable barrier to the implementation of environmental measures. Those measures that are passed are often fragmented and inconsistent, with several policies developed without national horizontal coordination or consistency (OECD, 2003). Further, the frequent reorganisation and restructuring of the Ministry's supporting bodies have led to some inconsistencies in environmental technical and scientific support. For example, ANPA has under gone two restructurings since its creation in 1994 and is now under the heading of the Institute for Environmental Protection and Research (ISPRA), while ENEA may see a return to nuclear energy research as a result of Berlusconi's recent decision to reinstate nuclear energy into the roster of national energy production (ENEA, interview). The poor coordination between these and other technical or research bodies have additionally led to fragmented knowledge production, as responsibilities come assigned in an *ad hoc* manner in the absence of a national research strategy or mandate (ENEA and ISPRA, interviews).

Italy's structure of sub-national bodies and responsibilities adds an additional dimension to environmental policy formation and implementation. Twenty regions in Italy make up the largest sub-national unit of Italian government, divided into two categories of 'ordinary' and 'special' statute. The 15 ordinary regions are granted legislative and administrative powers but have little to no financial autonomy. The remaining five regions of special statute (the two islands of Sicily and Sardinia, plus the regions of Trentino-Alto Adige, Friuli-Venezia Giulia and the Aosta Valley) are granted additional financial autonomy, though the differences between the two types of regions are becoming less distinct as increasing power is delegated to the regional level (ISPRA, interview). A more common distinction between regions is drawn geographically as opposed to structurally, dividing the richer and more advanced regions of the north from their poorer and often less developed counterparts in the south (cf. Putnam, 1993).

Beyond the distinction between ordinary and special regions, the relationship between central and regional governments is complex. Italy's system of regional administrative federalism has transferred policy jurisdiction in several sectors to the regional level (Cotta and Verzichelli, 2007). Since a Constitutional Reform in 2001, ordinary regions have retained legislative power over all matters not expressly reserved for the State, and share legislative power in matters concerning health, food, land-use regulation and planning, transportation, and energy. This structure allows strategic planning and legal coordination to be maintained within the Ministry while regions adapt national legislation to their particular circumstances (OECD, 2003). The creation of several ARPA in 1994 provided additional technical and scientific support to the regions' environmental administrations, though agency capacities now vary considerably from region to region and according to different regional environmental foci. While this form of decentralised control over resources and planning is encouraged by the EU, the regional differences in leadership and capacity and a lack of central coordination have led to considerable differentiation in the implementation of environmental measures across the regional scale (Gualini, 2004; OECD, 2003).

The dynamics between national and sub-national scales are rendered even more convoluted by the addition of provinces, municipalities and large metropolitan areas, each with differing roles and abilities. These bodies generally carry out the administrative functions of the regional level, but maintain some degree of autonomy in terms of the ways in which national and regional priorities are fulfilled. Municipalities, or *comuni*, additionally represent strong centres of local autonomy, retaining jurisdiction over matters pertaining to local planning including public transportation, water provision and infrastructure. The establishment of the LA21 bodies across the local level has increased the engagement of local authorities in environmental sustainability issues, particularly in the metropolitan areas of Rome, Florence, Venice, Bologna and Genoa, as well as in a number of northern regions and provinces (A21Italia, 2008a; Ramieri et al., 2001). The development of the Italian Local Agenda 21 Association in 2000 created a much-needed LA21 coordination network that now provides a forum for the exchange of sustainable development knowledge and practice between LA21 offices throughout Italy (A21Italia, 2008b).

The relationship between regions and local authorities and the State with regards to environmental policy is somewhat reflexive, in that national policies require sub-national scales of government to adopt certain legislation but have also been influenced by proactive regional or local environmental actions. While (typically northern) regions and local authorities have had a positive role in stimulating national environmental policy, conflicts over jurisdiction and overlap in responsibilities between the State, regions and local authorities have also led to considerable confusion and policy impasses in several sectors (Lewanski, 1998). According to Lewanski (1998), this has created a policy arena ‘crowded with an increasing number of actors who are more concerned with trying to stake out their areas of influence in this new field than with coping with substantive problems. . .Notwithstanding the increase of functions and tasks required by the growing demand for environmental quality, the distribution of powers is perceived by actors as a zero-sum game’ (p. 145). Though constitutional reform in 2001 has to some extent clarified the division of responsibilities between regions and the State (cf. Cotta and Verzichelli, 2007), overlapping jurisdiction still exists in several issue areas (ARPA and ENEA, interviews).

Similarly, in the absence of any coherent form of national planning, responsibilities are shared between the various levels of government, creating confusion and problems of implementation and accountability. Regions and provinces issue territorial plans, which are in turn used as guidelines for *comuni* to create master planning plans updated on a case by case basis (OECD, 2003). The degree to which concerns of the environment are incorporated into such plans has largely been a function of the cascading priorities from region to local authorities and the existence of local concern for specific and relevant environmental issues, resulting in considerable differences in the extent and quality of plans and policies between regions. Increasingly, EU mandates are playing an important role in bringing regions to a common level, despite differences in the extent to which environmental issues are addressed.

Thus, while Italy has seen a significant improvement in environmental policy over the last decade, the complexity of the administrative system and inadequacies in the ability of actors at different scales to coordinate environmental plans and programmes has left much room for improvement. Frequent changes in national administration, in combination with discrepancies in capacity and overlaps in jurisdiction between scales, have resulted in considerable horizontal and vertical fragmentation and the discontinuity of environmental policies. As a result, different parties have ascribed more or less importance to the environment sector and thus to the general capacity to design and enforce environmental measures needed to bring Italy up to EU standards. According to the OECD review of Italian environmental policy in 2003, Italian environmental policy would benefit considerably from an improvement in its environmental infrastructure, the efficiency of its environmental policies, the integration of environmental concerns into economic and social decisions, and the reinforcement of international cooperation (OECD, 2003). However, Italy’s progress towards the implementation of EU directives and sustainable development measures should be recognised as an achievement considering the country’s institutional context and political history.

6.3 Entry into the Climate Change Arena: Mitigation Policies and Programmes

Both the developments and challenges faced in Italy's present environmental policy sector are reflected in the country's more recent engagement in the climate change arena. As in other industrialised countries, Italy's entry into climate change politics began with the need for mitigation of greenhouse gas (GHG) emission reductions under international agreements. However, Italy has seen a slower development of climate change policies than several EU Member States. Italy's first explicit and integrated climate policy was developed upon the country's ratification of the United Nations Framework Convention on Climate Change (UNFCCC) in January of 1994. The same year, CIPE approved the 1994 *National Programme for the Stabilization of Carbon Dioxide Emissions at the levels of 1990 by the year 2000*, the first significant commitment to greenhouse gas reduction on the part of the Italian government (Marchetti, 1996). Under the first Prodi government, CIPE was also given the responsibility for the creation of a framework through which a programme to achieve Italy's GHG emission reduction targets could be developed, to be coordinated by the Ministry of Environment (MATTM, 2006). The resultant *Guidelines for National Policies and Measures regarding the Reduction of GHG Emissions* were created in 1998 and outlined the specific methods and deadlines for achieving the voluntary reduction targets set by the National Programme (ISPRA, 2004; Massetti, Pinton, & Zanoni, 2007; MATTM, 2006). Activities were concentrated within six priority areas:

- Promotion of efficiency in the electricity sector;
- Reduction of energy consumption in the transport sector;
- Increased use of renewable energy sources;
- Reduction of energy consumption in the residential, commercial and industrial sectors;
- Reduction of emissions from non-energy sources; and
- Promotion of carbon sequestration through forest management (Ministry for the Environment Land and Sea (MATTM) 2006).

Between 1999 and 2000, various policies and regulatory measures were subsequently approved by CIPE to assist in meeting reduction targets, as well as the establishment of a fund for emission reduction and renewable energy promotion (Massetti et al., 2007). In June 2002, Italy ratified the Kyoto Protocol, transforming the targets outlined under the 1998 Guidelines from voluntary to binding. Under the burden-sharing agreement of the European Union, Italy became committed to reducing its greenhouse gas emissions by 6.5% of 1990 levels by the period 2008–2012, or the equivalent of 487 MtCO_2. The approval of the first National Action Plan (2003–2010) and the revised Guidelines for National Policies and Measures regarding the Reduction of Greenhouse Gas Emissions put Italy's strategy to achieve reduction targets under Kyoto into action. Among the mechanisms used to reduce emissions are the White Certificate programme through which end-use energy users are encouraged to improve energy efficiency; the Green Certificate programme that provides incentives for renewable energy production; and the reduction of energy consumption through the imposition of a carbon tax. The National Action Plan and Guidelines continue to be updated to account for changes in GHG projections and to add new policies and measures (MATTM, 2007a).

Despite the development of the variety of mechanisms for GHG reductions, national mitigation measures have had little success. While Italy may boast of low emission intensity, national energy efficiency also remains quite low and total emission rates have steadily risen (EEA, 2006; MATTM, 2006). According to work by Marchetti (1996), the reasons behind Italy's slow entry into mitigation are similar to those encountered in dealing with other environmental issues: low scientific capacity, the struggle to translate the information into concrete policies and measures, and the challenge of control over the private sector. The lack of success of mitigation measures has also been attributed to both the absence of a formal national energy policy and the exclusion of sub-national levels of government into emission reduction activities (WWF, interview). Local responsibilities and roles in mitigation remain undeveloped without a formal system of distributing national emission reduction targets across regions or local authorities. As a result, sub-national bodies to date receive no explicit financial or other forms of incentive from central government to address GHG emissions. Most importantly, perhaps, is the present administration's position that meeting Kyoto targets would reduce the country's GDP and increase unemployment, which has created difficulties in reaching an agreement on EU climate plans. As a result, several mitigation policies are receiving decreasing financial support from central government (ENEA and WWF, interviews).

However, regional and local governments have begun to engage with mitigation efforts as a result of EU regional funding requirements, encouragement from non-governmental organisations, and the participation in various environmental networks. For example, Italian regions signed the 2001 Turin Protocol and committed to mainstreaming mitigation into all regional policies, including Rural Development Plans and Transportation Plans. Under the EU, regions are additionally expected to create and implement Regional Energy Plans that outline measures for energy efficiency and production. Several national and international networks also provide support for and encourage mitigation at the sub-national scale: the Kyoto Club's *Local Authorities for Kyoto* programme, the Sustainable Energy Europe Campaign's *Covenant of Mayors* and the *Cities for Climate Protection* programme under ICLEI-Local Governments for Sustainability represent some of the most successful networks at engaging regions, provinces and municipalities in mitigation activities.

6.4 Adaptation and the Italian State

Having reviewed the evolution of Italy's engagement with environmental issues and entry into the climate change arena, this chapter now turns to the assessment of the ways in which adaptation to climate change in Italy has been addressed.

Though adaptation has been formally reported on since Italy's First National Communication to the UNFCCC in 1995, it is only recently that adaptation has begun to receive more serious attention. The particular framing of adaptation in Italy as a component of sustainable development demonstrates that in many areas and at various scales, adaptation is considered to be an extension of existing environmental resource and risk management activities, and thus ongoing.

This section begins at the national level, providing initial descriptions of climate change impacts and vulnerabilities as they have been reported in Italy's National Communications to the UNFCCC.[1] Though several difficulties have been encountered in the formation of concrete adaptation measures at the national scale, events such as the National Conference on Climate Change in 2007 and the creation of a consortium for climate change research indicate an increasing willingness and ability on the part of national administrations to transcend these difficulties and create effective adaptation measures. While the creation of a National Adaptation Strategy (NAS) has not been a priority for the current administration, initial steps towards the development of a NAS are reportedly underway. The approach to adaptation as an environmental issue and the appointment of the environmental sector as the principal actor in national adaptation measures have led to the emergence of similar challenges to those faced in environmental policy. As such, gaps in climate impacts and vulnerabilities research, an absence of coordinated adaptation measures and changes in government have all contributed to the relatively slow development of formal adaptation measures.

6.4.1 From Impacts to Adaptation: 1995–2007

Unique geographic and socioeconomic characteristics make Italy both a complex and highly vulnerable country to the impacts of climate change. Italy's diverse topography and varied landscape produce very different climate regimes, ranging from a Mediterranean climate in the south and along the coast, characterised by hot dry summers and mild winters, to the cold, wet and snowy winters of the humid subtropical and continental climates in the north and in the mountains. Nearly 59 million people are distributed over its area of roughly 300,000 km^2, giving Italy the fifth highest population density in Europe.[2] Such high density and high rates of urbanisation in several areas, combined with a varied physical and socioeconomic landscape, have rendered much of the Italian territory susceptible to climate-related risks.

---

[1]Each Annex-1 country under the UNFCCC is obliged to submit a National Communication detailing measures underway for the mitigation of GHG emissions, as well as data on impacts, vulnerability and adaptation activities. The reports represent the sum of climate change-related activities in each of the Annex-1 countries and are a primary source of information on the state of climate science in each country.

[2]196 ppl/km^2 (ISTAT, 2008).

Despite extensive climate monitoring networks, knowledge and information on the expected impacts of climate change and the vulnerabilities of Italian regions and sectors remain somewhat limited in Italy. Impacts of climate change and national vulnerabilities were initially reported on in very little detail in the first National Communication. Estimates of impacts and vulnerability are vague and unmeasured, using IPCC projections to assess the degree to which coastal areas and hydrological cycles could be potentially affected. However, the Second Communication represents an improvement in terms of climate data and vulnerability assessments, using external data sources from various studies to assess the potential for negative effects on seven different sectors. In both the First and Second National Communications, insufficient climate projections at the regional scale and the absence of national level data on changing conditions and vulnerabilities are presented as significant obstacles to the quality of the report, citing the need for 'greater human and financial resources than the ones available' (MATTM, 1998, p. 190). Throughout the first three Communications, information on climatic trends and assessments of vulnerability derived from external or EU-funded research activities is added as it becomes available, but remains relatively undeveloped in terms of the quantification of impacts or vulnerability assessments at the national scale. Sections on impacts and vulnerability maintain a general focus on coastlines and the Po River delta until the latest and Fourth National Communication (2007), which provides information on impacts and vulnerability in relatively greater detail (see Table 6.1).

Conversely, the development of adaptation measures within the first three reports remains relatively stable. The discussion of adaptation is generally absent in the First Communication, while adaptation measures are described in the second as developments in general environmental protection legislation (such as the establishment of river Basin Authorities or the approval of flooding laws) or suggestions for the improvement in existing flooding and coastal defence measures along the Adriatic coast. The Third National Communication highlights measures that have been traditionally used to combat issues of sea level rise and flooding, as well as a general outline of response measures identified to combat desertification and risks to agriculture and forestry. The most recent National Communication, issued in 2007, acknowledges the need for national-scale adaptation strategies that involve 'strategic decisions regarding the management of the peculiarities and vulnerabilities of the territory, economic activities and resources' (MATTM, 2007a, p. 208). This last report indicates the intention of the national government to involve ministries and subnational levels of government in the design of an adaptation policy that would link adaptation strategies with existing or future mitigation policies. The Fourth National Communication additionally briefly outlines possible adaptation strategies in the areas of coastal management, water and health, agriculture, ecosystems and biodiversity, desertification, tourism and the energy sector. However, the majority of these actions had not yet been implemented at the time of the publication of the report, and are more reflective of a preliminary overview of possible adaptation strategies.

This relatively slow development of information on climate impacts and vulnerabilities may be partly explained by Italy's uncoordinated approach to climate data collection and elaboration, and the fragmented nature of the relationships

Table 6.1 Expected climate impacts in Italy as outlined in the Fourth National Communication to the UNFCCC (adapted from MATTM, 2007a)

Sector	Projected changes in climate	Expected climate impacts
Energy	Reduction in overall water availability Increase in summer and winter temperatures	Reduction in electricity generation from hydropower Decrease in energy consumption in winter months Increase in energy consumption in summer months
Agriculture	Increase in extreme climate events (e.g., heat waves, heavy precipitation) Increase in average temperatures	Reduction in summer crop yields Shift in agroecosystems to the north Increase in length of growing season by 10–15 days
Coastal zones	Sea level rise Increase in extreme storm events	Loss of humid zones near river estuaries Salt water intrusion into coastal fresh water sources Increase in flooding and erosion of coastlines
Soil and water resources	Increase in heavy precipitation events (north) and droughts (south) Reduction in precipitation (south) Shrinkage of glaciers	Increase in soil erosion Increase in incidence of landslide phenomena Reduction in water supplies Increase in risk of desertification
Biodiversity and terrestrial ecosystems	Increase in average temperatures	Migration of ecosystems north and into higher latitudes Reduction in species richness Increase risk of forest fire Loss of benthic species density and biomass Spread of invasive species
Tourism	Increased summer temperatures and incidence of heat waves Increased winter temperatures	Excessive heat in tourist season Decrease in water availability in tourist season Loss of snow cover in Alpine areas
Health	Increase in average temperatures Increase in extreme weather events	Increased risk of summer heat-related illness and mortality Increased risk of climate event-related accidents Changes in water and vector-borne disease distribution

between several of the primary research bodies responsible for putting together national climate data. Unlike several other EU countries, Italy has not instituted a national programme of climate data collection for the purposes of establishing climate impact trends. National climate data is collected primarily by the Italian National Meteorological and Aeronautic Climatology Centre, a body operating under the Italian Ministry of Defence, as well as by the National Hydrographic and Oceanographic Service and the National Agricultural Information Service. However, these focus principally on past and current climatic trends and are used for the compilation of climate records and short-term forecasting.

Outside these national climate networks, independent sections and projects by ENEA, ISPRA, the National Research Council, and the regional ARPA work on varying aspects of climate change and projected impacts. However, the absence of coordination between these institutions' efforts has resulted in fragmented pieces of information collected and elaborated using different methodologies (ENEA, interview). Institutions concerned with climate phenomena and their effects (such as ISPRA and ENEA) receive little national funding or direction for climate change work, and so instead seek funding and direction from EU projects. While these have been an important source of research funding and information, they have not been a substitute for a nationally-led and funded integrated climate research strategy, as much of the baseline data required for coherent studies of impacts and vulnerability is still piecemeal:

> Everything that is funded, in terms of knowledge that could end up in the family of impacts, comes from the [European] funds. A little comes from other funds, but the largest percentage comes from the European funds. But the provision of these funds, which anticipates an exhaustive knowledge of the country on a given theme, limits us in being able to use a portion of these funds to cover the gaps in knowledge. If ones says, "I'll create a national database", that is charged to the government. "Give me an innovative methodology, create a different indicator"...that is a different approach than Italy needs, but since we work with those funds we use them to do what we can. (ENEA, interview)

As described in Section 6.2, research institutions lack an institutionalised network for information sharing, relying on personal relationships and projects to bring work together. While ISPRA continues to publish an annual climate indicator report using data from various sources around the country, coordination and validation of climate data remains difficult.

In response to the perceived lack of climate change assessments and projections, in 2000 the CIPE approved a request for the National Research Plan to include the creation of the Strategic Programme for Sustainable Development and Climate Change. Funded by the Integrated Special Fund for Research, the Programme was initiated in 2005 and included the establishment of the Euro-Mediterranean Centre for Climate Change (CMCC), now the primary body for climate change research in Italy. Created from various existing research institutes including the National Institute of Geophysics and Vulcanology and the Eni Enrico Mattei Foundation, the aim of the CMCC is to obtain a better understanding of climate change in Italy and to provide technical support to MATTM in matters relevant to climate impacts and adaptation. Research performed by CMCC includes the creation and downscaling of global circulation models and scenarios, impacts of climate change on forestry, agriculture, ecosystems, soil and coastlines, and the economic evaluation of these impacts and climate policies. CMCC now represents a significant portion of the climate change research ongoing in Italy, acting as a coordinator between various associate climate research centres, a focal point for several EU and international projects and a disseminator of climate change information to the media.

6.4.2 The National Climate Change Conference: 2007

Progress toward a complete picture of climate impacts and the development of concrete adaptation policies received an enormous push with the National Conference on Climate Change (NCCC). Held in September 2007 under the direction of the Prodi government, the national conference was jointly held by ISPRA and its National System of Environmental Agencies (the collective ARPA), in which the primary focus was to assimilate knowledge of climate impacts and vulnerabilities and identify appropriate adaptation options. The conference also aimed to investigate the role of various policy actors and technical bodies at different scales and across sectors in order to improve coordination between the different actors involved in adaptation. Prompted by the release of documents such as the European Commission's Green Paper on adaptation (Commission of the European Communities, 2007) and WWF Italia's *Guidelines for a National Adaptation Plan* (2007), the overarching goal of the conference was to instigate the formulation of a national adaptation strategy (Carraro and Sgobbi, 2008).

The conference was preceded by several thematic workshops conducted over the three days prior to the conference on identified primary climate-related risks: (1) Erosion and coastal risk; (2) Desertification; (3) Glacier and snow cover loss; (4) Hydro-geological risk, and; (5) the Po River basin. Workshops were conducted separately on each theme and subsequently combined together over the two days of the conference in order to discuss their implications for the vulnerability of the health, agriculture, tourism, water management and biodiversity sectors. Participants included members from various research institutions, universities, ministries, private companies and environmental NGOs from around the country, drawing principally from national and regional level agencies.

Beyond a summary of impacts and possible adaptations in each sector, two short documents constituted the principal outputs of the conference. The first is *The Climate Manifesto – a New Deal for Sustainable Adaptation and Environmental Security* (MATTM, 2007c), a document that committed the Ministry of the Environment and Territory to developing a National Adaptation Strategy by 2008 to be implemented over the following three years (MATTM, 2007c). The Manifesto concluded that while mitigation and the attainment of Kyoto targets were to be prioritised and achieved by 2012, adaptation efforts were to be coordinated and integrated into existing policy and legislation. Under the proposed strategy, matters of soil protection, integrated coastal zone management, tourism and water resource management were to receive primary attention and to be linked to national plans developed in response to the UN Conventions on Biodiversity and Desertification. A second two-page document entitled *The First 13 Actions for Sustainable Adaptation* outlined priority areas for intervention to be the focus of efforts of the MATTM and other ministries to promote coordinated adaptation policies (MATTM, 2007b) (see Box 6.1).

Box 6.1 Priority areas for adaptation intervention identified by the NCCC (2007)

The first 13 actions for Sustainable Adaptation (adapted from CNCC, 2007)

1. Development of an extensive climate change research programme, including the preparation of an annual report on climatic changes and their impacts;
2. Expansion and support of energy conservation and renewables, integrating of emission reduction with adaptation activities;
3. Development of incentives for new forms of consumption compatible with climate adaptation needs;
4. Adaptation of water resource management to adaptation through improvements in conservation and distribution;
5. Protection of the agricultural sector from climate change impacts;
6. Protection of the Italian coastline through improvements in planning and remediation of natural areas;
7. Identification and improvement of responses to extreme events, particularly floodplains and areas at risk of landslides and desertification;
8. Sustainable management of marine resources through improved fishing practices, the protection of ecosystems and water management practices;
9. Encouragement of sustainable tourism in mountainous areas;
10. Insertion of new climatic risks into health sector strategies;
11. Creation of a more efficient early warning system in areas of high flood and landslide risk;
12. Increase in the level of civic participation in policies of mitigation and adaptation through participation initiatives including the establishment of a national 'Climate Day';
13. Realisation of forms of environmental incentives for private companies and workers related to new forms of environmental accountability

The conference remains something of a question mark in terms of its achievements in the minds of many of its participants and observers, as well as a delicate point of enquiry. Despite its success as the first concerted effort at formulating a coherent approach to adaptation, the initial deadline for the completion of a National Adaptation Strategy by the end of 2008 has come and gone unmet. After losing a vote of confidence in the Italian Senate, Prodi's second coalition government fell at the end of January 2008 and was replaced by Berlusconi's Freedom People Party, a right-wing coalition government in mid-April. Many climate change actors point to this change in government as the principal force in the loss of momentum for the creation of an adaptation strategy; however, the absence of provisions in the 2008 national budget for such a strategy (created at the end of 2007 while Prodi was still in power) indicates that perhaps momentum was being lost regardless. Competition between national technical bodies for responsibility over the creation of an adaptation strategy may have also played a role: while the conference concluded with the appointment of ISPRA as the coordinating body for the National Adaptation Strategy, other bodies allegedly contested the decision, causing it to be rescinded and subsequently placed on hold. What is more certain is that the disintegration of the original adaptation strategy committee and the absence of any central competence for the drafting of an adaptation strategy have affected the course of adaptation at the national level.

However, the 2007 National Climate Change Conference still played an important role in Italy's ongoing adaptation policy formation process. The conference confirmed the need for an improved understanding of climate impacts and vulnerabilities in the Italian context and highlighted a lack of basic understanding of the meaning and implications of adaptation. Conference organisers realised too late that the concept of adaptation was poorly understood by many participants who had difficulty separating out mitigation and adaptation:[3]

> The first difficulty we had was that the word 'adaptation' generated a series of confusion. There was a lack of preparation; we raced ahead. People and local administrators, even the scientists themselves, aren't prepared to understand what adaptation is. For many, adaptation means resignation; that is, to passively undergo what is happening. So the first thing to be made understood is that adaptation is the prevention of negative consequences, prevent and minimize the negative consequences and exploit, if there are any, new opportunities that may be created. (ENEA, interview)

Despite the recent creation of the CMCC, deficiencies of information in terms of climate monitoring and the development of various forms of scenarios were found to persist, as data was present or absent in certain areas and along select themes in significantly varying degrees. As one actor noted,

> At the 2007 Climate Change Conference, it emerged that in terms of knowledge in the country, there were monstrous shortages. . .it came out that we lacked monitoring networks, we lacked scenarios, we lacked impacts, everything. There are wonderful things, but they are being done only on certain parts of the territory and not in a general way. In the end everything is missing because we are still unable to produce a number that applies to the whole country. (ENEA, interview)

The conference thus presented an important opportunity for the dissemination of climate-related work undertaken by individual research institutes on potential impacts in Italy and the evaluation of adaptation options in a variety of sectors:

[3] *Mitigation of risk* is often used in Italy to describe adaptation efforts to reduce risks in, e.g. coastal areas, hence the confusion between mitigation (used in climate change circles to refer to GHG emission reductions) and adaptation.

> If nothing else, that conference. . .had a great value of bringing together many scholars who spoke about these problems of adaptation. So, water regimes, regeneration of soils, all these big themes were addressed in depth by competent people. . .we had a technical instrumentation of "summing-up", of development, that not everyone has. (WWF, interview)

The loose and informal network of actors that had existed prior to the conference was strengthened as various actors were brought together to pool information from different sectors and scales of governance, and were provided the technical instrumentation to sum up all the existing work on adaptation. Inspired in part by the collaborations forged from the conference, the publication of Italy's first economic evaluation of adaptation measures entitled *Climate Change and Strategies for Adaptation in Italy* (Carraro, 2008) was jointly published by the Eni Enrico Mattei Foundation, ISPRA, and the CMCC. Though the difficulties in coordinating different institutions still exist, the conference represented an important occasion for individual and isolated research to come together into a more coherent picture of climate change-related research throughout the country.

6.4.3 National Adaptation to Climate Change: The Present

Despite the loss of momentum in the creation of a national adaptation strategy, movement on adaptation at the national level has been reportedly reignited. In 2008, the Directorate for Environmental Research and Development (RAS) began the process of putting together a scientific committee to discuss and formulate a national adaptation strategy, made up of the relevant technical bodies and representatives from the Ministries of Environment, Forestry and Agriculture, Finance, and Research (CMCC and MATTM, interviews). The sensitisation of these and other ministries to the concept and importance of adaptation has been highlighted as an important prerequisite to the adoption of an adaptation strategy, particularly given the absence of any legal authority of the MATTM (MATTM, interview).

Indeed the need for such education on the importance of adaptation was frequently raised as a considerable challenge to the development and implementation of adaptation measures and to combat the low public and political perception of the importance of the climate change issue. As the national conference partly revealed, poor understanding of adaptation even in relevant research circles has led to confusion over its role and importance on political agendas. According to several Italian climate actors, strong media scepticism has additionally influenced both public and political perceptions of the need for climate change measures. Climate science is considered by climate adaptation actors at all scales to be either underreported or misrepresented in both national and local media that instead focus on presenting controversial stances on the existence of climate change, or sensationalising weather events as signs of imminent and catastrophic changes in climate:[4]

[4] It should be noted, however, that not all Italian media has put forth such a stance, and that more left-wing publications have in fact applied pressure on the current administration for greater action in the climate change arena.

> There isn't an accurate perception of climate change in Italy, and the mass media has a huge responsibility. . .they give extreme emphasis to details that are of no interest, they look for controversy instead of information, and the feeling is that they are culturally unprepared, people who interact between the results and the people. We have a huge problem of information. (ENEA, interview)

> The media doesn't do very good work, in Italy, on the climate. . . We have many tabloid articles on climate change. We also have a strong reaction of negationism in Italy, much more than in other countries. Compared with other [countries], I see that we are the ones that have the most, at least in Europe. And this is also a problem, these 'negationists', these sceptics. (CMCC, interview)

As such, bodies such as the CMCC and NGOs such as the WWF, *Italia Nostra* and *Legambiente* have taken on the role of disseminating information on climate change and the need for response measures to both civic and political communities. Notwithstanding the need for sensitisation, the early stages of work by the MATTM indicate that the creation of a strategy is still considered by the national authorities responsible for adaptation as the appropriate means to integrate adaptation across sectors and to continue to fill the gaps in climate change research. This idea is elaborated by a MATTM official:

> We have particularly vulnerable sectors in which we do activities: the organisation of water management, integrated strategies, etc. We do not call it a strategy; we haven't managed to converge them onto one piece of paper, one single strategy, and to say that this is the adaptation strategy. . .to try to manage the most vulnerable sectors, to seek an understanding of how vulnerabilities might develop over time. But this acknowledgement isn't lacking, in my opinion; what is lacking is an integrated management. We lack a level of coordination in practice. (MATTM, interview)

This view that the combination of these policies under a coherent framework oriented towards future planning would be the ideal method of addressing adaptation is shared by several actors at the national level. Different sectoral legislative frameworks, policies and actions are under consideration for inclusion within the strategy, several of which have been identified in the National Communications as starting points for adaptation. Existing national-level frameworks for various aspects of sustainable development are represented as actions that essentially constitute adaptation measures in that they address existing climate-related vulnerabilities. One example is the National Action Plan to Combat Drought and Desertification, approved by the CIPE in 1999 as a part of Italy's commitments under the UN Convention to Combat Desertification. The plan focuses on principle areas of soil and water conservation measures, land restoration and the reduction of environmental impact of economic activities on soil. National water protection measures such as the creation and decentralisation of power to the river Basin Authorities and various monitoring and prevention activities by the DPC are also considered adaptive measures under the current approach to adaptation taken by the MATTM (MATTM, 2007a, 2007b).

Other actions to be considered under the strategy include sectoral plans that address climate vulnerabilities and risk, such as the National Plan for the Prevention of the Effects of Heat on Human Health. The plan aims at reducing social vulnerability to heat waves through a joint effort by the National Centre for prevention and monitoring of disease (CCM) of the Italian Ministry of Health, the National

Department of Civil Protection (DPC) and the National Centre for Prevention of Heat Health Effects initiated in 2006 (Ministry of Health, 2005). The Plan was in part forged by work in 2004 between the DPC and the National Centre through which a 3-year programme for monitoring and preventing the effects of heat waves on human health was instigated (Ministry of Health, 2006). The main focus of the project was the coordination of a cross-country network to prevent heat wave health effects and mortality by monitoring the effects of heat on mortality and introducing national and local mortality prevention programmes within cities of over 200,000 inhabitants, beginning with the implementation of a Heat/Health Watch/Warning System.

The EU Framework Directives for Water, Birds and Habitat are additionally highlighted as priority areas for implementation as a part of the future adaptation strategy, and cut across several of the measures identified as relevant to adaptation (MATTM, interview). The importance of the EU to adaptation actions in Italy, as with environmental measures in general, cannot be overstated; though the EU has a less direct role in adaptation than perhaps in mitigation (see Chapter 2), its indirect role in shaping policies for resource management and planning will certainly have an impact on Italian adaptation measures. In addition to providing research and technical bodies with opportunities for information transfer and funding through EU projects, the EU presents an important influence over Italian environmental policies and thus over adaptation.

As may be noted from the short list of examples of national 'adaptation measures', several activities pertaining to resource management and planning that have been identified for inclusion under an adaptation strategy are managed not by central government, but at the regional scale. As described in Section 2, the regions constitute an important level of governance and administration, taking on several of the responsibilities over sectors such as agriculture, water management and coastal risk management. As such, the development of an adaptation strategy intends to incorporate regional level activities as well as national policies (MATTM, interview). In the next section, examples of adaptation practices at the regional and sub-regional levels are used to illustrate the types of adaptation activities that may occur at sub-national tiers of governance.

6.5 Adaptation at Sub-national Scales

The ability and necessity of local actors to respond to changing climatic conditions has been well-established in adaptation literature as examples of local actions continuously emerge throughout both the industrialised and developing world (e.g., Næss, Bang, Eriksen, & Vevatne, 2005; Tompkins, 2005). This is also the case in Italy, where regional and local authorities in select areas have moved forward on adaptation either simultaneously or prior to action on the national scale (see Box 6.2 for two prominent examples of local climate change adaptation action). Though no formal adaptation strategies yet exist at the regional scale, the structure of regional federalism allows regions in particular the sufficient autonomy to move on both mitigation and adaptation (given sufficient resources and political will), despite the lack of a coordinated national strategy. As important local planning bodies, cities and municipalities are allowed significant jurisdiction over spatial planning and development. However, despite regional and local autonomies and EU intervention, there are limitations to the extent to which successful planned adaptations have and may be incorporated into local activities. As with environmental policy implementation, fragmented and piecemeal activities at local scales across the country may additionally lose their total value without a coherent and centrally-guided direction.

Box 6.2 Examples of planned climate change adaptation activities in Italy

Beyond the case study areas, several other pockets of interest and action on adaptation have begun to emerge across Italy over the last few years. While the case study areas represent a limited application of adaptation measures, the appearance of climate impact assessments and adaptation measures at the local scale provide evidence of the opportunity for local governments to address climate change issues. Below are two examples of some of the more prominent and extensive adaptation efforts in Italy.

The region of Trento-Alto Aldige

Trentino-Alto Adige is an autonomous region in north-eastern Italy that consists of the two autonomous provinces of Trento and Bolzano. Located in the Italian Alps, Trentino-Alto Adige is bordered to the north by Austria and Switzerland, and to the south by the regions of Lombardy and Veneto. While the administrations of both provinces have been involved in climate change activities, holding separate joint conferences with various NGOs, the 2008 Climate Project of the Province of Trento has resulted in a comprehensive report on climate change forecasts and consequences in the province (Provincia Autonoma di Trento, 2008). The report offers information on ongoing monitoring and availability of information, potential impacts and existing or suggested means of adapting in the following sectors: water resources, tourism, energy and industry, and environment and planning. The document also concludes with the importance of information provision, particularly given the pervasive polarisation of climate change attitudes in both media and the public between climate negationism and 'catastrophism'.

The City of Venice

The City of Venice is the only area of Italy that has taken a particular climate scenario (sea level rise) into account in the development and implementation of coastal protection measures to date, and represents the most oft-cited example of climate change adaptation in Italy. Measures that address both existing and potential climate events under an integrated strategy for mitigation and adaptation have been developed, combining aspects of the environment, economy and city planning. Beyond a municipal energy plan and public energy awareness campaigns, the strategy includes adaptation efforts projects designed to reduce the effects of increases of summer temperatures and thus the effects of heat waves on health, while simultaneously reducing GHG emissions and ozone levels in the city. (Massetti et al., 2007). Additionally, the Centre for Forecast of Tide Level and High Water Alerting of Venice has implemented a sea level rise forecasting system under the coordination of the city's Department for Environment and Security (DAST). The newly-upgraded system (2007) alerts residents of high water levels through a variety of means, including an alarm that uses four different sounds according to the height of the water (110, 120, 130 and 140+cm). The system also offers an SMS high water alert through the Civil Protection Service, in collaboration wit Venice ARPA and the Fire Department (Città di Venezia, 2008). In 2003, a system designed to lower maximum water levels during high water events was approved and is now in phases of construction. The MOSE system includes mobile flood barriers that are designed to isolate the City of Venice from the sea to prevent flooding, complementing ongoing efforts to 'raise' lagoon banks and public walkways in low lying areas of the city in times of high water (EEA, 2006). The City of Venice was also a participant in the *Climate Alliance* and *Cities for Climate Protection* projects.

In the next sections, the way adaptation (and in its absence, mitigation and/or sustainable development) has been addressed across the three scales of administration in Italy will be examined drawing primarily on case study work conducted in the winter of 2009. The region selected for the study is Emilia-Romagna, a regional leader in Italian environmental issues that has recently begun to demonstrate a considerable interest in adaptation. The Province of Ferrara and its capital Municipality of Ferrara are also assessed in terms of the means and extent to which they have addressed adaptation to climate change.

6.5.1 Adaptation in the Region of Emilia-Romagna

Emilia-Romagna is a region of ordinary statute in north-eastern Italy, bordered to the north by the Po River that divides it from the region of Veneto, to the south by Apennine ridge and the regions of Tuscany and Marche, and to the east by the Adriatic Sea (see Fig. 6.1). The region's geography is dominated by flat, low-lying plains in the east that give way to first hills and then mountains in the west and south. This varied geography has given rise to a thriving economy based on agriculture, tourism, energy and industry, but also presents a diverse range of climate vulnerabilities given existing risks of coastal and low-land flooding, landslides and water scarcity and conflict. The addition of a high population density renders

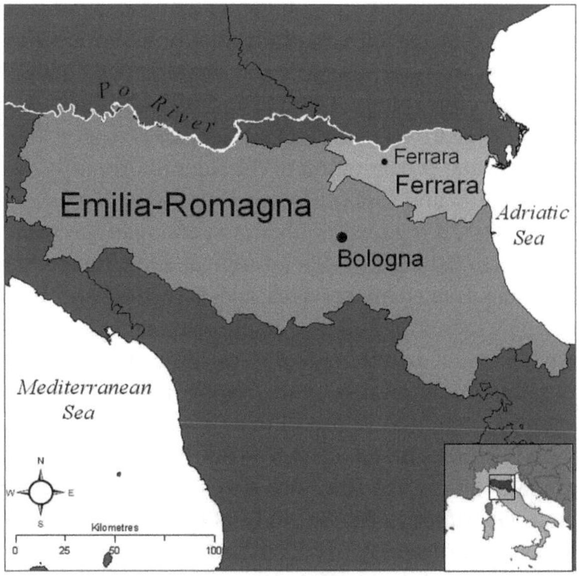

Fig 6.1 The Italian case study area

Emilia-Romagna a unique and complex region.[5] Now the third richest region in Italy by GDP per capita, Emilia-Romagna enjoys relatively great financial wealth compared with other regions, particularly in comparison with those of the south. The structure of government within the region is such that it has decentralised many of the administrative competences to the nine provinces, maintaining a principle of subsidiarity that favours the perceived increase in effective and relevant service provision, despite higher administrative costs.

Emilia-Romagna has shown considerable interest in adapting policies of sustainable development and resource management to projected climate impacts and vulnerabilities in comparison with other regions. Though no comprehensive adaptation strategy has been designed for the region, several regional plans that guide regional resource management and planning have taken climate change into consideration in varying, albeit limited, degrees. The major activities pertaining to adaptation in Emilia-Romagna largely stem from the strength of climate-related technical bodies and their engagement with climate impacts research and in regional policy formation, as well as existing water management and conservation activities across the various sectors within the region.

Emilia-Romagna's Regional Environmental Protection Agency (ARPA) was established in 1994, among the first to be created along with their coordinating national body, ANPA (now ISPRA). Having grown to include over 1100 people, the basic function of Emilia-Romagna's ARPA is to inform and provide technical

[5]Emilia-Romagna's population density is 189 ppl/km^2, increasing to 232 ppl/km^2 in the area of the Po River Basin (UNDP, 2008; ISTAT, 2009).

support to regional environmental and other policies. The ARPA contains various services that perform multiple functions within the regional environment sector, among which are research and development activities in the fields of meteorology and climatology. Created in 1985 and later incorporated into the ARPA, the region's Hydro-Meteorological Service is responsible for monitoring trends of temperature and precipitation in the region, primarily in response to local demands for information and support for agricultural activities.

In 2007, a regional deliberation changed the name of the Hydro-Meteorological Service to the Hydro-Meteorological and Climate Service (IMC), prompting a shift in orientation of its services and according to regional authorities, indicating the region's increasing interest in climate matters. With this shift, activities of the IMC have expanded to include downscaling of global circulation models to determine possible future climate in the region, and the participation in national projects that examine potential future climate impacts in the agricultural sector (ARPA, interview). While climate change work is still largely limited to the study of potential climate futures, the IMC also participates in a number of EU projects that have created opportunities for the study of climate mitigation and potential climate impacts in a variety of sectors. The IMC now represents the strongest regional climate centre in Italy (ARPA and ENEA, interviews).

Though the ability of the ICM and the ARPA to formally influence policy occurs through the use of Strategic Environmental Assessment of regional and provincial plans, the long histories and strong capacities of both the ARPA and the IMC have lent considerable weight to environmental and climatic issues in the region. According to one regional environmental authority, the strong link between the political and technical bodies of the regional government has been the foundation of political interest in climate adaptation in Emilia-Romagna:

> The region has always considered that the consequences of climate change weren't necessary only to be read as variations of environmental parameters, such as temperature; consequently, it has been very important to monitor and check that which could have repercussions at the economic and social levels. It is a historical vocation, the attention to the environment and thus to this "new factor" that is emerging. (Directorate General for Environment, Soil and Coastal Defense, interview)

A new proposal seeking additional funding for the creation of a Climate-Environment Centre is a further indication of the movement towards independent climate impacts and vulnerability research. If approved, the Centre would become the central body for assessing impacts in five regional sectors, moving beyond agriculture and into environment, health, transportation and productive activities (ARPA, interview). The approval of such a centre would also indicate an interest from the political sphere in climate impact information.

Beyond the study of impacts, the potential effects of climate change have been considered in regional policies including the regional Environmental Action Plan and the Rural Development Plan (Regione Emilia-Romagna, 2007). However, the strongest example of the inclusion of adaptation into resource management can be found in policies relevant to regional water management implemented by the regional ministries of the Environment and Agriculture. The management of water resources in Italy is a complex process, involving several bodies and interests, from the national and basin scales to local agricultural cooperatives. While the MATTM retains responsibility for water issues deemed of national importance, the lack of a system of national coordination of regional water policies has left the management of water largely up to individual regions. The establishment of River Basin Authorities in 1989 as the principal unit for basin land and water management transferred the responsibility of planning and programming of water and flood risk management to the basin scale, with regions acting as implementing and operational bodies of the larger basin plan (MATTM, 1998). Despite this hierarchy, regional authorities retain the ability to design regional water management plans.

In Emilia-Romagna, three documents pertain to the insertion of climate change considerations into the management of water resources in the region: the *Climate Change and Water Planning* report (Regione Emilia-Romagna 2003), the *Water Protection Plan* (2005) and the *Drought Management Plan* (under development). The first of these, the Climate Change and Water Planning report represents an effort by the region's General Directorate for the Environment, Soil and Coastal Defence to examine the potential changes in climate on water management in Emilia-Romagna. Using IPCC projections as well as national and regional climate data, the report provides an overview of potential changes in climate and their effects for the region, and highlights possible planning strategies to overcome changes in water availability, including the example of the UK's 'twin-track' water management strategy. Conclusions of the report include general recommendations for the conservation of water supplies and the development of specific short and long-term strategies to address periods of drought given the likelihood of diminished water availability, as well as the development of regional climate impact research capacities.

The 2003 report served as a basis for the development of sections within the 2005 Water Protection Plan on drought management and the need to incorporate potential future changes in the management of water resources. However, the extent to which the plan incorporates specific climate impacts is limited, citing the considerable uncertainties in available climate impact information on the agricultural and other sectors as grounds for retaining a 'neutral' position. As a result, the report maintains the need for the general conservation of water resources over the 12-year duration of the Plan, to be updated every 6 years as new information on climate impacts and other changes becomes available (Regione Emilia-Romagna, 2005). Ongoing activities include the promotion of water conservation through extensive educational programmes, the development of a regional water tariff system and the establishment of a multi-regional water conservation forum (Directorate General for Environment, Soil and Coastal Defense and Water Resource Protection Service, interviews). The Climate Change and Water Planning report also served as a basis for the development of the Regional Drought Management Plan, which will establish guidelines for drought management plans that will be decentralised to the provincial level. Drought monitoring and risk potential at local and regional scales are used in order to determine appropriate short and long-term responses to drought potential in the agricultural and water provision sectors. At the time of the case study, the Drought Management Plan was still under formation and had not yet been made publicly available.

In the agricultural sector, the Rural Development Plan (2007–2013) represents the central planning document through which EU, national and regional goals and targets are met. The plan outlines measures for economic competitiveness and environmental sustainability to be implemented within the sector and funded through the national Ministry of Forest and Agriculture and EU structural funds. According to a ministry representative, climate change is generally addressed through the plan:

> We do not have an adaptation plan to climate change. We try to adapt public policies to the climate change that is in action; so, climate change for us is a compass, an obligatory reference point. . .because the only true plan of intervention that we have is the Rural Development Plan. . .therefore, we have tried to steer the Rural Development Plan towards mitigation and adaptation to climate change (Regional Ministry of Agriculture, interview)

While mitigation measures are more explicit in the plan and in some cases directly incentivised and required by the EU or central government (e.g., through the national Green Certificate programme), adaptation is considered inherent within several measures used to counter existing problems of drought and high rates of irrigation. The plan outlines the adoption of a water consumption monitoring system by regional drainage consortia, private bodies contracted by the region to maintain drainage and water provision to agricultural cooperatives. As irrigation represents a significant proportion of water extraction from surface water sources, periods of low water availability are being addressed through the construction of reservoirs to be used in lieu of waters from the Po River (Regional Ministry of Agriculture, interview). Ongoing research on drought-resistant crop varieties and efficient irrigation technologies, and the provision of climatic and meteorological data on optimal irrigation timing complement active policies. Proposed activities include the use of economic incentives to shift the cost of water consumption from a land area-based tariff to a consumption-based tariff (Regional Ministry of Agriculture, interview).

Outside of the ministries, activities performed by the regional Department of Civil Protection (DPC) are also relevant to climate adaptation, in that its network of weather and risk forecasting is slowly beginning to include preventative measures as well as emergency response. Though Civil Protection is coordinated at the national level, regional departments have strong and important roles in the prevention and response to climate-related and other risks. In the case of Emilia-Romagna, considerable authority is allocated to the regional department, reportedly as a result of its capacity to coordinate effectively and quickly while maintaining an efficient use of financial resources (Regional DPC, interview). Working with the ARPA, one of the regional DPC's principal tasks is to provide 72 hour weather forecasts to determine potential areas of risk of flooding, heat waves, seismic risk and others. The DPC has additionally begun to initiate a number of preventative measures in order to address the greater frequency of climate events and the increased vulnerability of the territory as a result of increased urbanisation (Regional DPC, interview). The two-phase system of prevention and response includes the design of territorial safety plans and intervention strategies for improved response, as well as the design and construction of technical measures used to prevent potential damages from future events (the construction of land bridges over potential landslide areas, and the construction of coastal flood defences in important tourist areas are two examples). In 2003, the DPC and the Po River Basin Authority instituted a technical roundtable in response to the intensive droughts during the summer of the same year, and is now used on an ongoing basis.

In general, the Region of Emilia-Romagna's proactive interest in the sectoral impacts of climate change and the development of measures relevant to adaptation is attributed to a combination of factors, not the least of which has been the strength of regional environmental institutions and the region's subsequent long-standing involvement in matters of regional environment and safety:

> We [ARPA] are an agency that. . .expresses an important capacity in the process of evaluation of the impact of public plans on several aspects of life, among which are the climate and the economy. Therefore there is a capacity of the agency to carry out certain tasks and certain functions that whoever directs the agency, but also whoever works on the lower levels, would like to see evaluated. . ..Capacity in this case strongly influences will. (ARPA, interview)

Emilia-Romagna has a strong record of implementing measures in the agricultural and environmental sectors, even, according to regional actors, superseding national requirements and frameworks in the area of water conservation. Climate and environmental monitoring in the region of Emilia-Romagna are also well-established and well-funded in comparison to other regions, allowing for the creation and distribution of regional-level climate data.

The success of the implementation of environmental measures is also attributed to the long-standing history of participatory practices and stakeholder engagement applied by regional authorities. High stakeholder involvement in the development of policies such as the Water Protection Plan is considered to have improved acceptance and implementation by provincial and municipal governments. This is explained in further detail by two regional authorities:

> These action plans have always been done through sharing with a series of actors working on different spatial scales. For example, the provinces: there was never an act of imposition; this plan was proposed, discussed and truly shared. . .The various demands of the provinces and local stakeholders in some way were contributed, and everyone gave an opinion. So the end is a document that is not imposed from above, but shared. And it is certainly tiring to do so, but productive, because it is clear that it is then easier to impose a law and oblige [lower levels] to follow it. (ARPA, interview)

> The words 'sharing', 'participation' and 'cooperation' are a part of the history of the region of Emilia-Romagna. . .they have historical value. (Directorate General for Environment, Soil and Coastal Defense, interview)

Horizontal partnerships between bodies have also resulted in a strong network of actors working in environmental and civil protection at the regional and interregional levels, strengthening monitoring activities and improving the quality of baseline data both within the region and between regions sharing the Po River Basin. This spirit of cooperation has fostered extensive involvement in EU projects that have provided opportunities to engage in climate adaptation research and interregional projects on climate change and resource management, including the EU-funded WATERCORE project on climate change and water scarcity. EU influence has also reportedly been exercised through regional plans such as the Rural Development Plan, through which structural funds are allocated to regions that have met EU requirements. Whilst select other regions have lagged behind on their delivery, Emilia-Romagna has been timely in the development of such plans. The development of a proactive approach is also beginning to manifest in the activities of the DPC as efforts to coordinate longer-term data collection and the implementation of safety measures with longer-term benefits are increased.

Despite political interest in adaptation and relatively advanced climate information facilities, regional documents have still highlighted uncertainty in future climate impacts and the improvement of climate data as barriers to developing more concrete adaptation measures. As such, regional adaptation is still largely approached in terms of sustainable development and existing risks. Thus, while regional plans reference the need to address climate change, resource management and planning activities are only weakly linked with potential climate futures. Many ongoing regional activities are designed to respond to present climate-related risks to regional resources, particularly the regional water supply, and the effects of extreme climate events on the territory. While information and awareness of climate change scenarios and possible impacts are relatively high in the region, it appears the present tendency is to use climate information as an additional incentive for actions taken to resolve existing issues that are expected to increase in the future.

Finally, it should be noted that the region of Emilia-Romagna is among the most economically-rich regions of Italy which, in combination with supplemented funding from EU regional integration strategies, has provided relatively ample availability of funding for climate change-relevant programmes. However, while EU requirements have played a significant role in pushing environmental measures and mandating the inclusion of GHG emission mitigation measures into regional plans, adaptation is not yet a requirement for the receipt of regional funding. In the absence of national or EU-level economic incentives, regional action on adaptation to date thus remains largely a function of the individual capacity and political will of regional authorities. In Emilia-Romagna, reoccurring problems related to water scarcity, hydrogeological instability and other climate-relevant issues have incentivised action relevant to likely future climate vulnerabilities, but without national coordination or support and the uncertainty surrounding authority for matters of resource management and planning, adaptation measures have remained largely ad hoc and are not generally supported by a long-term risk management approach.

6.5.2 Adaptation Actions in the Province of Ferrara

One of Emilia-Romagna's nine provinces, the Province of Ferrara is bordered by the Adriatic coast and the Po River in the upper north-eastern corner of Emilia-Romagna (see Fig. 6.1). The province is rich in agricultural land located only a few metres above sea level, and is highly dependent on agriculture as an important source of revenue. As a result, the province's vulnerabilities lie principally in the susceptibility of coastal developments to flood risk and coastal erosion, and the risk of flooding and drought on interior agricultural and industrial activities and civic water needs. The Province has been active in environmental measures in general, administered by the provincial Ministry of Environment alongside the Department of Local Agenda 21 and the Service of Policies for Sustainability and International Cooperation. Relevant activities performed at the provincial level include environmental impact assessments, strategic environmental assessments and environmental education, through which energy efficiency and conservation requirements and programmes are introduced. The agricultural sector has also been involved in several environmental measures, the majority of which have been mandated by EU Directives through the regional Rural Development Plan. The recent institution of the Environmental Budget will provide baseline data from which to design and improve future environmental measures in the agricultural and other sectors.

As an administrative arm of the region, the Province of Ferrara naturally implements the policies and plans developed at the regional level, translating regional plans into provincial level programmes such as the Province's Rural Integrated Programme. Beyond the adoption of regional plans and programmes, however, the Province of Ferrara has been able to use discretion in the application of regional policies in order to address the specific needs and conditions of the province, and as such has made its own forays into climate change adaptation. This interest has manifested principally though the Province's participation in the Climate Alliance *Adaptation and Mitigation – an Integrated Climate Policy Approach* (AMICA) programme. Funded by the EU INTERREG IIIC and completed at the end of 2007, the aim of AMICA was to engage European local authorities in mitigation and adaptation activities to stimulate comprehensive approaches to climate change at the local scale. Local governments participating in the project included the Cities of Dresden, Stuttgart and Venice in addition to the Province of Ferrara, each bringing experiences of climate mitigation and risk management strategies of heat waves, floods and drought. Participants presented their individual means of addressing climate risks to share experiences and draw lessons for the application of various techniques and approaches in other contexts.

Provincial authorities involved in the project noted that before entering into AMICA, the concept of adaptation was unused and poorly understood, but that several ongoing technical and operational activities performed by the province were in line with what was considered 'adaptation':

> ...the adaptation concept was a bit foreign. We were much more concentrated on prevention [mitigation] than on adaptation, even if historically we've worked quite a lot on adaptation. This system of drainage is an adaptation. But this conceptual development in which actions of adaptation must be done at the same time as actions of prevention, we still had not thought of it. (Provincial Ministry of Environment, interview)

Thus, while provincial environmental authorities had previously been focused on mitigation as a primary activity in the climate arena, the project presented an opportunity to learn alternative means of adapting to climate events and to impart a means through which the two spheres could be combined in ongoing environmental management activities. Participants were shown a framework for dealing with episodic climate-related risks in a more integrated manner. Further, the project provided a methodology for bringing different actors together that built on the existing roundtables and means of stakeholder involvement used in different sectors during episodes of crisis or in general policy development.

> [The project brought] a new way of working together. Because in any case there were already, premises for us to work together. Many things have been done together. . .inside our structure, the Agriculture and Environment Sector, there were already systems of stakeholder involvement that have been active and operative for years. (Agriculture and Environment Sector, interview)

Among the Province's ongoing activities that corresponded with the goals and concerns of AMICA are the management of coastal erosion and storm events, and of drought and flooding in the province's agricultural lands. In the coastal sector, the Province is responsible for monitoring the level and quality of coastal waters and changes in sea level, as well as the maintenance of favourable conditions for the aquaculture that constitutes an important component of Ferrara agricultural activities. As a result of an agreement between the region, province, Municipality of Goro and local fishers, responsibilities of the provincial authorities mainly pertain to the maintenance of water circulation in the bay through the transfer of excavated sand from areas of sedimentation to areas of shellfish production. As storm and high precipitation events regularly carry sediments into the area of the bay, the maintenance is an ongoing process of dredging and depositing. Additional activities have included the provision of climate data to agriculturalists in the province, and the construction and maintenance of reservoirs to provide additional water drainage and reserves in periods of flooding and drought, respectively.

The final output of AMICA included the production of different tools with which to assess the potential for the development and integration of mitigation and adaptation within local governments.[6] However, the process of engagement in the project itself was highlighted by several Ferrara authorities as the central benefit to participation. The project provided an integrative understanding of ongoing risk management issues in the province under the lens of adaptation and allowed for the transfer of practices and models used to manage climate-related risks in other local governments in Europe.

Thus, the Province of Ferrara serves as an example where engagement in adaptation has occurred as a result of the participation in an external network that provided information and the opportunity to pool the experiences of different local bodies under a common framework. As is evident from the above activities, several of the measures used both before and after the project are in fact technical and/or operational activities carried out by various sectoral bodies that pertain to the management of current risks and issues of environmental management. Given the importance of

---

[6]Details on the tools developed through the project can be found at www.amica-climate.net

the agricultural sector to the economy of the province, it is not surprising that a significant proportion of the measures associated with adaptation in the province have concerned the provision and drainage of water in agricultural lands. However, several respondents indicated that the history of strong vertical and horizontal relationships between different stakeholders had contributed to the province's interest in adaptation. The use of networks and partnerships as a means of accessing information on best practices and improving provincial measures was emphasised in all sectors:

> We really believe in the exchange of good practices; it is for this we participate in networks. We are a part of the coordination of the Italian Local Agenda 21, which in turn is a part of. . .the European networks. We are also a part of the Sustainable Cities Campaign. We participate in international meetings, when we can. We participate in projects like AMICA, when we can, simply because we think that, to copy from who has already done [something] is an absolutely positive thing. (Provincial Ministry of Environment, interview)

The Province's participation in the project and thus its access to knowledge of climate impacts and possible adaptations was additionally fostered through the structure of the Emilia-Romagna region, in which general approaches to environmental and agricultural activities are set by the regional level but allow for provincial discretion in the design and implementation of specific activities. Provincial authorities spoke favourably of this extensive decentralisation to the provincial level:

> The benefits are that you have a management that is in contact with the territory and therefore with the stakeholders. . .My political reference point is no longer the regional minister, who I saw very little, but is now the provincial council, a municipal councillor who is here, on the territory. The physical contact is more frequent; the users, too, are closer to us. . .the knowledge that, either technically or politically, the government is here, for the citizen, the consumer, the final beneficiary. . .it is a concrete reference point. They can touch it. (Department of Agriculture and Environment, interview)

This structure has also been attributed the success of the Province's strong performance in EMAS registration and LA21; however, the absence of public funds through which to complete necessary measures was identified as a limitation to the province's ability to act on various aspects of sustainability and adaptation. Actors from different provincial sectors described the need to manage financial and human resources efficiently given the low availability of public funds:

> The resources are never sufficient. Among other things, we live in a phase in which an increase in our responsibilities corresponds to a reduction in resources. . .Now we are also in a moment of economic crisis and so the State is reducing, little by little, the resources that are transferred to the periphery. (Provincial Ministry of Environment, interview)

In both the agricultural and coastal sectors, an increasing number of activities are now completed by public bodies using alternative sources of funding in order to ensure the continued success of agricultural activities:

> In general terms, the tendency is to join all possible projects. It's not that we say yes to one and no to another. If we can, because these things are not always easy to follow particularly with such insufficient human and financial resources, we try to participate in all projects. (Service for Policies for Sustainability and International Cooperation, interview)

Though EU requirements for regional funding now play a significant role in shaping regional (and thus provincial) plans, both the Province of Ferrara and the other provinces in the region have been actively involved in sustainability over the last decade. Several actors identified this engagement in sustainability issues, as well as the interest in addressing themes of adaptation of one or more political figures, as the motivating factors for Ferrara's involvement in adaptation prior to other provinces. The participation in AMICA and subsequent awareness of the possibility of implementing adaptation measures occurred as the result of the existing partnership in the Climate Alliance and of the interest of authorities of the Environment and Agricultural sectors. Authorities also spoke of the general awareness of and support for sustainability activities in the province as an important factor the province's ability to implement environmental measures. However, adaptation to climate change remains a concept that has not yet been fully integrated into provincial activities by any means. As at the regional level, measures to counteract the effects of climate events and variability are still largely designed in response to current climate vulnerabilities.

6.5.3 Adaptation at the Municipal Level: the Municipality of Ferrara

The capital city of the province of the same name, the Municipality of Ferrara is home to a population of 135,000 and is a UNESCO World Heritage Site located along the Po River (see Fig. 6.1). The municipality was selected in the case study not as a result of its engagement in adaptation issues, but for its relatively extensive engagement in matters of energy efficiency and sustainability. Ferrara is acknowledged for its strong civic engagement and social capital that have contributed to its progress in achieving sustainable development goals (cf. Evans, Joas, Sundback, & Theobald, 2005) and received the European Sustainable City Award in 2003. Municipal actions have included the promotion of green and protected areas, sustainable transportation, urban regeneration and efficient waste management, as well as extensive civic environmental education provided by the IDEA Environmental Education Centre.

Together with the provincial Environment Ministry, the municipal Environment Service and Ministry of the Environment engage in ensuring energy efficiency measures are incorporated into municipal planning and building legislation. Through the system of decentralisation, the province is responsible for coordinating the inclusion of energy efficiency measures into the planning documents of all its municipalities. Thus, several activities have occurred as a result of regional requirements for energy conservation through the Regional Energy Plan (and subsequent creation of municipal energy plans) and regional planning legislation and instruments for the mitigation of GHG emissions. However, the municipality has demonstrated a proactive approach to the promotion of renewable energies and energy efficiency. For example, Ferrara has hosted a number of energy and climate change events, including the 1999 release of the Ferrara Declaration on Geothermal Energy by the European Geothermal Energy Council, and the *From Global to Local: Action Plans for the Climate* conference in 2007. Renewable energies and conservation have more recently been promoted through the provision of incentives for domestic solar panels, the implementation of maintenance standards in domestic heating, and the adoption of improvements in insulation requirements into building codes in 2006. While a successive change in regional legislation in 2008 mandated similar standards throughout the municipalities of the region, the Municipality of Ferrara's early adoption of more rigorous standards indicates some willingness and ability to act on such themes. Information on these measures is provided to the public via publications through the LA21 and the Sustainable Cities Service of the municipal Ministry of Environment, including a user-friendly *Energy Savings in Buildings* booklet.

While several of these municipal activities could be considered practical advancements in mitigation policy, the municipal approach to such measures is different. In reference to the energy measures taken in the municipality, one authority remarked: 'We can also call it mitigation, but in reality the objective is energy savings reflected in the expenses of families, and in the quality of the air' (Environment Service, interview). As in the examples at regional and provincial levels, such interest in such measures is generally attributed to a long-standing engagement with matters of sustainability and the strong presence of public and civic environmental organisations. The establishment of the municipality's Local Agenda 21 office in 1999 was among the first in Italy and has since been active in both environmental education and various energy projects, including the Climate Alliance's PRO-EE (Public Procurement Boosts Energy Efficiency) project for public energy efficiency.

However, the municipality has yet to engage in climate adaptation activities, in part attributed both the low incidence of climate impacts and low public perception of the need to engage in adaptation activities. Municipal authorities emphasised the prioritisation of mitigation measures, citing a lack of immediate effects of climate change in the area:

> The part of mitigation and compensation comes first. We haven't yet had this cultural leap [toward adaptation] because we probably still hope there won't be a need to. We are still in the phase of prevention. We put mitigation and compensation first because we've not yet seen any concrete events on the territory, even if we are planning various actions, such as rainwater management. These actions are a part of routine territorial planning procedures that we have considered as elements, but we are more concentrated on mitigation. (Municipal Ministry of Environment, interview)

Climate impacts considered to have immediate relevance to the municipal level were limited to the incidence of extreme climatic events and so far were not considered of particular concern to either public or political spheres, though some awareness-raising measures were undertaken:

> The issue of climate change is not perceived as an emergency on the part of the citizens...We are trying to do sensitisation work, pushing, and not only in moments of emergency, a kind of widespread knowledge, even using very commercial instruments [such as] the film by Al Gore... (Municipal Ministry of Environment, interview)

Two projects have additionally been implemented by municipal health authorities to address social vulnerabilities in periods of high summer temperatures, using climate forecasts issued by the regional ARPA.[7] Actions to incorporate improvements in targets for water quality and quantity within the strategic municipal structural plan have been advocated through the creation of the *Water as a Resource* working group. However, these activities are not considered within a framework of adaptation or incorporated under a unified approach or strategy, and have been largely reactive in nature. The absence of such a strategy is explained by one respondent as the result of a lack of systemic or comprehensive approaches to climate impacts:

> This is because issues are dealt with one at a time. For example, the *Uffa che afa* project that we spoke of earlier, the project against summer heat, is to help people in moments of suffering, but has no systemic thinking. This happened even with the problem of tiger mosquitoes, which was also probably related to climate change. . .where there was a lack of a systemic vision of all these problems, and where the ability to put them all together and understand what to do for everything was lacking. (LA21, interview)

Further, while certain actions with regards to energy efficiency have been mandated by the regional level, municipal decisions to take on additional tasks in the area of mitigation and sustainability have been voluntary and thus receive little financial support from either regional or central government. Funds for measures such as the creation of an Environmental Budget are beyond the scope of dedicated channels of financing tied to regionally- or centrally-mandated planning and management tasks. As such, municipal authorities indicated that they are obliged to use resources carefully in order to implement municipally-led tasks, incorporating measures within broader urban planning strategies. The participation in networks such as the Climate Alliance and the Kyoto Club have also been used as a means of overcoming financial constraints to the implementation of climate change measures, and to provide an opportunity to learn from the activities of larger, more well-funded municipalities within the region (e.g., Bologna), throughout Italy (e.g., Milan) or internationally (e.g., Barcelona). According to local authorities, these networks represent a means of collaboration with other municipalities and an important forum for the transfer of best practices:

> As a city we are fortunate to be able to compare experiences with others as part of several networks of "virtuous" cities, innovative cities. Effectively, networks are important, and fortunately we have this opportunity, on the part of staff and financial resources, to be engaged. (Municipal Ministry of Environment, interview)

However, the lack of an integrated and comprehensive national coordination between municipalities has ultimately led to local variations in terms of the implementation of energy efficiency or mitigation measures. Though the National Association of Italian Municipalities is designed to link local authorities across the country, national coordination of municipalities in the practical sense is still limited.

---

[7]The *Giuseppina* project provides low-cost transportation to elderly people in summer months, while the recently- implemented *Uffa che afa* project (2003) ensures the provision of assistance and support to high-risk populations during heat waves.

6.6 Conclusion: Multi-Level Governance in Italian Adaptation

This review of climate change adaptation actions in Italy and in select sub-national administrations presents an interesting example of adaptation governance in which various actors at different scales have entered into adaptation planning through different avenues and at slightly different times. At the national level, progress on adaptation has been delayed through recent governmental instability, but has also maintained some continuity in its approach through its use of existing legislation, policy and knowledge of vulnerable areas as the basis for a national adaptation strategy. The participation of the various relevant ministries in the creation of a strategy represents an important step in uniting individual, piecemeal activities that directly or indirectly address climate change risks and vulnerabilities. Given that adaptation has largely been approached as an environmental issue, collaborative efforts to address climate change may present an opportunity to reduce fragmentation and connect isolated legislation and policies under a common framework. This raises the question as to the role and importance of the CIPE, through which decisions on the environmental and climate change are made by several ministries. While such a decision-making structure has improved the coordination of policies, it is perhaps natural to question whether the input of other ministries has hampered the implementation of climate change measures (particularly mitigation) that are perceived to have detrimental effects on the Italian economy. However, adaptation will likely require extensive cooperation and cross-sectoral engagement in order to be effective in addressing climate change vulnerabilities, which may be fostered by such an institution.

The success of the Italian administration in pulling together such a strategy remains to be seen, and will likely hinge on the extent to which existing issues of capacity are addressed. While the creation of the CMCC in 2005 signified a valuable step in the advancement of comprehensive, nation-wide impacts and adaptation research, the improvement of coordination both between research institutes and between central and regional governments would considerably advance the progress of adaptation in Italy. This coordination is also lacking at the regional scale, where the devolution of power to the regions in several matters has both fostered the abilities of regional governments to engage in matters of sustainability and climate change, as well as discouraged the universal implementation of rigorous environmental standards (beyond those set by the region itself). Thus, while select regional and local administrations have begun their own explorations of impacts and adaptation through existing channels of climatic research, environmental policy and NGO networks, the absence or presence of certain characteristics and capacities has shaped the extent of their engagement. Such discrepancies also create considerable fragmentation in the ways in which resources are managed across the country, creating challenges for those areas interested in pursuing proactive measures particularly along the physical and administrative boundaries between areas.

However, the reluctance of the current Berlusconi administration to accept EU mitigation proposals and to form concrete or meaningful emission reduction targets is perhaps indicative of a deeper, political opposition to engage with adaptation.

Despite the ongoing efforts by various technical and administrative bodies at the national scale, an admission of the need to reduce emissions by the current administration required to significantly engage with adaptation is a likely hindrance. Thus, while the failure of the National Climate Change Conference led by the Prodi government in 2007 has been attributed to the inability of such left-wing parties to effectively command resources and implement policy, the lack of serious attention paid to the larger climate change issue and its associated requirements exhibited by current powers surely plays an important role. This partisan approach to climate change is further exhibited by the left-wing region of Emilia-Romagna and its efforts to incorporate climate impacts into regional policy, in the larger vein of close science-policy linkages fostered by regional authorities.

The results of the case study further reveal that the existence of political leadership and interest in adaptation issues, strong vertical and horizontal networks, and the engagement in international climate change networks all contributed to the development of planned adaptation measures in Emilia-Romagna. Regional action in the adaptation arena has been enabled through long-standing political and civic interest in environmental issues, and ties to local agricultural activities that have required quality environmental management and information in order for its success. This interest has been additionally fostered by the ARPA's strong research and technical capacities for climate monitoring and impacts studies and implemented successfully as a result of the strength of participatory networks and linkages between the science and policy realms. The development of a new climate impacts research programme within the ARPA demonstrates continuing interest in incorporating adaptation issues into regional plans, and would serve as a novel movement towards a more integrated study of impacts and vulnerabilities to fill gaps in regional knowledge. Coupled with strong stakeholder involvement in regional planning and the decentralisation of power from the regional to provincial levels, the success adaptation efforts in Emilia-Romagna are in some ways considered simply an extension of the region's positive environmental policy implementation record. The provinces themselves are additionally able to move beyond purely administrative duties, using regional funds and existing networks to take on projects of interest. In the case of the Province of Ferrara, this structure has enabled the interest of local authorities in climate change to translate into the participation in the exchange of information and practices relevant to adaptation. The province's involvement in the Climate Alliance's AMICA project provided an exceptional opportunity for learning on adaptation needs and practices, building on existing participatory structures between ministries and scales to develop an integrated framework for addressing adaptation.

However, despite movement forwards in the climate adaptation arena, the lack of an integrated central government framework on adaptation has to some extent constrained sub-national abilities to address adaptation. This is particularly reflected in the provincial and municipal case studies, in which the absence of financial and human resources presented a significant barrier to work on issues of interest, including adaptation and mitigation. Funding from the central to local governments via regional authorities are transferred according to legal frameworks and legislation to be implemented by sub-national levels of government; given that adaptation remains thus far undeveloped at the national level, the absence of economic incentives for adaptation presents a considerable challenge. As government cutbacks fall increasingly heavily on local authorities and funding for climate change-related activities is scarce, smaller and less financially-stable municipalities and even regions could encounter considerable difficulty in designing and implementing adaptation measures.

Both the experiences of the Province and the Municipality of Ferrara highlight the importance of networks in the development and establishment of issues such as adaptation. Authorities have drawn on new and existing networks between local authorities in Italy and across Europe to obtain information, share knowledge on practices, facilitate the implementation of policies, and access funding to overcome financial difficulties, including private-public arrangements. The EU represents a particularly important source of such financial and informational networks on sustainability, climate change and adaptation, both through regional integration programming and individual impact and vulnerability-related projects. Though adaptation is currently not a required component of regional plans to access EU funds, climate change mitigation actions are required and may set a precedent for additional climate-relevant activities (European Climate Change Programme, 2006). In the absence of national support, projects funded by INTERREG and orchestrated by NGOs such as the Climate Alliance have additionally served as important sources of information and resources for local authorities to engage in various environmentally-relevant activities, including adaptation planning. However, it should be noted here that only those local authorities with sufficient resources to engage in such networks could be able to benefit from such participation. As a result, while financial and human resources may be scarce in the case study areas, they were able to harness enough to engage in specific networks, indicating that access to certain resources (i.e. financial and human) in essence may beget further access to others (i.e. information and knowledge).

An additional way of overcoming barriers to action on adaptation highlighted by several authorities was the need for a mainstreaming approach, through which considerations of future climate change could be streamlined into existing policies as an additional issue or impetus for policies of sustainable resource management. Addressed this way, issues of funding and human resources may become less restrictive, while issues of information and education become more significant. However, as local climate change impacts had not yet been identified or considered particularly relevant in the case study municipality (and unlikely in less technically-proficient regions), improvements in the effective creation and transfer of climate impact information from the regional level may be warranted.

Finally, the approach to adaptation shared among the different Italian case study areas is worth noting. Actions that were considered as adaptations by the majority of actors at all scales have generally been designed and implemented to address current vulnerabilities to climate-related risks such as drought, flooding and geologic instability. Using this approach, authorities that are to contribute to the development of a national adaptation strategy have indicated the need to combine existing policies, legislative frameworks and activities that address ongoing concerns, identified as important sectors or themes given the likely characteristics of a future climate. Similarly, adaptations identified at the regional and local levels represent measures for the ongoing management of reoccurring issues of resource scarcity or risk, such as water management in the agricultural sector and the maintenance of coastlines.

While the use of current vulnerability as the starting point and basis for planned adaptation acknowledges the need to address well-known and rooted environmental and political problems, some actors raised concerns as to the lack of future considerations in adaptation activities, or what one respondent termed a 'planning for today'. Select administrations and bodies have engaged in cross-sectoral research and policy-making, but adaptation is still largely considered in terms of present risks and addressed by extending existing sectoral policies and measures to contend with the increasing frequency or magnitude of present issues. The tendency to consider adaptation as an extension of existing measures or policies based on past and present risk trends is somewhat reflective of traditional Italian approaches to environmental management. As described in Section 2, Italian environmental policy and risk management have been predominantly reactive in nature and focused on emergency relief in lieu of preventative planning methods. As such, the need to address shifting conditions posed by climate change through planned, anticipatory actions may present an interesting challenge to the prevalent methods of environmental governance and planning in Italy.

References

Interviews

Euro-Mediterraneo Centre for Climate Change (CMCC). Bologna, January 23, 2009.
Institute for Environmental Protection and Research (ISPRA) – Department of Environment and Meteorology. Rome, January 22, 2009/
Institute for Environmental Protection and Research (ISPRA) – Department of Soil Protection. Rome, January 17, 2009.
Italian National Agency for New Technologies, Energy and the Environment (ENEA). Bologna, January 23, 2009.
Italian National Agency for New Technologies, Energy and the Environment (ENEA). Cassacia, February 16, 2009.
Legambiente. Rome, January 22, 2009.
Ministry of the Environment and Territory (MATTM) – Department of Environmental Research and Development. Rome, March 5, 2009.
Municipality of Ferrara – Department of Environment. Ferrara, January 24, 2009.
Municipality of Ferrara – Department for Local Agenda 21. Ferrara, February 5, 2009.
Municipality of Ferrara – Environmental Service. Ferrara, February 5, 2009.
National Department for Civil Protection (DPC). Rome, March 4, 2009.
Po River Basin Authority. Parma, February 2, 2009.
Province of Ferrara – Agriculture and Environment Sector. Ferrara, January 27, 2009.

Province of Ferrara – Agriculture and Environment Sector, Agro-Environment and Technical Services. Ferrara, January 27, 2009.
Province of Ferrara – Department of Water Resources and Environmental Protection, Coastal Resources and Fisheries. Ferrara, January 26, 2009.
Province of Ferrara – Provincial Ministry of Environment. Ferrara, January 26, 2009.
Province of Ferrara – Service for Policies for Sustainability and International Cooperation. Ferrara, January 26, 2009.
Regional Environmental Protection Agency (ARPA) Emilia-Romagna. Bologna, February 3, 2009.
Regional Environmental Protection Agency (ARPA) Emilia-Romagna – Hydro-Meteorological Service. Bologna, January 20, 2009.
Regional Environmental Protection Agency (ARPA) – Hydro-Meteorological Service. Parma, February 2, 2009.
Region of Emilia-Romagna – Department of Civil Protection. Bologna, February 3, 2009.
Region of Emilia-Romagna –Ministry of Agriculture. Bologna, February 24, 2009.
Region of Emilia-Romagna – Directorate General for Environment, Soil and Coastal Defense. Bologna, January 29, 2009.
Region of Emilia-Romagna – Directorate General for Environment, Soil and Coastal Defense, Water Resource Protection Service. Bologna, January 20, 2009.
WWF Italia. Rome, February 18, 2009.

Secondary References

A21Italia. (2008a). Downloads-Piani d'Azione. Retrieved September 24, 2008, from http://www.a21italy.it/a21italy/modules.php?op=modload&name=Downloads&file=index
A21Italia. (2008b). The role of the Italian Local Agenda 21 Association. Retrieved September 24, 2008, from http://www.a21italy.net/association.php
Biorcio, R. (2002). Italy. In F. Müller-Rommel & T. Poguntke (Eds.), *Green parties in national governments* (pp. 39–62). London: Frank Cass Publishers.
Carraro, C. (Ed.). (2008). *Cambiamenti climatici e strategie di addatamento in Italia*. Bologna: Società editrice Il Mulino.
Carraro, C., & Sgobbi, A. (2008). *Climate change impacts and adaptation strategies in Italy: and economic assessment* (Working Paper 6.2008). Milan: Fondazione Eni Enrico Mattei.
Città di Venezia. (2008). Servizio di allerta SMS – Rischio Idraulico. Retrieved October 6, 2008, from http://www.comune.venezia.it/flex/cm/pages/ServeBLOB.php/L/IT/IDPagina/17078
Commission of the European Communities. (2007). *Adapting to climate change in Europe- options for EU action*. Brussels: Commission of the European Communities.
Cotta, M., & Verzichelli, L. (2007). Local government: the long-term oath towards a federal system. In *Political institutions in Italy* (pp. 171–201). New York: Oxford University Press.
European Climate Change Programme. (2006). Regional planning, energy and public infrastructure and structural funds sectoral report. Working Group II Impacts and Adaptation. Brussels: European Climate Change Programme.
European Environment Agency (EEA). (2006). *Vulnerability and adaptation to climate change in Europe*. Copenhagen: European Environment Agency.
Evans, B., Joas, M., Sundback, S., & Theobald, K. (2005). *Governing Sustainable Cities*. London: Earthscan.
Gualini, E. (2004). *Multilevel governance and institutional change: The Europeanization of regional policy in Italy*. Aldershot: Ashgate Publishing Limited.
ISTAT. (2008). *Italy in figures*. Rome: Italian National Institute of Statistics.
ISTAT. (2009). Ufficio Regionale Emilia-Romagna. Retrieved June 8th 2009, from http://www.istat.it/istat/attivita/sediregionali/emilia/

Istituto Superiore per la Protezione e la Ricerca Ambientale (ISPRA). (2004). Italia: politiche sul clima. Retrieved 15 September 2008, from http://www.apat.gov.it/site/it-IT/Temi/Protezione_dell%27atmosfera_a_livello_globale/Cambiamenti_climatici/Italia_-_politiche_sul_clima/

Lewanski, R. (1998). Italy: environmental policy in a fragmented state. In K. Hanf & A.-I. Jansen (Eds.), *Governance and Environment in Western Europe* (pp. 131–151). New York: Addison Wesley Longman Ltd.

Marchetti, A. (1996). Climate change politics in Italy. In T. O'Riordan & J. Jäger (Eds.), *Politics of climate change a European perspective* (pp. 298–329). London: Routledge.

Massetti, E., Pinton, S., & Zanoni, D. (2007). National through local climate policy in Italy. *Environmental Sciences, 4*(3), 149–158.

Ministry for the Environment Land and Sea (MATTM). (1998). Second National Communication under the UN Framework Convention on Climate Change Italy. Rome: Republic of Italy.

Ministry for the Environment Land and Sea (MATTM). (2006). Italian Report on Demonstrable Progress under Article 3.2 of the Kyoto Protocol. Rome: Republic of Italy.

Ministry for the Environment Land and Sea (MATTM). (2007a). *Fourth National Communication under the UN Framework Convention on Climate Change*. Rome: Republic of Italy.

Ministry for the Environment Land and Sea (MATTM). (2007b). Le prime 13 azioni per l'adattamento sostenibile. Cambiamenti Climatici Conferenza Nazionale, Rome, 12–13 September.

Ministry for the Environment Land and Sea (MATTM). (2007c). Manifesto per il clima – un new deal per l'adattamento sostenibile e la sicurreza ambientale. Cambiamenti Climatici Conferenza Nazionale, Rome, 12–13 September.

Ministry of Health. (2005). *Piano Operativo Nazionale per la Prevenzione degli Effetti del Caldo sulla Salute*. Rome: Department for Civil Protection.

Ministry of Health, Direzione Generale Prevenzione Sanitaria & Centro Nazionale Prevenzione e Controllo Malattie. (2006). Linee Guida per Preparae Piani di Sorveglianza e Risposta verso gli Effetti sulla Salute di Ondate di Calore Anomalo Rome: Republic of Italy.

Næss, L. O., Bang, G., Eriksen, S., & Vevatne, J. (2005). Institutional adaptation to climate change: Flood responses at the municipal level in Norway. *Global Environmental Change Part A, 15*(2), 125–138.

OECD. (2003). *OECD Environmental Performance Review of Italy*. France: OECD.

Provincia Autonoma di Trento. (2008). *Previsioni e conseguenze dei cambiamenti climatici in Trentino*. Trento: Giunta della Provincia autonoma di Trento.

Putnam, R. (1993). *Making democracy work: Civic tradtions in modern italy*. Princeton: Princeton University Press.

Ramieri, E., Wallace-Jones, J., & Lewanski, R. (2001). Italy Converging pathways between central and local levels of government? In W. M. Laflerty (Ed.), *Sustainable Communities in Europe* (pp. 225–244). London: Earthscan Publications Ltd.

Regione Emilia-Romagna. (2003). Cambiamenti climatici e pianificazione idrica. Bologna: Direzione Generale Ambiente e Difesa del Suolo e Della Costa, Regione Emilia-Romagna.

Regione Emilia-Romagna. (2005). Piano di Tutela della Acque. Bologna: Assessorato Ambiente e Svilupo Sostenibile, ARPA Regione Emilia-Romagna.

Regione Emilia-Romagna. (2007). Programma di Sviluppo Rurale 2007–2013. Bologna: Direzione Generale Agricoltura, Regione Emilia-Romagna.

Rosenbaum, A. (1987). Italy's Green Party. *Environment, 29*(10), 3/34.

Tompkins, E. L. (2005). Planning for climate change in small islands: Insights from national hurricane preparedness in the Cayman Islands. *Global Environmental Change Part A, 15*(2), 139–149.

United Nations Development Programme (UNDP). (2008). Po. Retrieved June 8th, 2009, from http://www.grid.unep.ch/product/publication/freshwater_europe/po.php

WWF Italia, W. W. F. (2007). *Per un piano di adattamento al cambiamento climatico in Italia: Primi indicazioni*. Rome: WWF Italia.

Chapter 7
Planned Adaptation Measures in Industrialised Countries: A Comparison of Select Countries Within and Outside the EU

Lisa Westerhoff, E. Carina H. Keskitalo, Heather McKay, Johanna Wolf, David Ellison, Iosif Botetzagias, and Bertrand Reysset

Abstract This chapter provides a context for the discussion on the emergence of adaptation measures in the four case studies presented in this volume through a brief discussion of the ways in which planned adaptations are being developed in eight European countries. Adaptation actions at different levels in Austria, France, Germany, Greece, Hungary, the Netherlands, Norway and Spain are described in order to provide an illustration of adaptation measures in a range of countries representing differing structural contexts, modes of decentralisation and histories of engagement in environmental policy. An additional section on planned adaptation as it has manifested in Canada and Australia is also presented to provide further ground for comparison with countries outside the European context. Results show a tenuous link between the extent to which countries have engaged in adaptation and their environmental policy implementation record, though some interesting links between environmental policy institutions and adaptation can be made. The extent to which unitary and federal nations have decentralised responsibility to sub-national tiers is shown to have an impact on the way in which regions and local authorities have engaged in adaptation. The role of both NGOs and the European Union is shown to be of significance to regional and local governments, as well as those countries who have not yet extensively engaged in adaptation at the national scale.

Keywords Adaptation · Climate change · Europe · National adaptation strategy · Planning

7.1 Introduction

As governmental and non-governmental actors alike increasingly recognise the value and importance of adaptation as a response to changing climatic conditions, adaptation policies and actions have begun to emerge across the European continent.

L.Westerhoff (✉)
Department of Social and Economic Geography, Umeå University, Umeå 90187, Sweden
e-mail: lisa.westerhoff@gmail.com

E.C.H. Keskitalo (ed.), *Developing Adaptation Policy and Practice in Europe: Multi-level Governance of Climate Change*, DOI 10.1007/978-90-481-9325-7_7,

Governments at national, regional and local scales, as well as non-governmental organisations and private businesses, have demonstrated a burgeoning interest in designing measures that ensure the future viability and safety of populations and economies. A corresponding increase in the number of surveys of adaptation actions and policies across Europe is also occurring as researchers endeavour to assess how and when different countries have embarked upon adaptation measures, and to identify differences between such measures, which range from individual, regional or sectoral plans that indirectly address climate impacts, to collaborative national adaptation plans (cf. Hulme et al., 2009; Massey & Bergsma, 2008; Swart et al., 2009).

The aim of this chapter is to provide a broader context for the deeper analysis provided in the four country cases that form the major contribution of this volume (see Chapter 1). Following a brief discussion of the various forms and stages of adaptation policy that have emerged in European countries, the chapter provides a selection of case studies that reflect different forms of decision-making structures and traditions, including Germany, Austria, Norway, the Netherlands, France, Spain, Greece and Hungary. In order to understand the ways in which structural factors and institutional arrangements have influenced the design of adaptation measures, the focus here is on those countries that have begun to embark on adaptation at the national scales (with the exception of Greece), and that represent a diversity of political, planning and environmental policy traditions. The chapter further provides a review of adaptation governance in Canada and Australia in order to provide some comparison with non-European countries. Final sections discuss the role of national environmental policy implementation and diverging steering abilities in federal and unitary states, decentralisation and the participation of sub-national authorities, and the role of NGOs and the EU.

7.2 Approach and Methods

In order to provide a basis for comparison with the more detailed chapters of this book, the framework for analysis of adaptation in this chapter draws on the concepts of state political and administrative systems (including planning systems) and environmental policy traditions described in the volume's introduction (see Chapter 1). Each selected country is described and assessed in terms of four principal themes: (1) the impact of a country's structure as either a federal or unitary state; (2) the effect of varying extents of decentralisation of planning and/or resource management competence on the emergence and nature of planned adaptation at various levels; (3) the relationship between environmental policy institutions and the development of planned adaptation and; (4) the role of NGOs, the EU and other stakeholders in existing processes of planned adaptation.

The material presented here draws on the results of recent programmes and studies of adaptation, as well as policy documents, National Communications to the United Nations Framework Convention on Climate Change (UNFCCC) and national and regional legislation in the selected ten countries. A total of eight interviews were additionally conducted with representatives of coordinating bodies at the national-scale involved in processes of adaptation policy design or information dissemination in each of the European countries and Australia.[1] The data presented in the Hungarian section is drawn from a policy review and one additional interview with select representatives of scientific bodies involved in Hungarian adaptation processes. Interviews have in some cases been supplemented with additional personal communications.

The material presented in the Canadian section draws on research conducted between 2004 and 2006 that examined climate change responses in two coastal communities in the province of British Columbia (cf. Wolf, 2006; Wolf et al., 2009). The study relied on 103 interviews with 42 key actors professionally involved in climate change at the federal, provincial and local level; 44 members of the public in southern coastal BC; and an extensive literature review specific to Canada's climate change policy landscape and the actors involved. The state of adaptation policy was updated in 2009 with a further four interviews with key actors involved in policy development at the federal, provincial and municipal level, and a current literature review.

7.3 Background: Adaptation Actions Across Europe

Before turning to the individual country cases in this chapter, the next section first presents a brief background on the forms of planned adaptation that have emerged across Europe. As one of the most prominent forms of national adaptation response, National Adaptation Strategies are increasingly being developed by a range of countries throughout Europe and have already been adopted by several northern and western European states (cf. Swart et al., 2009). Finland was the first to adopt a National Adaptation Strategy in 2005, followed by Spain (2006), France (2007), the Netherlands (2007), Denmark (2008), Hungary (2008), Romania (2008) and Germany (2008). The bulk of these strategies are founded on the basis of comprehensive impacts and vulnerability assessments conducted and funded at the national scale that identify projected impacts and vulnerabilities in relevant sectors. France's Climate change impacts and management programme serves as an example of such a national programme, the results of which served as the partial basis for the development of the French National Adaptation Strategy released in 2007 (see Section 7.4.2).

The strategies themselves follow broadly consistent structures, summarising expected impacts and highlighting necessary adaptation measures within different sectors and/or regions. Participatory mechanisms were used in the formation of

[1] Interviews were conducted by telephone with representatives from Austria, France, Germany, Spain, Norway, the Netherlands and Australia, and transcribed in full (for further information on methodology, see Chapter 1). Though solicitations for interviews were also sent to representatives from Greece, these did not receive a timely response.

strategies in all cases (albeit to varying degrees) to include input from stakeholders from different ministries, public and private institutions, NGOs, and/or regional and local governments. However, while earlier strategies have utilised a sectoral approach to adaptation measures (e.g., Finland, Spain), those developed in 2008 or later tend to be more comprehensive and integrated in their treatment of adaptation, highlighting cross-cutting measures across multiple sectors (ESPACE, 2008). Further, while some include specific recommendations for adaptation responsibilities for actors on different scales of governance (e.g., the Netherlands, Norway), others are more general in their recommendations for possible adaptation measures in each sector.

Notwithstanding these differences, NAS generally serve as guiding documents to be applied within all national (with implications for regional and local) sectors, particularly with regards to the environment and spatial planning. In Finland, the implementation of the strategy is currently underway, monitored through an evaluation of progress carried out at the national scale (see Chapter 4). The design of plans to follow up and implement the general recommendations presented in the national adaptation strategies is also underway in countries such as France, where a National Adaptation Plan is scheduled to be finalised by 2011.

A selection of other countries are in the process of developing adaptation strategies and policies, with a formal strategy expected by the end of 2009 or the beginning of 2010 in countries including Latvia, Portugal and Austria. In some cases, groundwork has been already completed and the design of a strategy is in its final stages. For example, a first report outlining adaptation needs and suggestions for adaptation in Latvia was prepared by the Ministry of Environment and accepted by the Latvian government in August of 2008, and constitutes the basis for the future adaptation strategy. The report outlines vulnerabilities and identifies current adaptation measures under use, as well as recommendations for the strategy. In Portugal, the development of a National Adaptation Strategy is also underway and is based on a comprehensive two-phase programme entitled *Climate Change in Portugal: Scenarios, Impacts and Adaptation Measures*. The development of a 'reference document' to serve as the basis for the strategy was completed in 2009. Other countries have elected to carry out a slightly different process than the design of a national adaptation strategy, such as the UK and Sweden, who have instead engaged in adaptation through legislative frameworks that assign tasks to various institutions and scales of administrations. While these are perhaps not adaptation strategies by definition, they perform many of the same functions in terms of multi-ministerial action and guiding national and sub-national actions (Swart et al., 2009). In the case of the UK, adaptation measures have gone beyond general recommendations to include specific requirements of different scales of government (see Chapter 3).

Still other countries are only in the beginning stages of the development of adaptation policies, including the Czech Republic where the inclusion of an adaptation component into the existing climate strategy was planned in 2009. Other countries that are projected to have adaptation strategies within the next few years are Estonia (planned for 2009) and Belgium (2012) (EEA, 2009). Several countries have yet to address adaptation in a comprehensive manner, including the southern European countries of Italy and Greece, as well as several from Central and Eastern Europe, including Poland, Slovenia and Slovakia. In several of these countries, activities outlined in the National Communications to the UNFCCC as adaptations tend to constitute an assortment of management plans or programmes in different sectors such as agriculture, water and forestry. In several instances, actions of compliance to EU Directives (specifically those relevant to the management of water, birds and habitat) are also noted as relevant to (or perhaps prioritised above) adaptation. This is also noted by Massey's (2009) report on adaptation in Central and Eastern European countries, which identifies EU Directives as an important driver of adaptation in those countries in the beginning stages of adaptation policy formation. In other cases, multilateral agreements such as the United National Convention to Combat Desertification are also highlighted as potential adaptation measures, as in the case of Greece and Italy. The absence of an established interministerial group on adaptation is also notable in these instances; climate impact and vulnerability research in several of these countries has been sectoral and without the guidance of a national climate change research programme.

7.4 Adaptation Across Scales: Examples from the European Context

This chapter now turns to a more detailed treatment of planned adaptations as they have emerged in the eight European countries outlined in the introduction: Hungary, Germany, Norway, Austria, France, the Netherlands, Spain and Greece. As outlined in the review of central state structures provided in the introductory chapter (Chapter 1), European countries may be divided into categories of state characteristics according to national political structures and planning system, as well as their degree of engagement with environmental policy. In this section, a review of a selection of countries representing such differences is provided to illustrate how planned adaptation has emerged within a range of structural contexts. The review thus aims to give a broad overview of the ways in which adaptation at different scales of governance is shaped by national political and planning systems. As such, this chapter focuses less on the differences in the extent to which countries have engaged with adaptation, and more on the forms that planned adaptation has taken. Wherever possible, the different resources that have enabled adaptation (i.e. adaptive capacities) at sub-national scales of administration will also be assessed.

Countries selected for inclusion in this chapter and their characteristics according to the criteria described in the introduction of this volume are presented in Table 7.1. The selection includes Germany, a federal state of a Germanic planning tradition often considered among the leaders in environmental policy implementation, and with strong regional decentralisation in contrast with Austria, a country of the same structure and record of engagement with environmental policy with slightly fewer powers delegated to the regional level (Andersen, 1999). France, Greece, Spain

Table 7.1 Country characteristics and state of adaptation policy

Country	State system	Planning tradition	Activity
Austria	Federal; somewhat centralised	Germanic	• Policy paper released (2008) • NAS in progress (due 2010–2011)
France	Unitary; centralised	Napoleonic	• National Adaptation Strategy adopted in 2007 • National Adaptation Plan anticipated in 2011
Germany	Federal; decentralised	Germanic	• NAS adopted in 2008 • National Adaptation Plan anticipated in 2011
Greece	Unitary; centralised	Napoleonic	• No adaptation strategy • Assorted existing frameworks and policies outlined in 2006 National Communication
Hungary	Unitary; somewhat decentralised	Eastern European	• National Adaptation Strategy adopted in 2008 • Some facets under implementation
Netherlands	Unitary; decentralised	Napoleonic	• National Strategy on Climate Adaptation and Spatial Planning adopted 2007 • National Adaptation Agenda under development • Space for Rivers Policy, National Water Plan (to include recommendations of Delta Committee)
Norway	Unitary; decentralised	Nordic	• Climate change adaptation report issued in 2008 • Ongoing implementation of recommendations
Spain	Unitary; decentralised	Napoleonic	• National Plan for Climate Change Adaptation adopted in 2006 • Second Work Programme for implementation of the Plan underway

and the Netherlands are also reviewed, all of which represent Napoleonic planning traditions and unitary states with varying degrees of centralisation. While France and Greece are centralised and generally considered followers in terms of environmental policy implementation, Spain and the Netherlands decentralise considerable authority to the regional level. Spain may additionally be considered ahead of its southern counterparts with regards to environmental policy (cf. Börzel, 2003) and is similarly a leader among southern European states in terms of adaptation, while countries such as Italy and Greece have yet to address adaptation in a meaningful or integrated manner. Finally, Hungary and Norway represent examples of typical

Eastern European and Nordic planning families respectively, each having engaged with adaptation on the national scale despite very different track records with respect to environmental policy implementation (see Chapter 1).

In the next sections, each country is described in terms of its impacts and adaptation activities at the national scale, as well as through select examples of adaptation actions at sub-national levels. Each country's work on adaptation is presented in terms of the timing and form of adaptation measures; the events or phenomena that spurred adaptation activities; the actors involved and their respective responsibilities; the involvement of NGOs and the EU; and the independent activities pursued by select regional and local actors.

7.4.1 Austria

Austria is a federal democratic republic of nine states called Länder, or *Bundesländer* (to differentiate from German Länder). Legislative and executive competences are shared between the Bundesländer and the federal state according to the Federal Constitution Act. Though Austrian Länder have been considered less powerful than German Länder, Austria's neo-corporatist style is characterised by consensus-seeking decision-making approaches to governance (Andersen, 1999). Bundesländer are thus included in state processes, including those relevant to climate change. With respect to mitigation, the Austrian Fourth National Communication to the UNFCCC provides examples of issues under the competence of the Bundesländer, including residential construction, road construction and public transportation, regional planning, industry and mining. According to the Communication:

> [. . .] Article 15a of the Federal Constitution Act leaves open the possibility to come to agreements among the Länder or between Länder and the Federation in order to harmonise policies under the respective legal areas of jurisdiction. No party can be forced to enter into an agreement. In some important climate change-related policies, jurisdiction is distributed among the Federation and the Länder, e.g. energy policy, waste management and agriculture. (BMLFUW, 2006, p. 44)

Climate vulnerabilities in Austria are the result of expected changes in climate, including an overall increase in temperature, increased winter precipitation and reduced summer precipitation (BMLFUW, 2006).

7.4.1.1 The Development of a National Adaptation Strategy

Until recently, climate change research in Austria has principally been sector-based, with a focus on climate impacts and vulnerabilities by various research institutes (CIRCLE, 2008). Among the major research programmes relevant to adaptation is *StartClim*, one of the most comprehensive programmes funded by various federal ministries and the Austrian Environment Agency. Allocated an average of €100–200,000 per year, the focus of the *StartClim* 'Climate Impacts and Adaptation' programme initiated in late 2009 was to contribute to the development of an adaptation strategy. Other research is additionally conducted by the Austrian Environment Agency, the Federal Ministry of Science and Research, the Klima and Energiefonds Programme and the Austrian Academy of Science's Global Change Programme (Environment Agency Austria, 2009).

To date, an Austrian National Adaptation Strategy has yet to be created and as a result, measures for adaptation have so far been sectoral in nature. The 2006 Austrian Fourth National Communication outlines various adaptation measures, differentiating between existing measures used to mitigate environmental risks and those motivated by climate change impacts. The Communication further outlines possible adaptation measures in the areas of agriculture, forestry, tourism and natural hazards (avalanches, erosion and floods). Preparatory actions for a National Adaptation Strategy began in 2007, spurred by the release of the EU Green Paper on adaptation and the identified need for a coordinated strategy by the Austrian research community (Environment Agency Austria, 2009). According to a representative of the Federal Ministry of Agriculture, Forestry, Environment and Water Management (BMLFUW), the European Commission's work helped to remove the perception of adaptation as a relinquishing of responsibility for mitigation:

> The ministry first put a focus on mitigation. Adaptation was not discussed intensively in order to prevent criticism ("getting fit on climate change by just adapting to it, but leaving mitigation activities behind"). The development on the EU level was a kind of motivation to broaden the agenda and establishing adaptation as second pillar of climate change policy. (BMLFUW, personal communication)

The process has since been led by the BMLFUW through three preliminary workshops on the need for an adaptation strategy. The Climate Change and Air Quality Management division of BMLFUW in particular has taken a coordinating role in the strategy's development, supported by the Austrian Environment Agency and complemented by work in the departments of agriculture, water management, forestry and natural hazards (BMLFUW, interview). Two committees provide additional support to the process: the Inter-ministerial Committee to Coordinate Measures to Protect Global Climate (IMC Climate Change) and the Kyoto Forum, which coordinates climate change mitigation (and since 2007, adaptation) policies between national, regional and local scales. The latter Kyoto Forum is a cooperation between the BMLFUW and representatives from the nine Bundesländer, and has been particularly engaged and supportive of the development of an NAS (BMLFUW, interview).

In 2008, a survey of current adaptation actions in research and practice through an email survey of 600 experts was conducted to construct a database of adaptation-relevant activities. Among those surveyed, the most prominent fields of action that emerged included flood protection (particularly through the Federal Ministry of Agriculture, Forestry, Environment and Water Management's 2003–2008 FloodRisk programme, allocated €4 million in funding), agriculture and forest management (BMLFUW, personal communication). Other activities included a 2008 vulnerability assessment to determine current and future vulnerability in Austria, followed by a second assessment to identify possible adaptation measures that led to the drafting of a policy paper in 2009 (Environment Agency Austria, 2009). The draft, entitled *Towards a national adaptation strategy*, included a general outline of the objectives of the NAS, current status of adaptation initiatives, impacts and vulnerability, suggestions for adaptation and next steps (Environment Agency Austria, 2009). Practical recommendations in the policy paper included measures for agriculture, planning, forestry, electricity production, tourism and water management, such as the establishment of water saving irrigation systems and drought-tolerant crops in regional management, considerations for future requirements for air conditioning given increasing temperatures, forest and soil conservation and regeneration, reductions in water consumption, and flood protection measures. Measures that ensure protection against natural hazards in areas of construction, health, spatial planning and infrastructure were also included (BMLFUW, 2009).

This draft has recently undergone a review process through which various federal, regional, and non-governmental institutions were able to amend comments, and is now undergoing further revision through a participatory process. The process is carried out by the Austrian Environment Agency in collaboration with the BMLFUW and the Bundesländer representatives. The process aims to include various private, environmental NGO and public stakeholders in order to raise awareness, discuss potential conflicts, clarify adaptation measures and increase acceptance of the strategy and is to be completed by September 2010. Following further assessments of vulnerability and adaptation measures in the biodiversity, health, building and construction, natural hazards and transportation sectors scheduled for completion in 2010, the strategy is to be finalised in 2011 (Environment Agency Austria, 2009). The finalised strategy is to provide a guiding document 'to prevent maladaptation' and to support adaptation through the provision of information (BLMFUW, interview).

7.4.1.2 Regional and Local Actions in Adaptation

The decision to adopt a National Adaptation Strategy in Austria was in part the result of ongoing collaboration between the Bundesländer and the BLMFUW and the decision that 'it wouldn't make any sense if all nine provinces had their own deliberations and tried to handle this problem [on their own] and separately from one another' (BLMFUW, interview). Bundesländer are strongly involved in the process of the strategy's development, and particularly within the Kyoto Forum (BMLFUW, interview). Specific funding arrangements and responsibilities for different relevant authorities have yet to be decided, and will be addressed through the ongoing participatory process over the course of 2010 (BMLFUW, interview). In the meantime, individual Bundesländer have begun to act on adaptation, albeit in relatively limited degrees. According to Austria's Third National Communication, the majority of Austria's Bundesländer 'ha[d] already adopted their own regional climate change programmes, taking into account specific regional circumstances, needs and areas of competence' (BMLFUW, 2001, p. 44).

While these programmes first focused on mitigation actions alone, the development of mitigation plans has expanded in all Bundesländer to include some provision of information on adaptation (BMLFUW, personal communication). Various project networks have additionally enabled the engagement of certain provincial administrative bodies in adaptation work. The Regional Offices for water management in the provinces of Carinthia, Tyrol and Lower Austria all participated in the ClimChAlp project, a programme partly funded by INTERREG III B on climate impacts and adaptation strategies in Alpine regions of the EU in Austria, Germany, Italy, France, Lichtenstein, Slovenia and Switzerland. Further, Upper Austria's engagement in the Climate Alliance's AMICA project has also spurred attention to incorporating both mitigation and adaptation policies into regional planning (see Chapter 6, for more details on the AMICA project). Carinthia and Vienna have additionally been involved in adaptation through ongoing work in the fields of water management as well as forestry (European Environment Agency, 2009).

At the municipal level, authorities have been similarly engaged in mitigation efforts, particularly the 700 municipalities that participate in the Climate Alliance Austria network. The Alliance is a part of the larger Climate Alliance of European Cities with Indigenous Rainforest Peoples, a non-governmental network that aims to reduce GHG emissions by 50% by 2030 and support sustainable forestry in participating towns, cities and provinces (Climate Alliance Austria, 2009). Currently, 819 Austrian municipalities as well as 488 businesses, 149 schools and all nine Bundesländer participate in the Alliance, though adaptation is not yet a significant component of the Alliance's activities. The Austrian Council on Climate Change (ACCC), an 11-member institute for the provision of information and recommendations on climate change research, additionally provides information to interested stakeholders at various levels on themes related principally to impacts and the need for adaptation through an online Climate Portal, but has not yet been involved in policy.

7.4.2 France

France is a centralist unitary state that retains legislative power at the national level and distributes funds to the 26 regions (22 on the European continent and the four islands of Guadeloupe, Martinique, French Guiana and Réunion) out of levied regional taxes. A complex sub-national administrative system further divides the regions into departments and communes, each representing various administrative duties of the central government and the three principal tiers of local administration. Administrative and budgetary tasks in France are divided between central and local governments, which designate local authorities as both administrative bodies of the national level and independent, autonomous bodies. Funding is transferred from the central government for allocated tasks, which may be complimented by local taxations. Regions are tasked with broad economic development and planning, while departments take care of issues such as health and social services, transportation and rural infrastructure. Smaller communes are responsible for local services and are encouraged to form cooperative associations through which resources may be pooled and tasks simplified.

Changes in climate projected for France include increases in overall temperature, and a particular risk of increased temperatures and reduced precipitation during the summer months (ONERC, 2007). Such changes are expected to have deleterious effects on water availability, agriculture, forestry, tourism, health and energy, and an increase in natural hazards (Interministerial Working Group on the Evaluation of Climate Change Impacts, 2009).

7.4.2.1 Climate Research and Policy Development

According to one actor, interest in adaptation in France began largely in reaction to growing public interest as a result of climate-related events:

> People got more concerned about climate with two events: two big storms in 1999 [and 2009], which destroyed a large part of the French forest. . .and then there was a bigger heat wave in 2003, which was quite a large event, and I think in France there was about 15,000 casualties. . .It was clear that after that, this type of heat wave would become more frequent with climate change. These are the two events that raised the interest of the population in adaptation. (Metéo-France, interview)

This interest in adaptation has also been fostered by strong political ambition to include adaptation on the political agenda, in particular by those active in government who perceived an urgent need to respond to growing climate impacts. A prominent example of such leadership is Senator Paul Vergès, who has promoted awareness as to the susceptibility of the French islands (and France as a whole) to climate change-risks and helped to focus attention on adaptation within French research institutions over the last decade. France has since been active in developing both individual policies (such as the National Heat Wave Strategy) as well as a comprehensive adaptation plan.

As the foundation for such activities, climate change research has been conducted by the GICC (*Gestion et Impacts des Changements du Climat*), the Ministry of Environment's centralised research programme. Created in 1999, the goal of the GICC is to 'promote and develop research on identifying National Impacts of Climate Change and associated physical mechanisms' (CIRCLE, 2008, p. 92). Within the GICC's programme, the intent is to provide a scientific basis from which to make adaptation decisions, steered by an interministerial committee of ministries and NGOs under the Ministry of Ecology and Sustainable Development (its funding agency) and supported by the French Scientific Council. Additional research has also been conducted by the National Research Agency under the supervision of the Ministry of Research. Both programmes receive between €3–5 million a year from their respective ministries.

Work on an official adaptation strategy began with the establishment of the French National Observatory of the Effects of Climate Warming (ONERC), a coordinating body created by Parliament in 2001 under the Ministry of Ecology and Sustainable Development for climate information collection and dissemination, the provision of recommendations for risk prevention and adaptation, and contribution to dialogue on climate change with developing countries (ONERC, 2009). In

2007, the French National Adaptation Strategy (*Stratégie nationale d'adaptation au changement climatique*) was submitted by ONERC to an interministerial committee on sustainable development and subsequently adopted by Parliament. The strategy was transformed into a formal government document following a brief consultation process through the provision of an online forum for the receipt of comments from interested bodies and individuals (though few comments were received) (Metéo-France, interview).

The strategy outlines four major areas of concern: (1) Public security and health; (2) Social aspects, including inequity in the face of risks; (3) Limiting the costs and taking advantage of changing conditions, and; (4) Protecting natural heritage (ONERC, 2007). The strategy is further divided into nine axes, or principal areas of concern, outlined in Box 7.1. Upon these, 22 recommendations are made, ranging from the need to provide improved information and linkages between science and policymakers and other actors, to the establishment of regional centres and research into possible funding mechanisms. Cross-cutting approaches to adaptation are also described within the areas of water, risks, health and biodiversity (in addition to the sectoral approaches are taken to cities, coastal and marine environments, mountains and forests), addressed by an additional 21 recommendations.

Box 7.1 The nine axes of the French National Adaptation strategy

Axis	Description
1. Develop knowledge on climate change impacts and adaptation	Addresses the need for research that characterises risks and benefits associated with climate change, defines vulnerabilities, improves knowledge on extreme events and informs local through international adaptation.
2. Strengthen observations	Calls for the structure and reinforcement of current observations through the update of observation systems and the development of indicators of change.
3. Inform, train and sensitise all relevant actors	Addresses the need to foster the provision and exchange of information between elected representatives, administrations, communities, public bodies, associations, citizens, consumers and the media are to
4. Promote an approach appropriate for the different territories	Addresses the need to facilitate the adoption of adaptation into departmental, regional and local plans and activities
5. Finance adaptation	Outlines the requirements for studies on possible methods of costing and financing adaptation measures, and methods of financially incentivising adaptation.

6. Utilise legislative and regulatory instruments	Calls for a review of existing laws and texts to ensure improved resource management under climate change, and the assurance that climate change is taken into account in sectoral legislation.
7. Support a voluntary approach and dialogue with private actors	Calls for the encouragement of innovation and initiative in private actors, and insuring the participation of the private sector in adaptation.
8. Take the specific circumstances of overseas territories into account	Calls for improved cooperation and exchange between the four overseas divisions and their surrounding countries, and improve research on their specific needs.
9. Contribute to international activities	Recommends increased participation in European and international research programmes.

While the Strategy provides a general basis for adaptation, specific actions or changes to legislation are not outlined. However, the absence of specific mention of adaptation in the first Bill issued by the *Grenelle de l'Environnement* (a multi-ministerial and stakeholder group for the formation of sustainable development strategies for France) in 2008 prompted a call for further measures. As a result, the implementation of the recommendations put forth by the French adaptation strategy will occur through a National Adaptation Plan in 2011, as determined by Article 42 of the first Grenelle Law passed in October 2009. The Plan is to include a broader and more extensive consultation process to include interested NGOs, private bodies, local authorities and the public (Metéo-France, interview). A list of concrete measures to be implemented on different scales of administration is currently under development by the five groups of stakeholders included under the Grenelle: elected members of government, administration, civil society, trade union members and scientists. The recommendations provided by each group will undergo further review by consultation with the public, the regions, and finally, the French Parliament. The first Grenelle Law additionally requires the inclusion of adaptation to climate change into the national priorities for international cooperation (Article 48), and recommends the adoption of such strategies within the overseas departments and territories (Article 56).

A report released in 2009 by the interministerial working group on the evaluation of climate change impacts headed by ONERC and the General Directorate for Energy and Climate on the expected costs of impacts and recommendations for adaptation measures will provide a preliminary basis for the Plan (Interministerial Working Group on the Evaluation of Climate Change Impacts, 2009). The report outlines proposals for adaptation strategies and measures in general and for nine sectors, including water resources, natural hazards and insurance, biodiversity, agriculture, forestry, health, energy, tourism, and transport and infrastructure. Cross-sectoral issues are also highlighted.

7.4.2.2 Engagement at Regional and Local Scales

The French National Adaptation Strategy notes the importance of a bottom-up approach to adaptation that involves local governments and encourages them to formulate their own adaptation policies, but recognises the need for consistency: 'there is a need for coordination at the national level, because if adaptation is left so that each local authority or each private stakeholder just does what is best for [itself], it can very easily go to maladaptation' (Metéo-France, interview). The establishment of regional centres is thus recommended in the strategy in order to facilitate the role of ONERC, improve regional information and provide a basis for the development of local adaptation plans. Such structures are also intended to ensure that adaptation is taken into consideration in all regional sectoral planning documents. The implementation of the National Adaptation Plan will therefore be passed down to regional and local scales through legislation and/or administrative duties, as is the norm for matters of national importance. However, regional and local decision-making processes are also to be central within the forthcoming National Adaptation Plan, based on the principal of subsidiarity.

In January 2009, the stage for individual regional adaptation measures was further set through the Grenelle's National Bill of Commitment to the Environment (the Second Grenelle Law), through which regional governments will be obliged to develop strategies for energy, air quality, mitigation and adaptation.[2] The goal of these regional climate strategies is to assist companies of more than 500 employees and communities of greater than 50,000 residents to adopt climate plans by one year from the law's projected adoption in the spring of 2010, as outlined in the French National Climate Plan. Under the National Climate Plan, large communities will thus be legally required to design and adopt local climate plans, including measures for adaptation, mitigation, energy savings and air quality. These plans are to be co-designed by local authorities and technical central government departments to avoid contradiction between national and local activities. While funding amounts to be allocated for the completion of such plans is not specified within the Bill, informational support and guidance is to be provided by national authorities without charge to the local level.

At the time of writing, a few regions had already begun to include adaptation into climate plans, including the regions of Aquitaine, Languedoc Roussillon, Nord-Pas de Calais, Réunion and Rhône-Alpes (ONERC, 2009). While several strategies tend to remain focused on mitigation measures, the Rhône-Alpes region in particular has begun to design a climate change strategy that emphasises adaptation elements, and is a lead partner in the INTERREG III B ClimChAlp project (one of three French partners including ONERC and the University of Claude Bernard de Lyon). The region's adaptation strategy includes a review of expected changes in emissions and climate scenarios and their implications at national and regional scales, information on expected impacts within each sector, possible resources for adaptation and

---

[2] As stated in Article 26 of the current draft of the Second Law currently under assessment.

suggestions for possible partners in adaptation activities, and possible sectoral adaptation strategies (Rhônalpénergie-Environnement, 2008). The Nord-Pas de Calais region is additionally recognised as one of the regions most advanced in terms of addressing the impacts of projected sea level rise.

The roughly one hundred local plans that have been voluntarily developed to date have so far concerned only emission reductions, though cities such as Paris have begun to include concerns of adaptation into their local climate plans. The first French city to launch a process for adopting a climate plan, the capital of Paris' current plan provides some discussion of adaptation measures, though these remain fairly vague (Ghorra-Gobin, n.d.). Once the Second Bill is adopted, however, these plans (and their inclusion of adaptation measures) will become mandatory.

7.4.3 Germany

Germany is a federal democratic republic that shares authority between the state level and 16 *Länder*, or federal states. Germany's brand of federalism has often been characterised as cooperative or administrative federalism, which emphasises cooperation across different levels of government (Peters & Pierre, 2005). At the same time it is characterised as a 'unitary federal state' (Abromeit, 1992; Benz, 1999), as a result of extensive taxation rights and a large majority of legislative competencies retained at the federal level (Burkhart, 2008). However, this is not necessarily a contradiction, in that the German system is organised along a functional division of power: while the great majority of legislative competencies lie at the federal level (despite an increase in legislative competencies transferred to the Länder through reform in 2006), the Länder have almost exclusive administrative competencies. Furthermore, the Bundesrat and Länder governments have direct participation and veto rights in national legislation (Burkhart, 2008). Environmental protection is a matter of concurrent legislation, giving Länder the right to legislate where the federal level has not, as well as deviate from federal regulations (cf. Peters & Pierre, 2005). Integrated planning policies therefore require intense cooperation between levels as well as sectoral departments (Reiter, 2008), which in turn may slow down implementation processes.

With regards to climate policy, the German Fourth National Communication to the UNFCCC outlines the policy activities of the Länder (generally associated with mitigation) in terms of three broad responsibilities: (1) implementing German and EU programmes and regulations; (2) organising state measures and projects using federal or EU funding, and (3) pursuing independent assistance programmes and regulations (Federal Republic of Germany, 2006). Local authorities are identified as important actors in climate protection activities (including adaptation), given their multiple roles as implementing authority of state and federal legislation, energy entrepreneur (where applicable), owner of public property and standard-setter for local community activities and actions.

Climate projections for Germany point to an increase in average annual temperatures, with particular increases in temperature and corresponding decreases in precipitation during the summer months that will lead to summer droughts and shortages (Schröter et al., 2005). Precipitation is expected to increase in the winter, resulting in flooding events. These combined changes will have considerable implications for several sectors, including water management, transport, health, forestry, agriculture and tourism (Schröter et al., 2005).

7.4.3.1 German Adaptation Policy: The 2008 National Adaptation Strategy

According to the Federal Ministry for the Environment, Nature Conservation and Nuclear Safety, a number of events and frameworks served as the impetus for work on adaptation in Germany:

> The starting point [was] the UNFCCC, where every member is called on to create some national measure or action plan to enhance adaptation. Then we have the German Climate Protection Programme, which also calls for a national context on adaptation, and thirdly, we have the federal Länder....and an established conference of the environmental ministers, which meet regularly, two times a year. In March 2007, they voiced a need for a national and agreed regional strategy for adaptation. (Federal Ministry for the Environment, Conservation and Nuclear Safety, interview)

Events such as the flooding of the Elbe River in 2002, the droughts of 2003 and a more general series of storm events additionally raised awareness of the additional risks and challenges posed by climate change (Federal Ministry for the Environment, Nature Conservation and Nuclear Safety, personal communication). The resultant *German Strategy for Adaptation to Climate Change* (*Deutsche Anpassungsstrategie an den Klimawandel,* or DAS) was prepared by an interministerial working group at the federal level under the lead of the Federal Ministry for the Environment, Nature Conservation and Nuclear Safety in cooperation with the Länder. The strategy was adopted by the federal Cabinet in December 2008, and represents a framework designed by the federal government in order to provide guidance and information for adaptation, and to set the basis for a mid- to long-term process of adaptation (Federal Ministry for the Environment, Nature Conservation, personal communication). The general aim of the strategy is to ensure strong dialogue on adaptation to climate change between federal, state and local government, the academic community, industry and civil society (Federal Republic of Germany, 2008a). The strategy process is supported by the Competence Centre on Climate Impacts and Adaptation (KomPass), established in 2006 by the Federal Environment Ministry to disseminate climate research and information on impacts, vulnerabilities and possible adaptation measures to political decision-makers, businesses and the public (KomPass, 2008).

More specifically, the strategy outlines measures to improve knowledge of climate change impacts within the various regions of Germany, to identify where action is needed, and to develop and implement adaptation measures following an integrated approach. Expected impacts in 13 principal sectors are described, including two cross-cutting areas of spatial planning and civil protection, as well as possible adaptation measures at both national and state levels. Proposed measures include the inclusion of a minimum flow rate for rivers and streams in the

Federal Water Act (to be revised for 2010), the development of a climate biomonitoring system between the Federal Environment Agency, the Federal Office for Nature Conservation and the German Weather Service, and the commission of sectoral research projects such as the *Impacts of Climate Change on Waterways and Shipping–Development of Adaptation Options* programme (KLIWAS) under the Federal Ministry of Transport, Building and Urban Affairs. The strategy also outlines existing national measures that will assist more explicit adaptation efforts, such as the German Federal Nature Conservation Act (March 2002), the National Strategy on Biological Diversity (2007), and the development of a National Strategy for the Sustainable Use and Protection of the Seas (under EU legislation).

In order to build on the framework and foundations set by the national adaptation strategy, the federal Ministry for the Environment, Conservation and Nuclear Safety is in the process of elaborating a National Adaptation Plan in cooperation with the Länder and other relevant institutions. Scheduled for completion by March 2011, the action plan will clarify certain aspects of the strategy, including:

1. Principles and criteria for prioritising adaptation action and measures;
2. Federal measures and measures to be undertaken in cooperation with the Länder;
3. Information on financing options, especially through the integration of adaptation in existing national and EU assistance programmes;
4. Information about financing, especially through integration of adaptation in existing assistance programmes;
5. Suggested concepts for progress review; and
6. Further development of the strategy (Federal Republic of Germany, 2008a)

The design of the Plan is to be completed through an interministerial working group with representatives from all ministries, headed by the Federal Ministry for the Environment, Conservation and Nuclear Safety. The implementation of these actions are to be supported by an ongoing Federal Government-Länder Dialogue on Adaptation to Climate Change, and increased stakeholder participation through the inclusion of the private sector, local authorities and other relevant actors. Currently, funding for participation in such activities is taken out of existing budgets.

7.4.3.2 Independent Länder Policy Development

As noted above, alongside UNFCCC requirements and burgeoning interest in adaptation at the federal level, environment ministries of the Länder collectively voiced their concern over the need for a national and agreed regional adaptation strategy (Federal Ministry for the Environment, Conservation and Nuclear Safety, interview). The resultant process of NAS development included a parallel process of two separate working groups at the federal and Länder level (mostly within environment ministries), between which the interests and concerns of the Länder were brought to the federal level:

It was a parallel process, and into this parallel process we tried to put as much transparency as possible. So it was clear that the German adaptation strategy in the first step would be a concept of the federal level, but we tried to bring in the ideas of the Länder as much as possible...they were involved throughout the whole process. (Federal Ministry for the Environment, Conservation and Nuclear Safety, interview)

As such, the website of the Federal Ministry for the Environment, Conservation and Nuclear Safety describes the strategy with respect to sub-national scales of government as 'intended to make it easier for the various levels of the Federation, Länder, local authorities and for individual citizens to identify impacts and adaptation needs, and to plan and implement measures' (Federal Ministry for the Environment Nature Conservation and Nuclear Safety, 2008). Länder are expected to contribute significantly to the Adaptation Plan under preparation (Federal Ministry for the Environment, Conservation and Nuclear Safety, interview).

However, while Länder have acknowledged the framework concepts and recommendations in the NAS, the actual strategy does not constitute any formal changes in legislation, obligations or funding for the regional scale. The NAS does recommend the need to ensure that regional policy maintains high-risk areas free from development in order to reduce flood damage, and for urban areas to be designed to prevent heat build-up. To assist the states in planning issues, a working group on Climate Change and Civil Protection was set up prior to the NAS in 2007.

The adjustment of existing legislation (e.g., in the German Federal Building Code or the Federal Spatial Planning Act)[3] in order to facilitate and to mainstream adaptation at the regional or local scale is under consideration through consultation with the Länder and other experts included in the elaboration of the Adaptation Action Plan (Federal Ministry for the Environment, Conservation and Nuclear Safety, personal communication). In the meantime, actions outlined in the NAS reflect the need for cooperation between levels, reflected in the stated need for a common approach to implementation of the EU Floods Directive and Water Framework Directive as existing instruments for adaptation. Other examples of cooperation include the development of targets and adaptation strategies for soil conservation with the cooperation of representatives from agriculture, forestry, water management and research bodies. Joint work is further outlined in the promotion of the conversion of forest monocultures to more resilient stands of mixed forests.

Several research programmes under the umbrella of the Federal Ministry of Education and Research's German Climate Research Programme also support regional activities. For example, the Klimazwei assistance programme serves (among several other activities) to develop further research networks to develop adaptation measures and provide regions with adequate information on climate impacts and scenarios through the established Service Group Adaptation. Another

[3]German states use the 2008 Federal Spatial Planning Act as a legally binding document to establish their own legislative structures and laws, though responsibility for implementation (and its enforcement) rests in the hands of the individual Länder.

integrated research programme is *KLIMZUG: Managing Climate Change in the Regions for the Future*, which aims to help regions plan for extreme weather events under climate change conditions (Federal Ministry of Education and Research, 2009b). Within this programme, seven pilot regions are to establish networks, share experiences, technologies and strategies and jointly take advantage of opportunities. The programme's funding provided by the Ministry of Education and Research spans 2008–2014 and totals roughly €80 million (Federal Ministry of Education and Research, 2009c).

Alongside the process of the NAS, Länder interest in adaptation has increased (Federal Ministry for the Environment, Conservation and Nuclear Safety, interview). Several Länder have either agreed to or have begun to develop their own integrated or sectoral adaptation strategies, or have included adaptation into climate plans and research programmes, including Bavaria, Brandenburg, Hamburg, North Rhine-Westphalia, Baden-Württemberg and Saxony (Ribeiro et al., 2009). As an example, the state of Bavaria has recently adopted the *Climate Programme Bavaria 2020*, led by the Bavarian Ministry of Environment and including a broad range of public, private and civic stakeholders (Bavarian Ministry for Environment and Health, 2009). Though a central aim of the programme is to limit greenhouse gas emissions and increase the use of renewable energy sources, the programme also includes adaptation measures for expected floods and drought in the areas of agriculture and forestry, sustainable human development, economy and tourism, nature conservation, health, soil and geohazards. The adaptation component of the programme was based on the Climate Adaptation Bavaria 2020 assessment of regional climate change impacts, conducted by the University of Bayreuth.

As another example, the Land of North Rhine-Westphalia has also prepared a supplement to its regional climate protection programme that aims to reduce the region's vulnerabilities to climate impacts and take advantage of new opportunities. The central goals of this adaptation strategy are to raise public awareness of climate impacts, develop research and knowledge on projected regional impacts, develop and implement adaptation measures; increase the region's overall adaptive capacity, and provide assistance to various sectors in implementing adaptations (Ministry of Environment and Conservation Agriculture and Consumer Protection of North Rhine-Westphalia, 2009). The federal adaptation strategy serves as an important basis and framework for many regional adaptation programmes, which in turn represent guiding documents for regional policy and sectoral activities.

7.4.3.3 The Role of Local Authorities

According to the German national adaptation strategy, '[t]he dialogue and participation processes set in motion during the preparation of the Adaptation Strategy, which have so far focused mainly on the federal and regional authorities and academic circles, are to be put on a broader footing by increasingly integrating industry, local authorities and other actors from the various fields of activities' (Federal Republic of Germany, 2008b, p. 4). As a result, future discussions and exchanges with local authorities are to occur through online fora and conferences in close cooperation with the Länder in order to determine the ideal nature of central government support and guidance or the implementation of adaptations measures at the local level. So far, little engagement with climate adaptation policy is yet notable within smaller and less financially or otherwise well-resources municipalities (Federal Ministry for the Environment, Conservation and Nuclear Safety, personal communication). Existing programmes for support for local authorities remain focused on mitigation, while funding arrangements for 'institutional or longer-lasting support' will be addressed in the upcoming Adaptation Plan (Federal Ministry for the Environment, Conservation and Nuclear Safety, interview). However, cities are able to request funding for pilot projects from the Environment Ministry, including for adaptation (Federal Ministry for the Environment, Conservation and Nuclear Safety, interview).

Regardless, select cities that have engaged in climate change research or policy networks have exhibited some activity. An example of such a city is Dresden, the capital city of the state of Saxony that straddles the Elbe River. Dresden was an active participant in the Climate Alliance's *Adaptation and Mitigation: an Integrated Climate Policy Approach* (AMICA) project (funded by INTERREG III C), and has since become involved as a test site for the Integrated Regional Climate Adaptation Programme (IRCAP) for the Model Region of Dresden (under the project title of REGKLAM). The transdisciplinary regional pilot, funded by the KLIMZUG programme, aims to support adaptation decision-making in the state of Saxony and specifically the region of Dresden in the identification and implementation of possible adaptation measures, as well as the development of a regional actor network on adaptation (Federal Ministry of Education and Research, 2009d).

Downstream the Elbe, the city-state of Hamburg has also been involved in similar research projects, including the Ministry of Education and Research's KLIMZUG-NORD – *Strategic approaches to climate change adaptation in the Hamburg Metropolitan Region*. The project has five central aims:

1. To develop technologies and methods concepts for the prevention and reduction of the impacts of climate change;
2. To create strategies and approaches that allow the integration of these measures into regional planning and development processes;
3. To demonstrate the importance of the cost, effectiveness and efficiency of such measures to the environment and economy;
4. To create a meaningful dialogue between decision-makers and the public to promote the use of such measures; and
5. To create a 'master plan' for the management of climate impacts to 2050 (Federal Ministry of Education and Research, 2009a).

Within the broad aim of the project is the goal to unite political, scientific, economic, administrative and other groups in the process of identifying priority areas and climate solutions. Up to €3 million is provided to each region under the KLIMZUG programme for a total of five years, to be complemented by regional funds covering at least 50% of the total cost (Federal Ministry of Education and Research, 2007).

7.4.4 Greece

In contrast to Italy's movement towards regional federalism, Greece is a highly centralist and unitary state (Featherstone, 2005). Thirteen Greek regions (or 'peripheries') make up the largest administrative division of government, further subdivided into 54 prefectures. Both these levels of government carry out administrative tasks appointed by central government, while individual municipalities may exercise independent administration of local matters, including a range of local social, financial and cultural affairs. While central government does not engage with municipal activities directly, supervision over activities and the allocation of their funding are both centred within the State.

The literature on Greek politics has often described Greece as an ill-coordinated, inefficient state with conflicting inter-party relations and weak civil society (Featherstone, 2005). As noted by Featherstone (2005), systematic weaknesses in relation to state institutional capacity, 'disjointed corporatism' and clientelism and rent-seeking practices have been central characteristics of Greek administrations, and have had implications for the design and implementation of environmental policies.

Climate change impacts in Greece are expected as the result of changes similar to those noted in other areas of southern Europe, where increases in temperature and reduced precipitation in summer months are projected to cause widespread drought and reduced water availability. According to a study of the European Islands System of Links and Exchanges (EURISLES) network, the Greek islands (compromising roughly 5% of the population and 4% of the national GDP) are considered 'particularly exposed to risk from sea level rise' (EURISLES, 2002, as cited in Kizos et al., 2009, p. 97).

7.4.4.1 Sectoral Policies and EU Incentives in Greek Adaptation

Coordinated research programmes on climate change impacts or adaptation have yet to be instituted in Greece. Two major institutions are instead responsible for climate-related research in Greece: the National Observatory of Athens (NOA) and the National Centre for Marine Research (NCMR). Additional climatological research is carried out by select universities and a research group at the Academy of Athens, as well as the National Foundation for Agricultural Research of the Ministry of Rural Development and Food. Programmes such as *Climate Change and Impact Research: the Mediterranean* (CIRCE) have begun to complement national research and have produced assessments of Mediterranean impacts and examples of adaptation from different Mediterranean countries.

Greece also has yet to embark on formal, comprehensive adaptation measures at the national scale. The current National Climate Change Programme, prepared by NOA for the Ministry for the Environment, Physical Planning, and Public

Works, addresses mitigation actions until 2010, but so far has not included adaptation. Instead, existing policy frameworks relevant to reducing vulnerability to climate change impacts are noted in the 2006 National Communication to the UNFCCC, including the Draft Programme for the Management of Water Resources in Greece completed in 2003 and the Observation Network for Water Quality. The National Action Plan for Combating Desertification (2001) and the related ministerial decision for its implementation are also highlighted as important, as well as the Operational Programme *Agricultural Development of the Countryside* (2000–2006), funded by the 3rd Community Support Framework. Finally, the Communication lists several ongoing actions to enhance and conserve water resources, biodiversity and forestry. In this context, the EU Framework Directive for community action in the field of water policy is noted as particularly important, as are the EU Directives on conservation of natural habitats and wild birds. Since 2003, Greece has also been involved in the Mediterranean European Union Water Initiative (EUWI), a collaboration between the Barcelona Process countries on 'good practices' and adaptation in the water sector (EUWI, 2009).

Some sectoral initiatives have also occurred, principally in the agricultural sector and in the form of research projects funded by the EU. First, ADAGIO (Adaptation of Agriculture in European Regions at Environmental Risk under Climate Change) is a project funded under the Sixth Framework Programme (2007–2009) and with Greek representation by NOA. The aim of the ADAGIO project is to evaluate potential and actual adaptation measures in agriculture for different climatic and agroecosystem regions under risk in Europe (ADAGIO, 2009). In order to continue the work under ADAGIO, a proposal has been submitted for a new collaborative scheme under the Seventh Framework Programme which will aim to introduce climate change adaptation measures into EU farms. Second, a limited number of adaptation measures are outlined within the *Programme for the Rural Development of Greece 2007–2013* created by the Ministry for Rural Development and Food and funded under the European Agricultural Fund for Rural Development, including water management measures and the introduction of new crops (Ministry of Rural Development and Food, 2007). Third, a Special Framework for Spatial Planning and Sustainable Development of Coastal Areas and Islands was recently introduced into public consultation in August 2009 (Ministry for the Urban Planning Environment and Public Works, 2009). The last in a series of spatial development frameworks, this most recent version is the first to make explicit reference to climate change adaptation measures for vulnerable areas, and outlines a number of pilot studies to determine appropriate courses of action. The pilots are to be jointly undertaken by the Ministry of Environment and the Greek regions, and funded by the *Environment and Sustainable Development Action Plan 2007–2013*.

While comprehensive planned adaptation has thus been absent so far, recent developments suggest that the new socialist government elected in October 2009 may reverse this trend. Following the election, a specialised Ministry for the Environment, Energy and Climate Change was created, indicating a higher prioritisation of climate change issues. At her first appearance before the Greek Parliament, the new Environment Minister further indicated that climate change was to receive a high priority, and that following the UNFCCC meeting in Copenhagen in December 2009, the Ministry would 'take the initiative to develop, in cooperation with other Mediterranean countries, the necessary adaptation policies' (Birbily, 2009, p. 109). However, in the hitherto absence of either comprehensive or sectoral adaptation policies, other sectors have taken up the cause. The most important among these has been NGO engagement, including the joint project between WWF Greece and the International Union for Conservation of Nature. In 2008, a workshop on Mediterranean Forest Conservation and Management was held with the goal of developing a programme on climate change adaptation and forest management for the Mediterranean, while an additional programme on Water Resource Planning and Climate Change Adaptation by the same coalition is also underway. At the beginning of 2009, the Bank of Greece additionally commissioned a panel of experts to report in two years' time on the envisaged cost of adaptation measures to climate change (Bank of Greece, 2009).

7.4.4.2 Limited Regional and Local Action

Little action on adaptation is apparent at the sub-national scale in Greece. In a survey of existing literature, only one example of adaptation activity at the municipal scale was found. The Municipality of Kalamaria is a partner in the *Green and Blue Space: adaptation for urban areas and eco towns* (GRaBS) project initiated by the Town and Country Planning Association (a UK-based NGO). Supported by EU INTERREG IVC funds, the project aims to foster communication and the exchange of best practices on adaptation strategies in local and regional governments. According to the project website, 'the Municipality aims to develop awareness and knowledge amongst local and regional planners, developers, urban designers and architects, about the important role and multi-functionality of green and blue space infrastructure in creating climate change resilient development' (GRaBS, 2009).

7.4.5 Hungary

Hungarian political structure is highly centralised and has increased in centralisation since 1992, despite steps taken toward decentralisation from national to local governments (Fowler, 2001). In particular, the introduction of NUTS-II level regions[4] in 1996 and 1999 has provided the central government the opportunity to further strengthen its grip over local and county governance. As the NUTS II regional administrative unit was created for EU-related regional development planning and decisions concerning the distribution and allocation of EU funding, this political structure provides significant opportunities for the central government to control

[4]Nomenclature of Territorial Units for Statistics. For European Regional Development Funding (ERDF), the principle administrative and planning unit is the NUTS II level.

regional policy, as national government typically controls the decision-making process in the Regional Development Councils (Ellison, 2008a).

Public policy in Hungary is thus very strongly driven by central government strategies. With regard to climate change adaptation, Massey (2009) has noted that the bulk of strategy and policy has been centrally-determined, as the government has simply allocated tasks to lower level administrations. At the same time, however, specific regions in Hungary have contributed strongly to the development of strategies suited to local environments and needs. Expected climate change impacts in Hungary include an increase in the number of extreme precipitation events, decreased overall precipitation, and increased temperatures, all with potential implications for water availability, forestry and agriculture (Ministry of Environment and Water, 2005).

7.4.5.1 The Hungarian National Adaptation Strategy

As an established leader among Central and East European states in democratic and market transition processes, the integration of environmental issues, and progress toward EU membership (cf. Ellison, 2006, 2008b), Hungary has likewise moved comparatively rapidly on adaptation. The first and to date the only Central and Eastern European country to do so,[5] Hungary issued its first National Adaptation Strategy as part of its *National Climate Change Strategy 2008–2025* (NCCS) in February 2008. The NCCS was unanimously adopted by the Hungarian parliament on March 17th, 2008. Hungary's adaptation strategy represents the culmination of several years of research on the potential impacts of climate change, and the VAHAVA (Weather and Climate: Changes-Impacts-Answers) project in particular. Headed by the Hungarian Academy of Sciences in collaboration with the Hungarian Ministry for Water and the Environment, VAHAVA was conducted over 2003–2006 and played a central role in defining the Hungarian position on adaptation, significantly impacting the development of the Hungarian Adaptation Strategy. Published in 2006, the results of the VAHAVA study contributed a substantial share of the initial groundwork for the completed adaptation strategy (Hungarian Academy of Sciences/Budapest Corvinus University, interview). However, one weakness of the VAHAVA study however is its failure to incorporate climate-based modelling scenarios, weakening. The theoretical foundations for the construction of adaptation strategies.

One impetus for the development of adaptation policy in Hungary began with the National Communications required by the UNFCCC that include an assessment of impacts, vulnerability and adaptation measures. While Hungary's Second National Communication (Ministry for Environment and Regional Policy, 1997) does not directly address adaptation, the Third and Fourth National Communications

[5]There is some controversy here, however. According to Czako and Mnatsakanian (2008), Romania published a general National Strategy on Climate Change in 2005 and has been working on individual action plans on climate change and adaptation. A draft action plan on adaptation was reportedly published in March 2008.

(Ministry for Environment and Regional Policy, 2002; 2005) each offer individual chapters that specifically address adaptation and specific adaptation problems. These early contributions to the eventual Hungarian adaptation strategy note both considerable uncertainty regarding climate impacts, as well as perceptions that the potential impact of climate change may not be that severe. For example, the Third National Communication to the UNFCCC (2002) states that 'Hungarian agriculture will not suffer a major impact' (p. 83); however, this same report notes several potential impacts of climate change, including increased frequency, severity and consecutive reoccurrence of droughts and an overall drying tendency.

The Third National Communication further notes that a National Drought Commission is to be formed to work on the development of a National Drought Mitigation Strategy. The recent Drought and Increasing Aridity in Hungary conference (2009) additionally led to the publication of a position statement encouraging the government, among other things, to 'adopt a national drought strategy, action program, enacting legislation as well as financial support mechanisms' (cf. Kecskemét Conference Statement, 2009). Following contributions to the Hungarian Adaptation Strategy – the Fourth National Communication sent to the UNFCCC in particular – suggest that potential climate impacts on Hungary could be much greater, particularly with regard to water availability, extreme events and their impacts on economic sectors such as agriculture.

The official 2008 Hungarian Adaptation Strategy builds and extends upon these observations of the potential impacts of climate change and creates a basic framework for action, including an outline of the basic areas where adaptation strategies could be developed and mainstreamed into the existing policy framework. The strategy outlines a broad range of areas where policy strategies need to be elaborated, and proposes the development of strategies in five major areas: the natural environment and environmental protection, human health, water management, agriculture/forestry and rural development. Moreover, the strategy proposes a number of specific guidelines for:

1. The development and formulation of policy strategies for biodiversity, nature and water protection strategies;
2. Extending the New Vásárhelyi Plan (discussed below);
3. Building upon the EU Afforestation strategy; extending the existing knowledge-base and observations of future climate change;
4. Harmonising horizontal and cross-sectoral strategies (particularly across the agricultural, energy and water management sectors); and
5. Improving the effective dissemination of public information (KvVM, 2008).

Notable among the elements introduced by the strategy is its attention to the concept of 'ecosystem services', the importance of nature protection and the addition of a discussion of the potential impact of climate change on human health and well-being. The report also makes a connection between increasing dry spells and declining water quality for the first time, and gives attention to the increasing incidence of extreme weather events.

However, the adaptation strategy is little more than a framework for action, and lacks enacting legislation (such as an Action Plan) with clearly defined targets and funding mechanisms. A draft action plan, the *National Climate Change Program 2009–2010*, has been created and contains a subsection on adaptation, but has not yet been formally approved. Though it is not clear why this is the case, upheavals in the government (such as declining parliamentary support for the former Prime Minister Gyurcsany Ferenc and the installation of a new Prime Minister, Bajnai Gordon), repeated disputes over the Hungarian budget and proposed reductions of the deficit, the economic crisis, and the approach of new elections (to be held in Spring 2010) have presumably all weighed heavily on the ability to move ahead with policy. Though the draft action plan for 2009–2010 contains more clearly defined strategies as well as reference to specific funding mechanisms, such details as specific amounts, the total area covered by specific projects, or the number of related projects so far remain undefined. Most of the funding is to come out of EU structural and cohesion and rural development funding (KvVM, 2008).

Meanwhile, implementation of the Hungarian adaptation strategy is ongoing. In addition to work on the Vásárhelyi Plan, the Hungarian government has also been working on the revision of other sectoral legislation. Current efforts are focused primarily on collecting relevant data and information in order to prepare for the next round of efforts from 2010 onwards (Hungarian Academy of Sciences/Budapest Corvinus University, interview). The NCCS requires the government to review its strategy every two years.

7.4.5.2 The Regional and Local Level

Early efforts to address adaptation-related issues have in part been driven by local climate-related phenomena. According to a WWF report (WWF n.d.), extreme flooding events in 1999 and 2000 led to high reparation costs and put increasing pressure on the Ministry of Environment and Water to prevent further events in the future. As a result, a team of experts was formed in 2000 to elaborate a flood prevention plan, submitting a draft framework for 32 emergency reservoirs in 2001. Though the initial government plan was met with considerable local resistance, ensuing negotiations between local and central government representatives and NGOs led to the abandonment of the government's centrally-defined plan and ultimately gave rise to what is now called the New Vásárhelyi Plan (VTT). The VTT is highlighted in the National Adaptation Strategy as an example of the success of regional planning. Moreover, the VTT demonstrates an interest in developing integrated approaches to adaptation by attempting to address not only flood protection along the Tisza River, but also multi-functional interests and uses such as nature conservation, environmental protection, rural development and ecotourism (Matczak et al., 2008).

As the 'poster child' of more locally-driven and integrated strategies, the VTT project has attracted considerable attention and is currently the subject of both national and international research interest (e.g., ADAM, 2009). Despite suggestions that adaptation strategies in Hungary have been primarily centrally-determined

(Massey, 2009), the VTT example suggests that other more regionally-based actors and interests can determine policy outcomes. At the same time, it is important to note that overall progress on the VTT has been quite slow. Although the final decision was to build only seven reservoirs, only one has been completed and two further reservoirs are under negotiation (Hungarian Academy of Sciences/Budapest Corvinus University, interview). Negotiations have been slowed by land and water use conflicts, unclear property rights, inadequate local resources, tensions across interest lobbies and the relatively weak organisation of interests at the local level. Thus, current implementation has tended to favour large infrastructure projects primarily targeting flood protection (Matczak et al., 2008). Current research efforts in Hungary also focus on involving local governments in the adaptation planning process. According to reports, the Hungarian National Association of Local Governments (TÖOSZ) is actively and enthusiastically engaged in discussions on the development of the local development of adaptation strategies in Hungary (Hungarian Academy of Sciences/Budapest Corvinus University, interview).

7.4.6 Netherlands

The Netherlands have been classified as a corporatist country that, together with Belgium, serves as an example of 'democratic corporatism' in which policy change is dependent on agreement between coalition parties and their social partners (Hemerijck & Visser, 2000). Its consideration as both a decentralised unitary state as well as a consensus state signals the importance of the Netherlands' balance between the central state and local autonomy, and the role of consensus-building (Newman & Thornley, 1996). As with other countries with neo-corporatist styles (e.g., Sweden and Austria), the Netherlands has been seen as relatively open to new ideas and interests, reaching high scores on environmental performance as a result of consensus-seeking decision-making approaches (Andersen, 1999). The development of environmental policies and the definition of appropriate targets and independent environmental assessments therefore have a long history (Langeweg, 1989; Planbureau voor de Leefomgeving, 2009).

Power in the Netherlands is shared between the national government, the 12 provinces that form the principal sub-national scale of government, the water boards and the municipalities. Provinces receive funding primarily from the central government and share a number of responsibilities with the municipalities for such areas as transportation, agriculture and economy, infrastructure and planning and conservation. Municipalities themselves execute policy formed at the national or provincial scale and are provided funds by the province as well as directly from the central government. Responsibilities for planning at the municipal scale are increasing.

The Netherlands' approach to climate change adaptation has focused on the implications of climate impacts, particularly sea level rise, on various dimensions of spatial planning. Higher winter temperatures and increases in precipitation, combined with projected increases in sea level are expected to significantly impact water management and planning sectors, with additional implications for various sectors of the economy (Swart et al., 2009).

7.4.6.1 Climate-Proofing Spatial Planning in the Netherlands

Discussions of the need for adaptation in the Netherlands began in response to growing awareness of climate impacts and the need for action:

> At the time in the Netherlands, as in most of Europe, it wasn't politically correct to talk about adaptation at all; the emphasis was completely on mitigation. There was also some fear that if you started talking about adaptation, that the pressure would be taken off mitigation. I think it was only in the beginning of the century basically, say, 2002, 2003, that there were a number of new publications. Also, for instance, that climate change impacts were already observed in Europe; that generally built up pressure. And I think in our country, in 2005, there were questions in parliament about the need for adaptation, and that basically triggered a lot of activities and more emphasis on adaptation. (Knowledge for Climate, interview)

In response, the Ministry of Housing, Spatial Planning and Environment (VROM) organised a national congress in 2006 to begin the development of a comprehensive national programme formed of various government ministries, and organisations representing provincial authorities, municipalities and water boards, respectively. Titled the *National Programme on Climate Adaptation and Spatial Planning* (ARK), the purpose of the programme was to compile the results from a broad range of research programmes to form the two documents that comprise the 2007 National Adaptation Strategy. The first and shorter document was agreed upon by all relevant ministries and the umbrella organisations of the provinces, municipalities and water boards, while the second comprises more detailed background information on climate impacts, vulnerabilities, leading principles and possible adaptation strategies.

The main objective of the programme is to 'climate-proof spatial planning in the Netherlands' over the next 100 years, making adaptation 'a mainstream policy component by 2015' (VROM, 2008, p. 5). The strategy takes a two-fold approach, focusing primarily on amassing knowledge of climate change impacts and vulnerabilities in order to eventually promote the use of spatial planning as the basis for improving risk management practices for preventing and minimising damage and improving the resilience and flexibility of existing infrastructure. At the time of this study, programme activities focused primary on knowledge development and assessments from which to form a basis for necessary legislative or other changes. Though focused primarily on mainstreaming adaptation into water and flood risk management activities (e.g., water assessments), the strategy adopts an integrated approach involving several sectors including nature conservation, agriculture and tourism. An Adaptation Agenda is in the process of being drafted to extend the activities of the strategy and to begin the 'institutionalisation of adaptation' into Dutch policy (VROM, interview). The Agenda is to address several facets of adaptation needs, including broader adaptation processes in various sectors (e.g., health), improved networks between science, policy and practice, the role and impact of EU policies, and the enhanced participation of NGOs and the private sector (VROM, interview).

The adaptation strategy also draws on several additional plans and programmes that concern various facets of water management and coastal protection, including the Flood Protection Strategy, the National Flood Protection Programme and the National Water Plan. The National Spatial Plan will provide a further basis from which to develop concrete adaptation actions. The strategy lays the foundation for the eventual mainstreaming of climate adaptation into all planning aspects, scheduled to commence in 2015. Adaptation actions under the National Adaptation Strategy are to be carried out at the national scale by relevant national bodies and ministries according to the issue area, including the Ministry of Transport, Public Works and Water Management, the Ministry of Housing, Spatial Planning and the Environment, and the Ministry of Agriculture, Nature and Food Quality.

In addition to the National Adaptation Strategy, the Ministry of Transport, Public Works and Water Management has developed adaptation measures relevant to water management safety and policy. A principle component of existing Dutch water policy has been to retain open spaces alongside the rivers undeveloped. This is particularly evident in the Space for the River policy, through which three emergency overflow areas have been suggested at the low settlement areas of Rijnstrangen, Ooijpolder and the eastern part of the Beerse Overlaat. In order to address the inadequacy of existing coastal defence strategies in the face of climate change and to address the perception that 'the current approach to coastal defense may no longer be viable in the future' (VROM, 2008, p. 27), a new visionary Delta Project was embarked upon to identify possible means of improving the resilience of coastal areas to expected impacts. Twelve recommendations issued by the Delta Committee in September 2008 outline necessary actions for the protection of the country's coastlines from flooding, fresh water shortages and subsidence over the next 100 years, divided into measures required before and after 2050 (MVW, 2009). The recommendations include:

1. An increase in flood protection levels of all dyked areas 'by a factor of 10';
2. Cost-benefit analyses for the development of low-lying areas;
3. The prevention of new development from impeding river discharge outside dyke areas;
4. The immediate implementation of the Room for the River programme, and;
5. Ongoing measures for beach nourishment, storm surge barriers, flood management and excess water storage along the Dutch rivers and lakes (Delta Commission, 2008).

Recommendations are to be incorporated into the Dutch National Water Plan and the Flood Protection Strategy. Preparations for the formulation of the new Delta Act by a multi-ministerial steering group coordinated by the Ministry of Transport, Public Works and Water Management are now underway. A formal commitment of €1 billion has been additionally set aside for the improvement in water safety measures (Knowledge for Climate, interview).

7.4.6.2 Regional and Local Actions and Responsibilities

Though the adaptation programme represents an effort by the national government (with agreements from lower tiers), actions and related water and flooding policies are intended to be implemented at regional and local scales through the mainstreaming of adaptation into various plans and regulations. The programme's focus on providing climate impact and adaptation knowledge and the mainstreaming approach together form the basis from which adaptation needs are to 'trickle-down' to regional and local governments (VROM, interview). With the Adaptation Agenda, adaptation will be incorporated into e.g. the assessments of water management and risk required in the development of provincial and local plans (VROM, interview). Thus, while explicit funds are not yet provided for adaptation at lower levels of government, extensive support in the form of information and eventual requirements through legislation and regulations are to guide adaptation at regional and local levels.

The regional level is particularly highlighted as an important level at which the transition to climate-proof planning is to be enabled and the changes made in the spatial network to be planned and executed (VROM, 2008). Provincial authorities are thus encouraged to 'take the lead' in spearheading changes to spatial planning and enabling the cooperation between water management actors and levels of government. The responsibilities of the provinces, water boards and municipalities as outlined by the NAS are shown in Box 7.2; however, the NAS is intended to provide a framework for sub-national action and does not legally require authorities to engage in adaptation actions.

Box 7.2 Sub-national responsibilities under the Dutch National Adaptation Strategy (VROM, 2008)

Regional

- Create the administrative foundations for the spatial reserves for water along rivers and the coastal zone (as set by the Space for the River policy).
- Invest in regional water systems and rural/urban areas according to updated climate scenarios.
- Produce 'climate atlases' to serve as the basis for climate-proofing district policy and future plans and projects.
- Incorporate climate change-related activities into area programmes, including the launch of 'hotspot projects' that serve to produce and disseminate knowledge on climate change.
- Work with municipalities and water boards to determine what and how adaptation measures should be taken using a number of pilot projects, providing a basis for developing timelines for municipal targets and funding needs.

Municipal

- In accordance with a four-year agreement with the national government, identify types of adaptation measures that combine spatial/urban planning, water management and health care at the municipal level.
- Take greater account of climate change in planning and local regulation, including zoning plans and building regulations, to be supported through the permit process.
- In select municipalities, serve as pilot areas for preliminary adaptation measures and knowledge development to provide a basis for the development of scenarios through which other municipalities may meet government targets. The pilots must also render the costs of measures transparent so as to clarify the funding arrangements that will be decided between the national government, municipal authorities and the water boards.
- Promote innovation in the climate-proof design of streets and districts.
- Promote information on energy saving in conjunction with prospects for adaptation
- Promote the consideration of issues of heat stress, flooding and energy consumption in areas where the framework of the Budget for Urban Renewal is being applied,∗ alongside plans for the incorporation of water and land into ecological infrastructure and recreational areas.
- Encourage all actors to develop innovative adaptations in housing and building sites.

Water boards

- Ensure regional water systems comply with the new National Administrative Agreement on Water.
- Cooperate with public, scientific and private actors to contribute to climate objectives and provide knowledge on climate-proofing
- Promote the consideration of issues of heat stress, flooding and energy consumption in areas where the framework of the Budget for Urban Renewal is being applied,* alongside plans for the incorporation of water and land into ecological infrastructure and recreational areas.
- Encourage all actors to develop innovative adaptations in housing and building sites.

*The Budget for Urban Renewal is a pool of financial support available to cities that have taken on a government-promoted integrated approach to urban restructuring under the new urban renewal policy instigated in 2000 (KEI, 2009).

Regardless, several provinces have begun to develop their own adaptation strategies. Several such activities have been supported by the *Climate changes Spatial Planning* (CcSP), a national programme designed to support knowledge infrastructure on 'the interface between climate change and spatial planning' and to 'engage a dialogue between stakeholders and scientists in order to support the development of spatially explicit adaptation and mitigation' (CcSP, 2009b). Running from 2004 to 2011, the programme focuses on identifying gaps in current knowledge of impacts and is to be complimented by the *Knowledge for Climate* programme, a €100 million research programme spanning 2008–2014 and designed to apply knowledge generated by the CcSP programme to spatial planning (Swart et al., 2009). Though a formal body for climate knowledge dissemination was to be created within the programme, a 'virtual' facility has taken its place, using expertise in knowledge transfer to host seminars, workshops and a website (Knowledge for Climate, interview).[6]

[6]The success of such informal communications were considered possible as a result of the small size of the country and the lack of competition between institutions; however, a formal institution

Through the programme, eight ‘hotspots’ representing particularly sensitive and ecologically or economically significant areas of the country are targeted for the development of climate knowledge and in select areas, of adaptation strategies. Selected hotspots include Schiphol airport and the regions of Haaglanden and Rotterdam, as well as geographic areas covering major river areas, the South-West Netherlands Delta, shallow water and peat meadow areas, dry rural areas and the Wadden Sea (KFC, 2009). The Province of Gronigen is in the process of developing climate-proofing and energy measures, and is to share the results of its work with the municipality of Tilburg in the province of Noord-Brabant (CcSP, 2009a). The provinces of Zuid-Holland, Noord-Holland and Zeeland are also working on their respective assessment frameworks in order to adapt spatial planning to climate change impacts, while work on the development of climate-proof and sustainable neighbourhoods in underway in the city of Utrecht (Cramer, 2007).

The 2007 Climate and Energy Agreement between the national and municipal governments further sets a basis from which to cooperatively develop sustainable energy production, emission reductions and climate-proofed planning. Under the agreement, the Stimulating Local Climate Initiatives project provides funding to municipalities for the engagement in mitigation projects, though no funds have as yet been earmarked for adaptation. National funding is also being provided for climate-related projects through the Innovation Agenda, the Sustainable Energy Production subsidy scheme and through the promotion of local climate initiatives (Cramer, 2007). An adaptation handbook/guidance framework is currently under development by VROM to assist local and regional governments in incorporating adaptation into planning processes; however, the use of this tool remains voluntary. Over the course of 2010, VROM will decide as to whether sections of the existing guidance will become a part of national spatial regulations and as a result be a requirement for implementation in regional and local planning processes.

7.4.7 Norway

Norway is a northern European country within the Nordic planning tradition that allocates responsibility from the state level through the 19 regional county bodies and down to the 430 municipalities. As in other Nordic countries, municipalities in Norway are allocated local autonomy under national guidance, where the majority of planning and decision-making activities occur at the municipal scale. However, counties in Norway are somewhat more independent than in countries such as Sweden, and act partly as extensions of the State (County Governors appointed by the Government) and partly as autonomous, elected governmental bodies. Historically, Norway has been considered a state with strong peripheral regions, though some authors have noted its potential progression towards a more centralised

may still be created, building on the successes and failures of institutions such as the UKCIP (Knowledge for Climate, interview).

state (rather than towards a more a regionalised system, cf. Baldersheim & Fimreite, 2005).

Increased precipitation overall and an additional increase in heavy precipitation events are projected for Norway, as well as an increase in average annual temperatures. These changes are expected to have implications for several sectors, including transportation, energy, land use and planning, water management, fisheries, forestry and agriculture, health and others (Ministry of the Environment, 2008).

7.4.7.1 Development of National Adaptation Initiatives

While mitigation may remain the emphasis of climate change action in Norway, planned adaptation has been under development over the last few years. Policy literature and personal communications from actors involved in the process suggest that this development may to some extent be attributed to significant climate-related events that increased political focus on vulnerabilities in social planning and local crisis management (Steen, 2003). For example, the National Vulnerability and Preparedness Report published 2007 emphasises that

> perhaps the largest challenge for society's security in the future will be to deal with the consequences of climate change ... the landslide in Hatlestad ... in Bergen 2005 and the extreme precipitation in Vågå and Lom ... summer 2006 are only two examples of such events. (DSB, 2007, preamble, author′s translation)

Earlier events, such as a extensive flooding in southeast Norway in 1995, may have also served as a wake up call with regard to the increased need for planning to take flood risk into account (DSB, interview).

Thus, Norway's current stated approach to climate change adaptation prioritises mitigation while aiming to reduce Norway's vulnerability, increase adaptability and take advantage of possible benefits (Ministry of the Environment, 2008). Formal climate change adaptation policy at the national scale is based on coordinated climate research and the transfer of adaptation-relevant information between scales. The first development of regional climate scenarios through the Regional Climate Development under Global Warming programme (RegClim) in 1997 has since led to comprehensive impact and vulnerability assessments, among which NorACIA constitutes a domestic follow-up of the international Arctic Climate Impact Assessment (cf. Ministry of the Environment, 2008). The current central climate change research programme is NORKLIMA: *Climate change and its impacts in Norway*, a 10 year programme spanning 2004–2013 with a total budget of €11 million funded by the Research Council of Norway (The Research Council of Norway, 2008).

Early steps towards the development of an adaptation approach in Norway included a governmental report on societal security and civil military cooperation in 2004. The same year, a pre-study conducted by the Centre for International Climate and Environmental Research in Oslo (CICERO) indicated different paths in Norway towards adaptation to climate change: as administration-, sector- or research-led developments (DSB, interview). In 2008, a governmental report on vulnerability and adaptation to climate change within different sectors was published, entitled *Climate Change Adaptation in Norway: A government initiative on climate change adaptation (Klimatilpasning i Norge: Regjeringens arbeid med tilpasning til klimaendringene)* (Ministry of the Environment, 2008). The report represents a joint effort between different ministries, local governments and private actors, led and coordinated by the Ministry of Environment (DSB, personal communication). The report outlines three principal goals under implementation (Ministry of the Environment, 2008).

The first of these goals focuses on identifying Norway's climate change vulnerability through the development of a state commission to investigate exposures and adaptive capacities, with the aim to incorporate climate change considerations into planning. The resultant Commission on Climate Adaptation is based at the Ministry of the Environment and expected to report in November 2010 (Adaptation Program for the Norwegian Energy Sector, personal communication). Work within the Commission is mainly undertaken by a working group of 17 experts, including one person from the insurance industry (DSB, interview). In order to integrate adaptation into planning, various departments have reported on their work within their areas of responsibility in relation to the national budget for 2009. The report also highlights the need for special focus to be placed on the incorporation of climate change considerations into regulation and other planning frameworks for larger investment projects (Ministry of the Environment, 2008). To integrate adaptation within regional and municipal levels, modifications to existing legislation have been undertaken. In 2009, an amendment to the Planning and Building Act mandated the inclusion of climate change risk and vulnerability analysis into local level planning and construction, with ongoing considerations as to the need for more concrete guidelines (Ministry of the Environment, 2008). The Ministry of the Environment report (2008) also suggests that future changes in the building code (*byggeforskriften*) will need to address new requirements posed by climate change, some of which have already been acknowledged in the *Klima 2000* building industry research programme (e.g., the need for geographically-differentiated design).

Secondly, the report outlines the need to obtain greater knowledge on climate change impacts and potential adaptations through comprehensive vulnerability mapping and research programmes. The result has been an increase in funding for climate research since 2008, in accordance with the suggestions of the national action plan for climate research (Norwegian Research Council, 2006). The report notes the need for an improved understanding of potential impacts through risk and vulnerability analyses and improved monitoring in several sectors. In some sectors, this is being addressed by relevant sectoral research programmes led by such departments as the Ministry of Transport and the Norwegian Water Resources and Energy Directorate. The report further recommends the initiation of monitoring activities to define particularly vulnerable areas to climate change (Ministry of the Environment, 2008).

Thirdly, the report calls for the promotion of coordination of information and activities between administrative bodies and sectors. This has been undertaken through the establishment of an interministerial adaptation coordination group led by the Ministry of the Environment in 2007, to coordinate adaptation across the departments for an initial period of five years. An executive Secretariat has additionally been appointed within the Directorate for Civil Protection and Emergency Planning (DSB) for the purposes of practical coordination and providing assistance to the interministerial group. The Secretariat is to annually assess the development of adaptation to climate change in Norway, and develop and maintain the Climate Adaptation Norway (*Klimatilpasning Norge*) internet portal (as recommended in the initial report). Launched in March 2009, the portal provides 'information on the effect and consequences of climate change, through specific advice and examples of climate change adaptation' (Ministry of the Environment, 2009), particularly to local authorities and other institutions at the local and regional scale. In addition, the executive secretariat is to cooperate with the Climate 21 (Klima 21) forum for strategic cooperation on climate and environmental research, as well as the Norwegian Research Council (and possibly within the NORKLIMA programme) (Ministry of the Environment, 2008). Funding is thus attributed to the development of the state commission report, and the development and ongoing coordination of the adaptation executive secretariat at the DSB.

Other departments have also been engaged in adaptation, particularly with regard to water issues. A report issued by the Norwegian Pollution Control Authority (2007) suggested the need to revise municipal planning activities and their regulation following an analysis of requirements for limiting water pollution. A study by the Directorate for Nature Management (DN report 2006-2) noted the potential need for regulative change to include adaptation with regard to protected areas and cultural landscapes, freshwater, marine areas, wildlife and nature recreation/tourism (cf. Ministry of the Environment, 2008).[7] An interviewee further noted that the implementation of EU Directives (which Norway has voluntarily agreed to implement despite not being a member of the EU) may also impact Norwegian legislation with relation to water (DSB, interview).

7.4.7.2 The Allocation of Adaptation Responsibilities

Under Norway's adaptation strategy, governance of climate change adaptation is considered the responsibility of all sectors, scales and individuals, including both public and private actors. While responsibility for the coordination of adaptation activities on the national level lies with the Ministry of the Environment, '[t]he responsibility for societal adaptation to climate change lies both with the public sector, the private sector and with individuals' (Ministry of the Environment, 2008, p. 13; author's translation). Adaptation activities are therefore to be designed and implemented according to sector and the existing administrative system:

> [a]dapting to climate change does not involve a new distribution of responsibility. The individual sector or administration level still has a separate responsibility to reduce consequences of climate change within its sector. This means the individual player must map its own climate vulnerability, plan to handle climate change and implement measures (Ministry of the Environment, 2009).

---

[7]Specific reports have also been developed with regards to adaptation concerns in foreign development aid (Ministry of Foreign Affairs, 2007).

Thus, national and local responsibilities in adaptation are divided according to existing authorities. Ministries coordinate national research and guide local responses through national policy, such as the Agriculture Agreement and the Planning and Building Act mentioned above. The Planning and Building Act provides the municipal level the responsibility for municipal planning, while the county is responsible for regional planning with special emphasis on regional and national interests as well as cross-municipal issues (DSB, 2007). As expressed by one interviewee, '[i]t is easier to work with 18 counties than with 431 municipalities, and the counties also know their municipalities better than we know all the municipalities' (DSB, interview).

Given these responsibilities and the geographic distribution of climate change impacts and vulnerabilities, the Offices of the County Governors have a central role in coordinating adaptation. State adaptation requirements and expectations are to be developed through a joint project set up by the DSB, the Norwegian Water Resources and Energy Directorate and the Norwegian Pollution Control Authority (Ministry of the Environment, 2008). County governors held seminars and other events on climate change adaptation for the county's municipalities over the course of 2008, and are encouraged to support the exchange of practices and experiences both within and between counties (DSB, interview). So far, industry has not been a target group for adaptation, despite the participation of a member of the insurance industry in the NOU commission. However, court cases between municipalities and the insurance industry in one southern municipality indicate that in the future, the insurance industry may review requirements for the insurance of basements in flood and landslide prone areas (DSB, interview).

7.4.7.3 Local Engagement with Adaptation

A DSB representative noted that in general, Norway has placed significant focus on the development of cross-level approaches and on drawing lessons from experiences in other countries. In late 2007-spring 2008, participants from the DSB adaptation secretariat, the DSB at large, and the Ministry of the Environment visited Danish, Swedish (Commission on Climate and Vulnerability) and UK (DEFRA and UKCIP) adaptation initiatives, noting among other things the value of the UKCIP network approach to adaptation (DSB, interview). In 2009, different Norwegian state representatives (as well as municipal representatives funded by the state to participate) took part in a private initiative for an adaptation-focused 'Norwegian pavilion' side-event at the UNFCCC, giving 'different municipalities the chance to take part in adaptation seminars' (DSB, interview). Local engagement with adaptation has also occurred through national and international adaptation research networks, including the NORADAPT research project at CICERO (CICERO, 2007).

The most significant local example of action on adaptation is the Future Cities *(Framtidens Byer)* project (2008–2014), a collaboration between the Norwegian State, The Norwegian Association of Local and Regional Authorities (KS) and the municipalities involved in the *storbyforum* cities cooperation, including some of the country's largest cities in Norway, such as Oslo, Bergen, Trondheim, Stavanger,

Fredikstad, Tromso and Drammen. Developed in early 2007 following a pilot project on environmentally-friendly urban development, the current aim of the project is to support mitigation and adaptation to climate change in these 13 municipalities, together representing nearly 50% of the Norwegian population. Within the project, the Norwegian State is represented by four ministries: the Ministry of Transport and Communications, the Ministry of Local Government and Regional Development, the Ministry of Petroleum and Energy and the Ministry of the Environment. The cooperation focuses on four themes: transport, waste and energy, spatial planning, and adaptation. The adaptation group is coordinated interministerially through the adaptation secretariat at the DSB (DSB, interview).

While the collaboration was initially framed mainly in terms of mitigation activities and increasing pollution reduction (Ministry of the Environment, 2007), the project has since been elaborated through a letter of intent between the State and municipalities issued in 2008 (and subsequently modified at the request of municipalities) to better take into account local planning perspectives (DSB, interview). Beyond the original seven municipalities, other cities in highly populated regions have been subsequently invited to apply to the initiative. The initiative is also considered to have supported municipal awareness of climate change issues:

> We see very clearly ... the change between the applications [for Future Cities] from one and a half years ago with regard to adaptation, and what the municipalities are doing now. There has been a rather quick increase in the awareness of adaptation. (DSB, interview)

During the spring of 2009, each municipality further developed action plans that built on vulnerability assessments conducted at the municipal scale. Participating municipalities are able to apply for funding to develop specific projects under the initiative, such as the coordination of an adaptation network (as undertaken by one municipality) (DSB, interview). However, some municipalities have indicated a further need for national guidelines on the integration of risks associated with sea level rise into planning and in surface water management, prompting the creation of a working group on sea level rise under the adaptation secretariat at DSB. The group will consider guidelines for adaptation in relation to sea level rise and requirements on national as well as EU levels (such as in the Floods Directive) (DSB, interview).

Of the municipalities participating in the Future Cities project, Bergen has been particularly engaged in adaptation work. One of Norway's 'rainiest cities', Bergen is a coastal city of approximately 250,000 residents in Hordaland County. Risks and vulnerability assessments for different climate-related phenomena have been conducted as a part of the municipal master land use plan and according to the guidelines set by the climate adaptation secretariat, and is currently supporting the development of Hordaland County's adaptation strategy in the areas of land use, transportation, agriculture and aquaculture, waste and economic development (BaltCICA, 2009). Much of this activity has come through Bergen's partnership in several projects, including the *Climate Change: Impacts, Costs and Adaptation in the Baltic Sea Region* (BaltCICA) project, in part funded by the EU's European Regional Development Fund and European Neighbourhood and Partnership Instrument.

7.4.8 Spain

Though Spain is characterised as a unitary state, strong regional powers among the 17 autonomous communities that make up the primary sub-national tier of government have essentially rendered Spain a country with strong regional governments (Osterud, 2005). As a result of issues of decentralisation and differing resource bases, significant differences persist between regions, known in Spain as Autonomous Communities. However, these constitute an important and powerful level of competence. As such, provinces constitute the administrative level of regional governments, while municipalities represent the principal planning authorities under the rule of the regions.

Spain has also been recognised as a among the leaders in environmental policy implementation among the southern European countries (although it has not in particular held a strong role with regard to EU environmental policy). With regard to climate change, regions have been heavily involved in the formation of national strategies for both mitigation and adaptation, while local authorities have been less engaged.

Principal changes in climate of relevance to Spain include an overall increase in temperature and decrease in precipitation, particularly during the summer months, leading to risks of drought and water scarcity that will impact sectors such as tourism, health, water management and agriculture (Oficina Española de Cambio Climático et al., 2006). Other risks from climate change include an increase in flooding events as a result of a concentration of precipitation in shorter timeframes (Oficina Española de Cambio Climático et al., 2006).

7.4.8.1 The National Climate Change Adaptation Plan (NCCAP)

Development of formal national adaptation policy in Spain was prompted by a general increase in knowledge of climate change impacts (OECC, interview), particularly through the release of the *Effects of Climate Change in Spain* (ECCE), a project led by the Ministry of Environment between 2003 and 2004. Under the project, 15 sectors[8] were assessed in terms of climate change impacts with the joint contributions of over 400 experts (both national and international), resulting in the publication of the *Principal Conclusions of the Preliminary Evaluation of the Impacts of Climate Change in Spain* in 2005 (ECCE, 2005). The work comprised the first integrated assessment of impacts in Spain, and provided a basis for both further investigation into sectoral vulnerability and for the recommendations for private and public sector adaptation measures taken up by the national adaptation strategy.

The National Climate Change Adaptation Plan (NCCAP, or *Plan Nacionál de Adaptación al Cambio Climático*) was submitted to the Council of Ministers in

[8]The 15 sectors are: biodiversity, fishing and marine ecosystems, water resources, transport, forests, human health, agriculture, industry and energy, coastal zones, tourism, hunting and fishing, business and insurance, mountain areas, urban planning and infrastructure, and soils.

October of 2006, representing a cooperative effort between several central and regional bodies. The Spanish Climate Change Office (OECC), a General Directorate of the Ministry of Environment and Rural and Marine Affairs, is primarily responsible for the coordination, management and follow-up of the NCCAP. Of the three sections within the Office, one is dedicated to adaptation and is given relatively free reign in the design of adaptation measures (OECC, interview). Beyond the Office, two additional bodies are central to the NCCAP process: first, the National Climate Council is an assembly of relevant national departments, the Autonomous Communities, the Spanish Federation of Municipalities and Provinces, and representatives from research institutions, social actors and nongovernmental organisations. The function of the Council is to draw proposals and recommendations for the areas of climate change science, impacts and adaptation strategies, in addition to the identification of strategies to limit GHG emissions. Secondly, the Coordination Commission of Climate Change Policies (CCPCC) represents a cooperative body between the national and regional administrations for matters related to climate change, ensuring that both emission reduction targets and adaptation actions are integrated across national and regional sectors. The Interministerial Group on Climate Change and the Environment Sectoral Conference both represent additional fora for political cooperation among the environment ministry and other sections of the state administration.

The overall aim of the NPACC is to 'mainstream adaptation to climate change in the planning processes of all the relevant sectors or systems' (Oficina Española de Cambio Climático, 2006, p. 7) through (1) the provision of information and guidance to national and regional institutions; (2) the collection of information on regional and sectoral impacts; (3) the determination of pressing research and development needs and; (4) the design of adaptation measures. The Plan highlights the most likely climate change impacts and vulnerabilities for the 15 chosen sectors, and identifies next steps in research and vulnerability mapping in order to begin to define adaptation options. The plan identifies preliminary steps towards adaptation for select sectors, outlined in Table 7.2.

Implementation of the plan is occurring through a series of work programmes that identify activities to be implemented and the timing of their implementation under the coordination of the OECC. The First Work Programme, approved in 2006, focused on the development of regional climate scenarios and the assessment of climate change impacts on water resources, biodiversity and coastal areas. Financial resources required for the implementation of activities within each work programme were to be provided by 'those agencies, institutions and associations with responsibility in the sectors and/or systems to be evaluated' (Oficina Española de Cambio Climático et al., 2006, p. 43, author's translation).

The Second Work Programme was subsequently adopted in July 2009 and is to follow the results of the First Programme over a course of four years. The Programme is organised along four main axes: (1) the continuation of sectoral assessments of impacts, vulnerability and adaptation initiated in the First Work Programme, extending the sectors tackled to agriculture, forestry, health and soils; (2) the integration of adaptation into national sectoral legislation; (3) the

Table 7.2 Preliminary adaptation options identified in the national climate change adaptation plan (Oficina Española de Cambio Climático, 2006)

Sector	Preliminary adaptation options
Water resources	• Develop guidelines and regulations to incorporate the foreseen impacts of climate change into the processes of Environmental Impact Assessment and Strategic Environmental Assessment of Plans and Programmes within the hydrological sector
Forests	• Draft guidelines for an adaptive forest management • Develop and apply forest growth models under different climate change scenarios
Agriculture	• Draft guidelines for the management of agricultural systems • Identify long term and minimal cost adaptation strategies for fruit trees, olive trees and vineyards • Evaluate of the need to reduce livestock farms' stocking rates, change the management of grazing systems, and other adaptations for the sector
Finance-insurance policies	• Review the legal framework for construction and design, land planning and land use • Develop specific models for the insurance sector • Promote the insurance market as a vulnerability reduction instrument • Analyse the economic viability of the agricultural policy under the different climate scenarios
Urban planning and construction	• Incorporate information on the new climate scenarios into technical planning • Promote bio-construction techniques, particularly for public buildings

mobilisation of key public and private stakeholders, and; (4) the establishment of a system of impacts and adaptation indicators in all national sectors (Oficina Española de Cambio Climático et al., 2009). With regard to the second axis, legislative changes are to be promoted so as to ensure a coordinated approach between the respective areas of competence of the central and regional authorities, and will also be promoted at the municipal scale (i.e. for municipalities to incorporate adaptation into relevant municipal bylaws within their areas of competence). However, at the national level, initial emphasis for legislative change will be in areas under the competence of the Ministry of Environment and Rural and Marine Affairs, and eventually extended to cover all relevant sectors addressed in the NPACC.

7.4.8.2 Regional and Local Adaptation Actions

Given the autonomy of Spanish regions, the NCCAP can not delineate specific responsibilities to the different levels of administration but instead provides a guiding framework for the implementation of the plan to regional and local bodies:

> We [the OECC] are going to develop some guidelines to achieve some kind of homogeneous level in the development of the regional adaptation strategies . . . We don't have the powers to intervene in the way [the regions] develop their strategies, so what we do is to try to reach consensus on what the strategies consist of. But we can't impose any content in this regard. (OECC, interview)

However, the process of formulating the NCCAP and its work programmes involved strong engagement with regional authorities and local bodies, facilitated by the Working Group on Impacts and Adaptation of the CCPCC. The improved coordination between the central and regional governments further forms an important aspect of the Second Work Programme, established to avoid duplication and guarantee complementarity and synergy between central and regional actions (Oficina Española de Cambio Climático et al., 2009). As such, knowledge of impacts and adaptation is to be transferred bi-directionally between central and regional governments to support the development of guidelines for a common baseline for regional adaptation strategies.

A number of regions have begun formulating adaptation plans of their own since the adoption of the NCCAP. For example, the regional government of Aragón's Climate Change and Clean Energy Strategy proposes objectives and potential courses of action for mitigation, adaptation and information dissemination in ten regional sectors (Gobierno de Aragón, 2008). The proposal for the strategy was developed through the creation of the region's Climate Change and Environmental Education Service, a body under the broader Department for Environmental Quality and Climate Change created in 2007 to promote and integrate mitigation and adaptation measures within regional sectors. Other institutions created to address regional climate change in the region of Aragón include the regional Interdepartmental Climate Change Commission, a body responsible for ensuring the coordination and implementation of mitigation and adaptation actions across different departments of the region. The region also created a Climate Change Office responsible for the management, monitoring and organisation of information relevant to climate change, including methods and results of climate impact and vulnerability research (Gobierno de Aragón, 2007).

Similar structures have been created in other regions, such as the Basque Climate Change Office, working under the direction of the Department of Environment and Spatial Planning to promote and coordinate climate change mitigation and adaptation policies in the Basque region. The Basque Plan to Combat Climate Change aims to both reduce greenhouse gas emissions and progressively adapt the regional population, natural resources and economic activities to the impacts of climate change (Gobierno Vasco, 2008). Other Autonomous Communities that have embarked on climate and/or adaptation strategies include the regions of Andalucia, Murcia, Valencia and Cantabria (cf. Ribeiro et al., 2009).

Beyond the eventual incorporation of adaptation into municipal legislation, little work at the municipal scale has yet occurred. In November 2009, a first specific evaluation of impacts and vulnerability to climate change in the main Spanish cities was produced by the Federation of Municipalities and Provinces (FEMP). Guidelines for local adaptation strategies have also been produced by the Spanish Network of Cities for Climate (a sub-section of FEMP) with assistance from the Ministry of Environment to provide support to local governments. The resultant *Local Climate Change Strategy* report aims to help local municipalities to develop integrated and coordinated mitigation and adaptation actions through the production of a series of plans, programmes and other documents that serve to achieve the main objectives of the plan (Red Española de Ciudades por el Clima, 2008). The 670-page document contains detailed outlines for the implementation of plans and measures, as well as a system of indicators for select sectors. While some cities such as Seville and Murcia have completed mitigation-focused strategies, the relative novelty of the guidelines has limited their use so far (OECC, interview).

Both FEMP and the Ministry of Environment have also encouraged engagement with climate change issues through the issuance of best practice awards for work on climate change, including both mitigation and adaptation practices and the commissioning of a report on the role of local governments in mitigation and adaptation in the future (FEMP, 2006; FEMP, 2008). The Spanish Network of Cities for Climate additionally provides information and support for municipalities and was a leading partner in the development of the local guidelines, but remains focused on mitigation.

7.5 Beyond Europe: Adaptation Actions in Canada and Australia

Though a large proportion of adaptation work in industrialised countries has occurred in European countries, adaptation plans and measures have been an important component in the context of other countries as well. In order to provide some contrast between the ways in which adaptation has emerged in and outside Europe, the following sections review adaptation actions in two additional countries: Australia and Canada. Australia has conducted extensive research into the sensitivity of several of its sectors to past and future climate-related events, and has initiated a comprehensive national adaptation programme that has fostered several sectoral plans. Conversely, Canada was one of the first countries to develop regional impact assessment but has yet to form a framework for national adaptation, leaving much of the action on adaptation to lower tiers of government. Both countries are federal states that may be used to typify multi-level governance approaches outside the EU context.

7.5.1 Australia

Australia is a federal state made up of six largely self-governing states and two territories. States may thus create and implement legislation in all areas not expressly controlled by the federal level by Sections 51 and 52 in the Australian Constitution, which include matters of national interest such as foreign relations, fisheries, communications and defense. While states are conferred decision-making authority through the Australian constitution, Australian territories instead receive authority through national legislation. Within these second tier governments, Local Government Areas form the local level of administration and are allocated powers according to each state. Local governments receive funding from both state and federal governments, and in some states are enjoying increasing devolution of authority from the state level.

Increases in temperature and changes in precipitation patterns are expected to create reductions in water availability in some parts of Australia and increases in others, with an overall increase in the number of extreme weather-related events such as droughts, floods and fires (Australian Government, 2005). These changes will have particular implications for agriculture, settlements and infrastructure and water management, in addition to challenges along the coastal zone associated with sea level rise.

7.5.1.1 National Sectoral Research and the Development of Adaptation Plans

Considerable Australian engagement in adaptation began in 2004 with the Australian Government's allocation of AUS $14.2 million for the first National Climate Change Adaptation Programme, developed to determine the extent of impacts on Australian sectors and populations and the information required by decision-makers to manage projected risks, and develop tools and products to support adaptation measures (Australian Government, 2005). There is anecdotal evidence that the perceived need for action on adaptation was spurred by recent water shortages and droughts in Southern Australia that elevated general awareness of climate impacts:

> There was a decade that is still continuing, of reduced runoff....periods of low rainfall, and elevated temperatures in most of the major cities in Australia... there were some serious bleaching events on the Great Barrier Reef. And these were all things that, even if they couldn't be directly attributed to climate change, were consistent with the sort of things that we were expecting from climate change, and I think helped to raise the awareness of the need to adapt to the impacts of climate change. (Adaptation Policy Team, interview)

Among the programme's major outputs is a country-wide vulnerability assessment produced by the existing environment ministry that has since served as the foundation for further action and analysis (Adaptation Policy Team, interview). In 2007, action on adaptation was further developed through the Council of Australian Governments' release (and subsequent government endorsement) of the *National Climate Change Adaptation Framework* as the basis for government action on adaptation (Australian Government, 2007). Federal funding allocated toward the implementation of the measures outlined in the framework totals AUS $126 million over the next five to seven years (Australian Government, 2007; Adaptation Policy Team, interview).

The Framework provides the basis for current adaptation actions at multiple levels in Australia, and aims to reduce risks posed by climate change and to take advantage of opportunities by providing support, guidance and information to decision-makers of various scales in order to reduce vulnerability in eight of Australia's key sectors and regions[9] (Australian Government, 2007).

[9]These are: (1) water resource; (2) coastal regions; (3) biodiversity; (4) agriculture, fisheries and forestry; (5) human health; (6) tourism; (7) settlements, infrastructure and planning, and (8) natural disaster management.

Within each regional or sectoral area outlined in the framework, potential areas of action are highlighted that include recommendations for vulnerability assessments and improvements in knowledge, as well as reviews of existing legislation, the incorporation of climate change adaptation considerations into existing plans and programmes, and the provision of guidance to various sectors. Responsibility for implementation of activities to achieve the framework's outlined goals rests with relevant authorities at federal and state levels according to existing management and cost-sharing arrangements (Adaptation Policy Team, interview).

A significant part of the Framework focuses on the need for climate information and a greater understanding of climate impacts and vulnerabilities among decision-makers. To fulfil this need, a comprehensive research program oriented towards the needs of decision-makers was established. Out of the total funding, a large percentage has gone toward establishing a number of principal research initiatives, including the development of two broad research facilities: (1) the Climate Change Adaptation Flagship (as a part of the existing Commonwealth Scientific and Research Organization, or CSIRO) and (2) the Climate Change Adaptation Research Facility (NCCARF). Created in 2008, the Research Facility supports the development of National Adaptation Research Plans (NARP) to address key vulnerabilities in the Australian territory and to provide decision-makers with information to manage climate change risks. The Facility's program focuses on eight priority areas: (1) primary industries; (2) terrestrial biodiversity; (3) water resources and freshwater biodiversity; (4) marine biodiversity and resources; (5) disaster management and emergency services; (6) settlements and infrastructure; (7) human health, and; (8) social, economic and institution dimensions (Australian Government, 2007).

Progress on the Framework has been variable; some sectors have initiated measures prior to the adoption of the National Climate Change Adaptation Framework, while others are still under development. In most areas, tools and legislation have not yet been finalised but a considerable amount of research has been commissioned. Table 7.3 outlines the extent of select measures in different Australian sectors to date, including the development of NARP and networks. Direct support for the implementation of the individual NARP is further delivered by the CSIRO Flagship, which assists in the development of sectoral National Adaptation Research Plans (Australian Government, 2009c). NARP development is also supported by the Department of Climate Change, established in 2007 for the purposes of spearheading the development and coordination of Australia's climate change policies. Beyond this support, the principal responsibilities of the Department include the provision of policy advice, implementation and program delivery in the areas of mitigation policy and domestic emission reductions, adaptation to climate impacts and participation in international climate change strategies.

Several major policy activities with regards to adaptation are also ongoing within several federal sectors, including the development of water, agriculture and coastal plans that aim to ensure the long-term viability of these important sectors. The *Water for the Future* framework is a long-term national engagement to secure water availability in the face of climate change impacts and rising demand that promotes the

Table 7.3 Progress on sectoral adaptation research and measures

Sector	Developments
Agriculture and forestry	The Agriculture and Climate Change Action Plan developed in 2006 for a three-year period (2006–2009) provides the overarching framework for climate change policy for Australian governments and the agricultural sector, including agricultural forestry.
Emergency management	An Action Plan is in the final stages of development and is pending release. In addition to its National Adaptation Research Plan for Emergency Management, an Adaptation Research Network has been established.
Human health	A National Adaptation Research Plan for Human Health has been finalised out of the Adaptation Research Network for Human Health and complemented by the development of a climate adaptation human health research network.
Marine biodiversity and resources	A National Climate Change and Fisheries Action Plan is under development by the Department of Agriculture, Fisheries and Forestry (DAFF), through the Marine and Coastal Committee of the Natural Resource Management Ministerial Council. A Marine Biodiversity and Resources Research Plan is pending ministerial approval.
Settlements and infrastructure	Policy has yet to be determined in relation to adaptation responses; current research concerns adapting building and planning codes, assessing risk and vulnerability, and costing the impacts of climate change. Various small-scale projects are also ongoing.
Terrestrial biodiversity	A National Biodiversity and Climate Change Action Plan released in 2004 for the period spanning 2004–2007 lays out adaptation strategies and actions relevant to the biodiversity sector. A second draft is currently under public consultation.
Water resources and freshwater biodiversity	Principal work is being done through the ongoing implementation of incremental water system adjustments through the Water for the Future programme, designed to reduce civic and agricultural reliance on rainfall. A National Adaptation Research Plan for Water Resources is under preparation for public consultation

need to address climate change and foster sustainable water use while securing water supplies and strengthening the health of Australia's rivers (Australian Government, 2009d). AUS $12.9 billion has been invested in the framework through various strategic programmes, water policy reforms and water management arrangements. *Australia's Farming Future* similarly provides a framework for the agricultural sector to manage the impacts of climate change through the provision of information and capacity building, financial support to farmers for the management of climate impacts, short-term income support and training for farmers in severe difficulty, and the establishment of networks (Department of Agriculture Fisheries and Forestry, 2009). Finally, the *Caring for Our Coasts* policy addresses challenges of climate change and coastal growth through the provision of AUS $25 million for a national coastal risk assessment, AUS $100 million for a Community Coast Care Programme, and an additional AUS $200 million for the Great Barrier Reef Rescue Plan (Department of Climate Change, 2009).

The Department of Climate Change is also in the process of developing a national adaptation policy 'to enhance Australia's national capacity to manage climate change impacts, promote the use of market-based instruments to facilitate adaptation, and deliver the information and tools to enable decision-makers to manage risks from climate change impacts' (Australian Government, 2009a, p. 9). The policy will build upon the established National Climate Change Adaptation Framework. Further, the Council of Australian Governments has indicated that an increased focus on climate impacts and adaptation will be taken within the Council's activities, beginning with the first official meeting in 2010. As a result, policy development is underway both at the level federal level, as well as at the level of the sates and territories.

The Department of Climate Change and the CSIRO Adaptation Flagship are jointly advised by a multi-stakeholder group including members with experience and expertise in agriculture, tourism, mineral and energy, infrastructure, finance, biodiversity and local government and planning (Adaptation Policy Team, interview). The stakeholders derive from governmental, private sector and NGO areas, and provide input on the development of current and future adaptation-relevant activities.

7.5.1.2 Australian State Engagement with Adaptation

According to the Framework, 'risks should be managed by those best equipped to understand the context and likely consequences of action' (Australian Government, 2007, p. 4); thus, responsibility for the implementation of adaptation measures is allocated to the appropriate level. As a federation, Australian state governments in particular are expected to engage with adaptation and have been able to do so independently from the federal government.

The State of Queensland has been particularly engaged with adaptation through the Queensland Office for Climate Change and the support of its Climate Change Centre of Excellence. The office released its *ClimateSmart Adaptation 2007–2012* action plan in 2007 with the contributions of several different state bodies and agencies (Queensland Government, 2007). The plan builds on the earlier Queensland Greenhouse Strategy 2004 and outlines principles, strategies and actions to enhance the state's resilience to climate change through building and sharing information, incorporating climate change in decision-making, reducing vulnerability and enhancing resilience. The plan outlines seven priority sectors and establishes actions to be taken within each one, as well as cross-cutting measures to be implemented across all sectors.

Other states have also engaged in adaptation to a lesser degree. Both New South Wales and Southern Australia have begun to embark on adaptation research and the inclusion of adaptation into their general climate change action plans (which focus on mitigation). Southern Australia's *Tackling Climate Change* strategy includes a section on adaptation that outlines the need to include adaptation into planning, provide information to communities and local authorities, mainstream adaptation into resource management activities, reduce vulnerabilities and improve hazard management (Government of South Australia, 2007). Victoria has outlined a series of sectoral actions and projects that address various impacts, while Western Australia and the Northern Territory have only begun to address adaptation.

The Council of Australian Governments has also indicated that national coordination with regard to select adaptation issues may be necessary in order to achieve the best outcomes for adaptation, and that cooperative action between state and local governments in particular will occur where necessary (Adaptation Policy Team, interview). However, coordination will be used in areas where it serves to improve adaptation, and not to hinder independent activities within certain states:

> The objective would be to get good coordination in the areas where it matters, and the areas where it's going to be more effective to allow jurisdictions to work things out according to local conditions and so forth, it's best not to try to slow that down (Adaptation Policy Team, interview).

As a result, a collaborative approach is taken between the state and federal levels where necessary, where a primary federal level role is to provide basic climate information to facilitate adaptation at the state level. There are no specific funding arrangements between the two levels and individual states can engage in adaptation measures independently (Adaptation Policy Team, interview).

7.5.1.3 Support for Adaptation at the Local Level

One of the principal aims of the Australian National Climate Change Adaptation Framework is to provide support and guidance to decision-makers at different scales and help build adaptive capacity in order to reduce vulnerability and enhance adaptation (Australian Government, 2007). As such, the national government has allocated funding towards two programmes intended to assist local capacity for adaptation within the Framework. The first is the Local Adaptation Pathways Program (LAPP), an AUS $2 million fund for local governments to identify climate change impacts and implement appropriate adaptation measures. Through the program, funding is provided to local governments to assist in the completion of climate change risk assessments and adaptation action plans. Two rounds of applicants to the program have been allocated funding to date; successful applicants to so far total 33 projects in 2008, with an additional seven larger-scale projects in 2009 (Department of Climate Change, 2007). Project funding is used to assist local governments in assessing climate change risks and adaptation strategies at council, city, borough and shire levels of administration.

Interestingly, the LAPP additionally makes uses the Local Government Climate Change Adaptation toolkit developed by ICLEI Oceania, a non-governmental organisation oriented toward the support of local governments in achieving sustainability. This partnership evolved as a result of the strong collaboration between the (now defunct) Australian Greenhouse Office and the Cities for Climate Protection Campaign, designed to reduce emissions from local governments. The strong involvement and expertise of ICLEI in local government affairs was thus considered an important and useful means for designing adaptation guidance. Funded by the Australian Department of Climate Change, the resultant toolkit is intended to be used alongside the government's *Climate Change Impacts and Risk Assessment* guide for private and public institutions (ICLEI Oceania, 2008). The toolkit provides an adaptive management process for identifying and assessing risks and opportunities presented by climate change, and the development of adaptation options and plans and their implementation. Five pilot local councils around Australia have used the toolkit thus far. The Australian Government's release of the 2009 *Climate Change Adaptation Actions for Local Government* publication provides additional resources to local governments in terms of possible adaptation actions that may be suited for Australian communities. The report provides information on possible impacts that may be anticipated in a range of sectors, as well as information on the regulatory framework developed for Australian climate change adaptation and a brief framework for climate change risk management.

A second set of funding totalling AUS $2 million has been made available for the *Climate Change Adaptation Skills for Professionals Program*, which focuses on the professional development of architects, planners, resource managers and other similar occupations (Australian Government, 2009b). Founded on the perceived need to improve the capacity of key groups to make use of climate-related information, the programme provides education and training for professionals so as to improve their skill set to incorporate climate change impacts into ongoing activities (Adaptation Policy Team, interview). Though not directly targeted at local levels of administration, the program is designed to improve the skill set of individuals working in both public and private spheres often at the local scale.

Specific local activities with regard to adaptation have been taken on in a variety of areas, primarily through coalitions between city councils. The Sydney Coastal Councils Group, the Western Port Greenhouse Alliance and the Gold Coast City Council have all begun to undertake local climate change vulnerability assessments. The Gold Coast City Council has additionally begun to incorporate increased flood risks into planning, while the Clarence City Council has similarly begun to incorporate the implications of sea level rise into local activities. In some cases, financial assistance from the national government has been provided for analytical work (Adaptation Policy Team, interview).

7.5.2 Canada

Canada is a federation of ten provinces and three territories, in which the provinces have their own legislative assemblies while power to the territories is devolved from the federal government. While Canadian federalism relies on an overarching federal jurisdiction enshrined in the Canadian constitution, it gives certain powers exclusively to the provincial legislatures. It has been described both as 'executive federalism' (Skogstad, 1996) and more recently as 'collaborative federalism' characterised by the co-determination of national policies rather than by traditional federal leadership (Cameron & Simeon, 2002). However, the exact distribution of power between the federal and the provincial level can be ambiguous, and there is a substantial overlap between the powers of the federal and those of the provincial government (Harrison, 1996). As a result, both federal departments and provincial ministries are concerned with the management of natural resources, the environment, health, agriculture and others.

At the federal level, Natural Resources Canada (NRCan, primarily impacts and adaptation assessment) and Environment Canada (primarily science, including model development and scenarios) have been involved in adaptation efforts since the late 1990s. At the provincial level, environment ministries have taken the lead on adaptation in many cases.

7.5.2.1 National Research and (in)Action

To date, federal departments, in collaboration with provincial civil servants and university researchers, have prepared three reports that examine the impacts of climate change in Canada. Among the first countries to publish a national climate change impact study in the late 1990s, Canada released the Canada Country Study in 1997, which included an assessment of social, biological and economic impacts (Environment Canada, 1997). It was followed by a report titled *Climate Change Impacts and Adaptation: a Canadian perspective* published in 2004 (Lemmen & Warren, 2004) that summarised research conducted between 1997 and 2002. In 2007, Canada conducted an in-depth national assessment (Lemmen et al., 2008) of regional and sectoral climate change impacts, adaptive capacity and potential adaptation options. The federal health department, Health Canada, has followed with a report that examines the human health impacts of the changing climate in Canada (Health Canada, 2008). Together, these publications provide insights into the regional and sectoral impacts of climate change in Canada, while the social and cultural dimensions are less well covered.

In addition to these efforts, a core research initiative funded by the federal government, the Climate Change Impacts and Adaptation Research Network (C-CIARN) network was established in 2001 with a mandate to promote and encourage research on climate change adaptation and to promote interaction between researchers and stakeholders. In 2007, the network was closed after successfully meeting its mandate. A new funding programme called Regional Adaptation Collaboratives (RAC) now follows the efforts of C-CIARN out of the recognition that funding is necessary to enable decision-making on adaptation. Announced in 2008, the goal of RAC funding is to 'to catalyze coordinated and sustained action to reduce vulnerability to a changing climate by advancing adaptation planning and decision-making' (Natural Resources Canada, 2009).

As a result of nearly three years of intergovernmental collaboration, the National Climate Change Adaptation Framework was released in 2005 (Intergovernmental Climate Change Impacts and Adaptation Working Group, 2005) by a working group of representatives from federal and provincial departments and ministries. The framework highlights six areas in which action should be taken: (1) raising awareness; (2) strengthening capacity for coordinated action; (3) incorporating adaptation into policies and operations; (4) promoting and coordinating research;

(5) supporting knowledge-sharing networks, and; (6) providing tools and methods for adaptation planning. However, the framework has not been approved by the federal government, nor has there been any official follow-up (Brooks et al., 2009). As a result, there is no strategy for adaptation that coordinates federal and provincial initiatives. A review by the Auditor General of Canada notes that this framework document 'is the only systematic effort by the federal government to work with provinces and territories on a shared approach to adaptation across the country' (Minister of Public Works and Government Services, 2006, p. 8). The report highlights that the framework does not identify expected results, timelines or roles and responsibilities for adaptation initiatives in Canada.

7.5.2.2 Provincial Governments

Continued interest in the impacts of climate change and possible adaptation measures has been spurred at the national level by provincial ministers. The Council of Ministers of the Environment (CCME) initiated a report published in 2003 that aims to raise awareness of climate change impacts and adaptation among the Canadian public (Canadian Council of Ministers of the Environment, 2003). In February 2009, the provincial environment ministers agreed to continue to collaborate on climate change impacts and adaptation (Canadian Council of Ministers of the Environment, 2009). Some efforts were also undertaken by the Council of the Federation, a council of provincial and territorial premiers of Canada, to examine and summarise ongoing provincial initiatives on adaptation (The Council of the Federation, 2007).

Despite the absence of a national framework that lays out federal and provincial responsibilities, formal initiatives on adaptation are underway in select Canadian provinces. While the federal process that led to the development of the adaptation framework fostered networking among the provinces to exchange information, adaptation efforts are undertaken within multiple ministries throughout the provinces. Overall, the approach could be characterised as 'on-demand', as the efforts undertaken have depended on the urgency of the impact and the nature of the resource affected. Forest ministries in particular have generally been most active on adaptation due to the ongoing impacts of the mountain pine beetle on timber resources.

In British Columbia (BC), for example, the climate change branch of the provincial government published a report titled *Indicators of Climate Change for British Columbia 2002* (Ministry of Water Land and Air Protection, 2002) outlining key changes in climate in BC, and a climate change plan for BC titled *Weather, Climate and the Future: BC's Plan* (Ministry of Water Land and Air Protection, 2004). The latter has been replaced by the commitments made by Premier Campbell in 2007, which largely concern greenhouse gas reductions. However, adaptation in the province to date has been dominated by impact assessments, while attempts to identify vulnerabilities, decision making and policy implementation on adaptation at the provincial level have been limited.

In contrast, efforts in Nova Scotia are more recent than those in British Columbia, but have yielded an action plan. In early 2009, the provincial government released its climate change action plan, detailing 68 action pledges (Nova Scotia Department of Environment, 2009). Of these, fourteen are actions on adaptation, including an adaptation fund, a land use planning tool, a provincial vulnerability assessment and a progress report, as well as more detailed actions related to sustainability, water resource management and coastal development. Other provinces that have released climate change strategies have focused on mitigation, including Alberta, Saskatchewan and Ontario.

7.5.2.3 Municipal Governments

There are significant differences in approach and timing between municipalities in Canada that have addressed adaptation. Among smaller municipalities, those already directly affected by phenomena such as the mountain pine beetle or storm surges are developing strategies to adapt. However, many are only in the early stages or recognising the need for adaptation at the municipal level. A select few have developed and have already begun implementing climate change adaptation strategies. For example, in collaboration with a network of private sector companies in Nova Scotia, the Halifax regional municipality initiated a climate change strategy as early as 2003. The municipality identified four extreme events during 2003–2004, an ice storm, flash flooding, hurricane Juan and a severe snowstorm, that typify severe weather expected to be more frequent in the future. As a part of this effort, the municipality has begun to integrate both greenhouse gas reduction and adaptation into local decision-making processes. When launched in 2004, the strategy included explicit deliverables on vulnerability assessments, climate change risk management and an adaptation methodology (Halifax Regional Municipality, 2004). In the City of Toronto, work on adaptation is currently underway through the formation of the *Ahead of the Storm: Preparing Toronto for Climate Change* report. Released in 2008 by the Toronto Environment Office, the report provides a survey of expected impacts and outlines short- and long-term actions to improve the city's resilience to climate change (Toronto Environment Office, 2008).

In British Columbia, a study conducted in 2004–2005 on perceptions of climate change in the greater Victoria region suggested that at the time, public awareness of local climate change impacts was sparse. While some municipal managers, especially in the water sector, were aware of potential effects of climate change, the general public perceived the solution to lie principally in greenhouse gas reduction (Wolf, 2006; Wolf et al., 2009). Local and regional climate change impacts were not perceived as likely, and adaptation initiatives such as C-CIARN were not known among participants who were not professionally involved in some aspect of climate change. Notably, local NGOs were involved only in mitigation efforts on climate change, but not in adaptation.

Since the release of the study, high profile publications such as the Arctic Climate Impact Assessment and the Fourth Assessment Report of the IPCC have received attention in the media and may have increased awareness among the public. However, adaptation remains fragmented at federal, provincial and local levels due to a lack of political leadership on adaptation at federal and provincial levels, a lack of political coordination at federal level to facilitate provincial processes, and political cycles that have caused uncertainty and instability.

7.6 Discussion: The Emergence of Adaptation

Though brief, the above review of adaptation policies in European countries provides some insight into the ways in which adaptation has emerged, and the role that different traditions and structures have played in shaping the process. Of the eight European countries assessed, six have already designed a national adaptation strategy, one is in the process of doing so, and one had yet to demonstrate concerted national interest in adaptation. This selection demonstrates that where comprehensive national strategies have been adopted, they have been created following the production of national assessments of potential climate change impacts and vulnerabilities along sectoral and regional lines. As presented in the preliminary review at the beginning of the chapter, the strategies themselves remain largely focused on the improvement, provision and dissemination of knowledge on climate change impacts and adaptation options, particularly to lower tiers of administration.

As such, significant resources are dedicated in several countries to the pursuit of improving climate information and the establishment of relevant bodies and networks. Adaptation strategies also generally represent preliminary strategic work on adaptation that in some countries is to be followed by more detailed and structured adaptation plans. As such, several remain fairly vague in their prescribed activities at this stage and provide only a guiding framework for adaptation as opposed to a binding plan. In Australia and Canada, the different extent to which adaptation has been addressed reflects both varying national engagement and interest in adaptation, though the general approaches taken to adaptation are broadly consistent with European countries.

The differences between countries also highlight the range of possible approaches to planned adaptation that can be taken. Generally, mainstreaming approaches such as those taken by Spain and the Netherlands follow a top-down structure through which changes in legislation that are to occur within the implementation of specific plans are intended to create a cascading or 'trickle-down' effect. Such an approach may still be accompanied by a general allocation of responsibilities within national and in some cases, regional and local bodies, where regional and local bodies are to eventually address adaptation needs within their respective areas. However, no additional funding is necessarily earmarked in these instances; instead, requirements to incorporate considerations of climate change impacts at different levels are mainstreamed through existing mechanisms such as strategic environmental impact assessments and development plans. An alternative approach has been taken by such countries as France, through which sub-national tiers of government are expected to create additional adaptation plans. Adaptation is instead treated as an issue or factor to be considered in addition to other ongoing management and planning issues. In these cases, however, multi-ministerial and sectoral stakeholders are still involved in the process, and responses are to be coordinated between the different levels.

Beyond these general observations, several more specific points that emerged from the comparison may be discussed. As indicated earlier in the chapter, the range of European countries presented was also intended to represent diverging examples of national political and planning structures and decentralisation of power, and traditional engagement with environmental policy. Thus, in the following sections, the differences between state adaptation activities in terms of these broad characteristics and their effect on the capacity to engage with planned adaptation are discussed in further detail, drawing on the material presented above as well as additional illustrative comments from various interviewees.

7.6.1 Diverging Steering Abilities in Federal and Unitary States

The content of each strategy and the extent to which means for adaptation are founded and extended throughout national and sub-national scales of governance are of clear importance. To some extent, the nature of a state as federal or unitary is reflected in the ways in which adaptation policy has emerged and been carried forward. In federal states, for example, national adaptation strategies do not include explicit roles or expectations of regional bodies, but instead serve as a guiding document providing recommendations and guidance to facilitate adaptation across national and sub-national scales. The creation and nature of adaptation strategies are thus in line with federal arrangements that disallow direct involvement of the federal government into state affairs. In the case of Germany, this arrangement was seen to negate the need for strong coordination between Länder or federal 'interference', especially given strong Länder engagement and interest in adaptation (Federal Ministry for the Environment, Conservation and Nuclear Safety, interview). Future changes to federal legislation that could occur under the German adaptation plan (and likely within future Austrian plans) will thus likely serve to facilitate adaptation at the Länder scale, as opposed to constrain them to act (Federal Ministry for the Environment, Conservation and Nuclear Safety, interview).

The existence of a National Adaptation Strategy and strong Länder interest in adaptation in Germany have resulted in the direct engagement by Länder with nationally-led processes of adaptation, and have ensured independent Länder action, in part perhaps attributable to their relative autonomy. While the relatively less-extensive engagement with adaptation by Austrian Bundesländer may only be a matter of growing interest in adaptation, it is thus perhaps also the result of the diminished authority conferred to the Bundesländer, as they have both spurred and been involved in the process of developing the NAS. In both cases, traditions of cooperation and consensus-building in general policy design have persisted in the design of adaptation strategies. In Germany, this has been achieved through the establishment of strong and effective fora for transparent collaboration and dual-track dialogue between the federal and Länder scales.

Adaptation measures in the federal states of Canada and Australia share broad similarities with the European cases. In both countries, the regional scale (provinces and states in Canada and Australia, respectively) is conferred the ability to design and implement independent policies and legislation in several matters. Adaptation measures at the regional scale have been undertaken in both cases, but to a greater extent in some states/provinces than others, resulting in an inconsistent degree to which adaptation is addressed across each country. This has also been the case in both Austria and Germany, where certain states have been more proactive on adaptation than others. However, Canada presents an interesting case where sectoral adaptation actions at the provincial scale have emerged despite inactivity at the national level through the collaboration between provinces and to some extent out of processes at the national level. Conversely, Australia demonstrates engagement with adaptation at both state and federal scales, with national support received from above. Finally, and perhaps significantly, adaptation activities at the national level in all federal cases were initiated in part at the request of the individual regions, indicating a strong tendency of federal states to be mobilised from below (though in the case of Canada, a national response has not yet occurred).

The remaining examples of planned adaptation selected for this chapter are all unitary states in which central governments, in principal, retain considerable authority over the activities of sub-national tiers. In practice, the selection reflects a range of countries representing varying degrees of decentralisation and thus varying degrees of authority over sub-national levels. As a result, national governments have not been the only level able to engage in adaptation, and in decentralised unitary countries have been joined by efforts at regional and local levels (discussed further in the next section). Further, the principal aims of information gathering and dissemination and the preliminary nature of adaptation proposals demonstrate that at this stage, adaptation strategies represent guiding documents for various sectors and scales. As such, the authority of adaptation strategies even in unitary states to date remains limited, akin to those designed in the federal states of Germany and Australia in terms of their ability to mandate action or confer responsibility to regional or local scales.

That said, the ability of national governments in unitary states to require specific actions from lower-tier governments is higher than in federal states, and will emerge more clearly with regard to adaptation once strategies are implemented or transformed into binding plans. The extent to which national governments will exercise such authority will likely depend on both the approach to adaptation that has been taken (i.e. a mainstreaming approach vs. additional/separately-funded adaptation plans and measures) and the extent to which authority has been decentralised in each country. Changes in national legislation that may facilitate adaptation at lower scales have occurred in Norway and are anticipated in countries including Spain, the Netherlands and France; however, the associated distribution of funding may differ considerably. For example, the largely mainstreaming approach taken by Spain and the Netherlands suggests that the consideration of climate impacts and vulnerabilities will be incorporated into existing plans and tools, and as a result will eventually be required of local governments within their normal activities. Additional funding will therefore not be provided to local governments in these contexts; conversely, French funding arrangements to local governments are determined according to specific activities required by the State. As such, local authorities that request funding from the central government must ensure they have followed requirements outlined in the Grenelle Bill for local climate strategies. Presumably, this will also be true for the highly centralised Greek state, should an adaptation strategy be created.

Given the differences noted above, the difficulty in separating the differences between federal and unitary states in terms of their ability to engage in adaptation is clear. The examples of federal countries provided here do demonstrate that without national engagement (as in Canada), regional actions can emerge but risk remaining fragmented. This is also true for unitary states with strong regional or local autonomy, and thus conflicts with any notion of a unitary state as a necessarily more centrally-guided entity. The extent to which power is decentralised to lower tiers of government is therefore of significance, and is examined in further detail in the next section.

7.6.2 Multi-Level Governance: Decentralisation and the Participation of Sub-national Authorities

The allocation of responsibility to sub-national scales of government or the designation of a specific scale (often the regional or county scale) as the primary coordinating body for adaptation is certainly relevant to the differing planning and decision making structures that exist in each country. Firstly, the extent to which lower tiers of government are addressed or included in the national adaptation strategies reviewed here varies considerably, with explicit allocation of responsibilities in some and less in others. In several unitary states, national strategies include reference to specific present or future tasks and responsibilities for regional and local governments, even where adaptation is to be mainstreamed through existing sectoral activities and bodies. In these cases, while the authority to act on adaptation may exist at regional or local scales, action at the national scale has served to foster activities at lower levels of government, either through the provision of a guiding framework or through the engagement in national processes. In the case of Norway, for example, adaptation tasks have been allocated along existing responsibilities, where the state may guide local policy through legislative change and the mediating and coordinating presence of the counties, while implementation is left to be decided by individual local authorities.

Notwithstanding the specific allocation of roles and competences among subnational tiers, the strong engagement of regional and local bodies has occurred in both federal and unitary states in the design of NAS, where representative institutions of local and regional governments have been represented in the National Adaptation Strategy process. Input and collaboration with local governments was sought in all cases, and is to be extended in the creation of future adaptation plans in France and the Netherlands. In the latter case, the traditional approach of seeking consensus with different scales of government was reflected in the process of creating the national adaptation strategy, which emphasises the regional level as an important scale of implementation to be supported by national programmes. As another example, the involvement of regional and local actors in the Spanish adaptation plan has resulted in both the national allocation of tasks within different sectors (to be carried out across different scales), as well as the emergence of independently-produced regional strategies. Ongoing efforts in Spain to produce a common structure for regional adaptation strategies are coordinated between the central and regional scale by means of a coordinating body, the OECC.

The existence of such a coordinating body has been important in several countries (both federal and unitary) and has served as an important link between different levels in the absence of direct control over sub-national activities. These national bodies all have an important role in disseminating information on impacts, vulnerabilities and possible adaptation strategies to various bodies at both central and sub-national levels, indicating a strong importance attached to updated and accurate climate-related information. In countries such as Norway, the Netherlands and Germany, nationally-led programmes have engaged select regional and local authorities in adaptation activities, providing information, incentive and resources to engage in vulnerability assessments and the design of preliminary local adaptation strategies. In the centralised country of France, coordination in general is also highlighted as an important component in order to ensure local governments are taking on their share and to avoid maladaptation. In Hungary, however, despite a highly-centralised structure and limited autonomy of regional governments, little work on adaptation has of yet been delegated to local authorities.

Further, the ability of sub-national states to make their own forays into planned adaptation also varies. In the federal states of Austria and Germany, significant competences conferred to the Länder has allowed for the engagement of Länder in individual adaptation strategies, though in both countries, some Länder are more proactive than others. In both centralised and decentralised states, national scale legislative changes may occur in order to facilitate or require adaptation actions at lower tiers. In highly decentralised states such as Spain, the ability of the national government to enforce regional actions is significantly restricted, as in federal arrangements. In these cases, cooperation between regional and national scales has been necessary to ensure that regions (or states, in the case of federal countries) are able to act autonomously using changes in national legislation to facilitate adaptation. In the Netherlands, changes to regulations that would oblige provinces and local authorities to include adaptation in their activities are currently under debate. Despite strong centralist tendencies in Hungary, plans to implement specific plans in the Tisza region were met with strong opposition from local actors and were eventually abandoned as a result, indicating the ability of local actors to influence political processes even in centralised countries.

Select local governments have also begun to act on adaptation in the absence of national incentives or nationally-appointed obligations. Such activities have often occurred in areas with significant populations, economies and financial resources, including capital cities and economically-productive regions. In Canada, France and others, larger cities with clear vulnerabilities to specific climate-related impacts have typically engaged in adaptation, indicating that the larger pool of financial, human and perhaps political capital available in large urban centres has likely played an important role in local adaptation. This is somewhat true of the regional scale as well, articulated in particular by interviewees in the Spanish case who explained the differences in regional engagement with adaptation as a result of the combination of political will and sufficient resources present in some of the more economically-advanced regions. In federal states or unitary states with considerable decentralisation, regional disparities in terms of resources and interest in given issues are accepted as a part of the varied political and economic landscape of the country (OECC; Federal Ministry for the Environment, Conservation and Nuclear Safety, interviews). Notwithstanding the differences in state structure or decentralisation, these findings do indicate the importance of the relative access to financial and other resources in local engagement with adaptation.

7.6.3 The Relevance of National Environmental Policy Implementation

A broad analysis indicates an absence of a clear relationship between those countries that have been traditional leaders in environmental policy and those that have engaged with adaptation. For example, while Germany has been considered a European leader in environmental policy, the adoption of a National Adaptation Strategy in 2008 occurred two years after Spain (an identified follower in environmental policy) had adopted its National Plan for Climate Change Adaptation. Austria, while often considered among the leaders in environmental policy in an EU context, has yet to adopt a national strategy; at the same time France had adopted its own comprehensive strategy in 2007 and is currently working on an implementation plan. Examples where a connection is perhaps more notable include Hungary, which adopted a National Adaptation Strategy in 2007 well in advance of other Eastern and Central European states and perhaps in line with its general advancements in environmental policy since EU membership. Greece has apparently lagged behind on adaptation (though, as in Italy, the recent election of a left-wing government may prompt further action over the next few years), while Spain (as in the case of environmental policy) has engaged with adaptation earlier than its southern counterparts.

Given the relative youth of adaptation as an issue in the context of developed countries, however, this difference of a few years in the adoption of strategies between countries is likely reflective of a growing interest in adaptation across Europe and the recognition of the risks posed by climate change (and the political will to address them), as well as the time required for processes of planned adaptation to begin. A large number of interviewees indicated that interest in adaptation was largely raised as a result of a general increase in knowledge of impacts and vulnerabilities, and was spurred by climate-related extreme events over the last several years. In the Netherlands, for example, the existence of national climate change sensitivities (i.e. low-lying geography) and vulnerabilities were ascribed greater importance than the country's record or interest in environmental policy (VROM, interview). Additionally, several cases indicated that involvement in UNFCCC processes or action at the EU scale also encouraged action on adaptation, indicating the importance of external factors in the decision to engage with adaptation as well.

That said, it may be noted that in the majority of cases, the process of creating a National Adaptation Strategy was led, either directly or indirectly, by ministries responsible for matters of the environment. While this may reflect the consideration of adaptation as a primarily environmental issue, the dynamic nature of several of these ministries (which cover various sectors, from water management and forestry to spatial planning) indicates that adaptation is addressed by bodies with responsibilities over a range of areas that touch upon several important national sectors. For example, while water management is often a sector under the authority of an Environment ministry, the Netherlands' particular circumstances and long-standing engagement with water issues has led to the separation of the Ministry of Transport, Public Works and Water Management from the Ministry of Housing, Spatial Planning and the Environment. As a result, the principal adaptation strategy produced by the ministry has been extensively complimented by actions in the water management sector. Actors in France and Germany also highlighted that extensive adaptation research and the development of measures at the national scale had been conducted in other ministries as well as within the environment sector.

The interministerial processes through which most adaptation strategies were created and adopted have additionally ensured that a variety of sectors and concerns has been integrated into adaptation activities. This finding is consistent with the PEER review in 2009, which concluded that '[i]nterdepartmental commissions led and supported by strong institutions have played a central role in the cross-sectoral development of the NAS and could function as a powerful policy integration catalyst in the implementation phase' (Swart et al., 2009, p.165). Moreover, the responsibility for the design of a National Adaptation Strategy and the implementation of its measures has often been transferred to a separate and often multi-stakeholder body (e.g., ONERC), ensuring the incorporation of a range of perceptions and needs in adaptation planning, both across sectors and administrative levels.

However, environment ministries have still had a strong leading role in several cases; thus, the strength of both these ministries and the independent adaptation bodies (in terms of e.g., financial resources, membership, decision-making authority, etc.) may still be an important factor in the weight and implications of adaptation strategies. For example, Spanish interviewees indicated that the Spanish Climate Change Office had been conferred a significant degree of authority and freedom to design its own adaptation measures, indicating considerable decision-making authority over national adaptation. However, the Office's work with regard to changes in legislation may be limited to those areas under the authority of the Ministry of Environment, given their inability to create binding measure for other Ministries (OECC, interview).

Further, countries that have engaged less with adaptation (e.g. Greece, or Hungary up until, 2008) have often signalled the need to address existing institutions, such as the EU directives, or international agreements such as the UN conventions. Insofar as these are considered matters related principally to the environment sector, it may be said that a relationship between adaptation and environmental policy does exist. However, these institutions have recognisably important implications for a number of sectors and are often a matter of broad national interest. As such, the treatment of adaptation as a matter of realising the commitments outlined under such institutions is again both related to environmental policy traditions, as well as track records with respect to fulfilling the broader responsibilities under the EU or other frameworks.

Thus, while these findings indicate that while adaptation is now recognised by several countries as an issue for consideration in all sectors and economies, national environmental policy traditions and the strength of the environmental sector in each country may still have some significance with regards to adaptation. It can perhaps be assumed that those countries (as well as regions or local authorities) with greater experience with addressing environmental (and cross-sectoral) issues and that have dedicated considerable resources towards achieving environmental goals and commitments will have some advantage in addressing the adaptation needs posed by climate change. However, as adaptation is mainstreamed and/or legislated into activities within a range of sectors and scales, environmental policy records and independent environmental institutions themselves may decrease in importance.

7.6.4 Stakeholder Participation and the Role of NGOs and the EU

In most of the cases assessed here, the civil sector, academic bodies and the private sector were either included in the design of national adaptation strategies, or have been identified as stakeholders to engage further. Such stakeholders have been included in processes of coordination or consultation of adaptation strategies or legislation, either through stakeholder advisory groups, participatory processes or consultation. One reoccurring means for including non-governmental actors in the strategies' design processes has been the use of online fora, through which individuals and organisations are able to comment on their relevant strategy and its contents. However, the use of such a forum in France was noted to have received little engagement from the public or other bodies, suggesting that the views of non-governmental actors were perhaps not strongly included in the draft of the strategy. That said, France as well as the Netherlands, Spain and Germany have all indicated the need to incorporate stakeholders into future adaptation plans, and have begun to do so under current processes of adaptation plan design and the use of institutions such as France's Grenelle. While scientific bodies have certainly played a large role and the civil sector has participated in several instances, the private sector has yet to be significantly engaged in any of the countries surveyed. However, Norway and Australia have each begun activities to include various private actors at the local scale.

Below the national level, the engagement of regional and local governments with adaptation has been facilitated as a result of their participation in national or international networks or projects that have addressed locally-relevant impacts. The

Spanish Network of Cities for Climate provides an example of such a network in the national context, while others such as BaltCICA, AMICA and ClimChAlp represent networks of European regions and cities that share common climate-related issues and vulnerabilities. The prevalence of such networks as means for engaging local authorities in adaptation highlights the important role of non-governmental organisations in providing local authorities with the necessary resources (in the form of information, networks and in some instances, finances) to begin the process of engaging with planned adaptation policies. Several such initiatives receive the support of the EU, including BaltCICA, ClimChAlp and others. Such INTERREG-funded initiatives have been important for the development of knowledge and action at the regional and sub-regional scale in several countries, and in some cases have served to assist local authorities in establishing connections with similarly-impacted regions across Europe. In the case of Greece, where little has been done on adaptation to date, research and action on adaptation has been to some extent taken on by NGOs (such as WWF) and internationally-led research (such as CIRCE).

Australia further represents a case in which NGO work has been supported and used for the facilitation of adaptation at local scales. The involvement of ICLEI Oceania in the Australian government's adaptation programme presents a clear example of adaptation governance and the ability of government to draw on NGO expertise and networks, and reinforces the findings in the European context where internal and external networks and projects have facilitated adaptation activities at the local scale. However, while several projects in European local cases have drawn on EU funding, this is clearly not an available source of funding outside the EU and thus constitutes a possible benefit to adaptation in EU countries. Further, as the EU increasingly takes on adaptation through its various directives, European regions will be obliged to implement guidelines set by the EU for the incorporation of adaptation concerns into regional activities.

The role of the EU was additionally highlighted by several countries in terms of the effect of EU policy on national abilities to engage with planned adaptation. EU directives and activities such as the Natura 2000 network were highlighted by actors in both the Netherlands and Spain (and in National Communications in Hungary) as areas requiring significant changes to adequately address the impacts of a changing climate. Both further acknowledged, however, the EU's burgeoning attempts to address the need to incorporate climate impacts and adaptation into relevant actions through its White Paper on adaptation; as a result, Spain has included reference to the White Paper in the implementation of their adaptation plan. Increasing action on adaptation by is also raised as somewhat of an incentive for action in the individual Member States. Actors from several countries (France, Netherlands and Austria) highlighted that the release of EU reports had helped to remove the 'taboo' on the discussion of adaptation. Prior to these, adaptation was originally considered an admission of defeat with regards to emission reduction targets; thus, impacts and vulnerability reports and the Green and White papers issued by the Commission's helped to encourage Member States to begin their own investigations into adaptation policy.

7.7 Conclusions

Overall, the comparison of the countries presented in this chapter indicates that adaptation has been addressed largely in line with existing structures and delegation of authority, with some new elements. Though activities have largely been coordinated by environmental ministries and their technical and/or research bodies, adaptation is not treated solely as an environmental issue and has thus received significant input from other ministries in the design of adaptation strategies. At this stage, countries are focused on filling gaps in knowledge of impacts, vulnerabilities and adaptation measures and the dissemination of relevant information to help inform decision-making at various levels. As efforts have so far been concentrated on developing institutions and networks for climate change-related knowledge (i.e. building adaptive capacity), the actual implementation of specific adaptation measures is largely yet to come. However, actions to date reflect a growing acknowledgment of the need for cross-sectoral engagement with adaptation, linking together ministries that previously may have not been in contact.

Those countries with a strategy in place are moving toward what may be termed the 'institutionalisation' of adaptation into legislation, policy and regulatory instruments, but the degree to which adaptation may actually be mandated at lower levels significantly depends on the structure of the country in question. Where regions have considerable authority, the State's primary function will likely be the facilitation of adaptation through the provision of information and changes in national legislation that enable action at lower levels. In more centralised countries, more explicit requirements may be issued to ensure adaptation is addressed; in both cases, however, coordination between levels and sectors has been and will continue to prove a vital component of adaptation measures. These differences are also reflected in the different approaches taken to adaptation, where some countries have elected for a mainstreaming approach through which adaptation is incorporated into existing plans, where others have instead addressed adaptation through specific adaptation plans. Both approaches have exhibited elements of the other, where mainstreaming efforts are often accompanied by regional or local adaptation plans, while specific plans may be joined by subsequent changes to existing legislation and regulation. The effectiveness of either approach may hinge less on their respective differences than on the rigour with which adaptation needs are addressed.

With regards to the differences between EU and non-EU countries, it is clear that despite the lack of clear and binding adaptation requirements from the EU to date, the EU has and will continue to serve as an important institution in the adaptation of both nations and local authorities (see Chapter 2, for a more in-depth treatment of the role of the EU). The EU has provided both incentive and funds for adaptation through its various programmes, a resource clearly not available to countries outside its umbrella. As adaptation is mainstreamed into funding requirements and directives, the EU will provide further resources and impetus for adaptation, even in those countries, regions or local authorities that have not yet addressed it.

However, several regions and local authorities in both EU and non-EU countries have been able to access resources for adaptation, either independently or through various networks. It is apparent that the local bodies that have been able to engage in adaptation are those who have been able to access such networks or draw on the ample internal resources that are often present in larger and more-economically well off cities and regions. In most EU countries, such processes have generally been complemented (or fostered) by top-down processes established through National Adaptation Strategies, indicating interest in adaptation by political leadership at various levels. Indeed, regional governments have in some cases even stimulated discussion on adaptation at the national level, as in Germany and Austria.

This is true also of Australia, where both states and the federal government have been mutually involved in adaptation; conversely, in Canada the process has largely been driven from the bottom-up in the absence of national action. The absence of clear guidelines in place that would coordinate efforts, allocate responsibility and ensure accountability in Canada and countries in the EU that have yet to demonstrate national leadership on adaptation (such as Greece) constrains the ability of local decision makers, planners and engineers to take adaptation on as part of their other agendas. As demonstrated by the above examples, national action may not necessarily be a requirement for local engagement in adaptation, but may certainly be an important element for ensuring its success. That said, the paths taken by countries such as Germany, Spain, France, the Netherlands, Norway, Hungary and Australia will certainly need to be followed by other states if climate impacts and vulnerabilities are to be effectively addressed at both national and sub-national levels. National governments may still provide significant guidance, structure and resources for adaptation at lower levels and, as with EU directives, may in fact hinder adaptation in more advanced areas if measures are not taken to facilitate adaptation.

References

Interviews and Personal Communications

Academy of Sciences/Budapest Corvinus University, Hungary. Telephone interview, November 26, 2009

Adaptation Policy Team, Department of Climate Change, Australia. Telephone interview, December 17, 2009.

Adaptation Program for the Norwegian Energy Sector, Norway. Personal communication, December 6, 2009.

BMLFUW, Austria. Telephone interview, December 3, 2009.

BMLFUW, Austria. Personal communication, December 11, 2009.

Climate Change Adaptation Unit, Spanish Climate Change Office, Spain. Telephone interview, November 18, 2009.

Directorate for Civil Protection and Emergency Planning (DSB), Norway. Telephone interview, December 3, 2009.

Directorate for Civil Protection and Emergency Planning (DSB), Norway. Personal communication, December 6, 2009.

Federal Ministry for the Environment, Nature Conservation and Nuclear Safety, Germany. Personal communication, December 19, 2009.

Federal Ministry for the Environment, Nature Conservation and Nuclear Safety, Germany. Telephone interview, November 13, 2009.
Knowledge for Climate, Netherlands. Telephone interview, November 12, 2009.
Metéo-France, France. Telephone interview, November 13, 2009.
Ministry of Housing, Spatial Planning and Environment, Netherlands. Telephone interview, November 13, 2009.
Ministry of the Environment, Norway. Personal communication, December 4, 2009.

Secondary References

Abromeit, H. (1992). *Der verkappte Einheitsstaat*. Opladen: Leske+Budrich.
ADAGIO. (2009). Project info. Retrieved November 13, 2009, from http://www.adagio-eu.org/project_info.html
ADAM. (2009). The Tisza River Basin: Adaptation to climate change in floodplain management. Retrieved December 2, 2009, from http://adamproject.info/index.php/June-2008/The-Tisza-River-Basin.html
Andersen, M. (1999). Ecological modernisation capacity: finding patterns in the mosaic of case studies. In M. Joas & A.-S. Hermanson (Eds.), *The Nordic environments. Comparing political, administrative, and policy aspects* (pp. 15–46). Aldershot: Ashgate.
Australian Government. (2005). National Climate Change Adaptation Programme: Department of the Environment and Heritage, Australian Greenhouse Office.
Australian Government. (2007). National Climate Change Adaptation Framework: Department of Climate Change, Australian Commonwealth.
Australian Government. (2009a). Department of Climate Change Corporate Plan 2009-10: Department of Climate Change, Commonwealth of Australia.
Australian Government. (2009b). Key initiatives. Retrieved October 12, 2009, from http://www.climatechange.gov.au/impacts/about.html
Australian Government. (2009c). Support for climate change adaptation research. Retrieved October 12, 2009, from http://www.climatechange.gov.au/impacts/research.html
Australian Government. (2009d). Water in Australia: Water for the Future. Retrieved November 25, 2009, from http://www.environment.gov.au/water/australia/index.html
BaltCICA. (2009). Norway – Bergen. Retrieved October 5, 2009, from http://www.baltcica.org/casestudies/bergen.html
Bank of Greece. (2009). Commission for the Study of Climate Change Impacts Retrieved November 13, 2009, from http://www.bankofgreece.gr/Pages/el/klima/default.aspx
Bavarian Ministry for Environment and Health. (2009). Klimaschutz in Bayern. Retrieved September 18, 2009, from http://www.stmug.bayern.de/umwelt/klimaschutz/index.htm
Benz, A. (1999). From unitary to asymmetric federalism in Germany. Taking stock after 50 years. *Publius: The Journal of Federalism, 29*(4), 55–78.
Birbily, T. (2009). Inaugural speech before the Greek Parliament, *Parliamentary minutes of October 18, 2009*. Athens.
Börzel, T. A. (2003). *Europe. Why there is (not) a 'Southern Problem'*. Hampshire: Ashgate Publishing Limited.
Brooks, M., Gagnon-Lebrun, F., Harvey, H., & Sauvé, C. (2009). *Prioritizing Climate Change Risks and Actions on Adaptation – A Review of Selected Institutions, Tools, and Approaches*. Ottawa: Government of Canada – Policy Research Initiative.
Burkhart, S. (2008). Reforming federalism in Germany: Incremental changes instead of the Big Deal. *Publius: The Journal of Federalism, 39*(2), 341–365.
Cameron, D., & Simeon, R. (2002). Intergovernmental Relations in Canada: The Emergence of Collaborative Federalism. *Publius, 32*(2), 49–72.
Canadian Council of Ministers of the Environment. (2003). *Climate, Nature, People – Indicators of Canada's Changing Climate*. Winnipeg: Canadian Council of Ministers of the Environment.

Canadian Council of Ministers of the Environment. (2009). Press Release: Environment Ministers Collaborate on Climate Change. from www.scics.gc.ca/cinfo09/830956004_e.html
CICERO. (2007). NORADAPT – Community Adaptation and Vulnerability in Norway. Retrieved October 5, 2009, from http://www.cicero.uio.no/
CIRCLE. (2008). *Report on the current state of National Research Programmes on climate change impacts and adaptation in Europe. Extended country report – Issue 2*: Climate Impact Research Coordination for a Larger Europe.
Climate Alliance Austria. (2009). Climate Alliance Austria [in German]. Retrieved October 30, 2009, from http://www.klimabuendnis.at/
Climate changes Spatial Planning (CcSP). (2009a). Adaptation. Retrieved October 5, 2009, from http://climatechangesspatialplanning.climateresearchnetherlands.nl/nl/25222852-Adaptation.html
Climate changes Spatial Planning (CcSP). (2009b). Programme description. Retrieved October 5, 2009, from http://climatechangesspatialplanning.climateresearchnetherlands.nl/nl/25222866-Programme.html
Cramer, J. (2007). *Speech presented at the signing of the Climate and Energy Agreement between the Dutch national government and the provinces*. Amsterdam: VROM.
Czako, V., & Mnatsakanian, R. (2008). Impacts of and adaptation to climate change in the Danube-Carpathian region In Department of Environmental Sciences and Policy (Ed.). Budapest: Central European University.
Delta Commission. (2008). *Working together with water – A living land builds for its future*.
Department of Agriculture Fisheries and Forestry. (2009). Australia's Farming Future. Retrieved November 25, 2009, from http://www.daff.gov.au/climatechange/australias-farming-future
Department of Climate Change. (2007). Local Adaptation Pathways Programme. Retrieved November 30, 2009, from http://www.climatechange.gov.au/government/initiatives/lapp.aspx
Department of Climate Change. (2009). Caring for Our Coasts. Retrieved November 25, 2009, from http://www.climatechange.gov.au/government/initiatives/caring-for-coasts.aspx
DSB. (2007). Nasjonal sårbarhets- og beredskapsrapport (NSBR) 2007. Samfunnets sårbarhet overfor naturutløste katastrofer. Oslo: DSB.
ECCE. (2005). *Principal conclusions of the preliminary evaluation of the impacts of climate change in Spain (in Spanish)*. Madrid: Ministerio de Medio Ambiente.
Ellison, D. (2008a). Subnational regionalism in a supranational context: The case of Hungary. *Romanian Journal of European Affairs, 8*(1), 60–87.
Ellison, D. L. (2006). *Weighting the politics of the environment in the new Europe* (Working Paper No. 169), Institute for World Economics. Budapest: Hungarian Academy of Sciences.
Ellison, D. L. (2008b). Competitiveness strategies, resource struggles and national interest in the new Europe. No. 1901. Pittsburgh: Carl Beck Papers in Russian and East European Studies, University of Pittsburgh.
Environment Agency Austria. (2009). *The Austrian approach towards a National Adaptation Strategy*. Vienna: Environment Agency Austria.
Environment Canada. (1997). *The Canada country study: climate impacts and adaptation*. Ottawa: Environment Canada.
ESPACE. (2008). *Adaptation to climate change (EU/Espace) Final report*. Nijmegen: Ministry of Housing, Spatial Planning and the Environment (Netherlands).
European Environment Agency. (2009). Regional climate change and adaptation. The Alps facing the challenge of changing water resources. EEA Report 8/2009. Copenhagen: EEA.
European Environment Agency (EEA). (2009). National adaptation strategies. Retrieved October 5, 2009, from http://www.eea.europa.eu/themes/climate/national-adaptation-strategies
European Union Water Initiative (EUWI). (2009). Mediterranean. Retrieved November 12, 2009, from http://www.euwi.net/wg/mediterranean
Featherstone, K. (2005). Introduction: 'modernisation' and the structural constraints of Greek politics. *West European Politics, 28*(2), 223–241.

Federal Ministry for the Environment Nature Conservation and Nuclear Safety. (2008). German Strategy for Adaptation to Climate Change. Retrieved October 3, 2009, from http://www.bmu.de/english/climate/downloads/doc/42841.php

Federal Ministry of Agriculture Forestry Environment and Water Management (BMLFUW). (2001). *Third National Climate Report of the Austrian Federal Government*. Vienna: Federal Ministry of Agriculture Forestry Environment and Water Management.

Federal Ministry of Agriculture Forestry Environment and Water Management (BMLFUW). (2006). *Fourth National Communication of the Austrian Federal Government in Compliance with the Obligations under the United Nations Framework Convention on Climate Change*. Vienna: Federal Ministry of Agriculture Forestry Environment and Water Management.

Federal Ministry of Agriculture Forestry Environment and Water Management (BMLFUW). (2009). *Towards a national adaptation strategy*: BMLFUW.

Federal Ministry of Education and Research. (2007). Announcement by the Federal Ministry of Education and Research of regulations to fund research projects Managing climate change in the regions for the future (KLIMZUG) Regions profit from competition and the world profits too! Retrieved December 15, 2009, from http://www.bmbf.de/en/furtherance/7764.php

Federal Ministry of Education and Research. (2009a). KLIMZUG-NORD – Strategic approaches to climate change adaptation in the Hamburg Metropolitan Region Retrieved September 30, 2009, from http://www.klimzug.de/en/185.php

Federal Ministry of Education and Research. (2009b). Klimzug climate change in regions. Retrieved October 5 2009, from http://www.klimzug.de/en/index.php

Federal Ministry of Education and Research. (2009c). *KLIMZUG Managing climate change in the regions for the future*. Bonn: Federal Ministry of Education and Research.

Federal Ministry of Education and Research. (2009d). REGKLAM – Development of an integrated regional climate change adaptation programme for the model region of Dresden Retrieved September 30, 2009, from http://www.klimzug.de/en/98.php

Federal Republic of Germany. (2006). *Fourth National Report by the Government of the Federal Republic of Germany (4th National Communication).*

Federal Republic of Germany. (2008a). German Strategy for Adaptation to Climate Change (p. 73): Federal Ministry for the Environment, Nature Conservation and Nuclear Safety.

Federal Republic of Germany. (2008b). German Strategy for Adaptation to Climate Change – Executive Summary (p. 4): Federal Ministry for the Environment, Nature Conservation and Nuclear Safety.

Federation of Municipalities and Provinces (FEMP). (2006). La Ministra de Medio Ambiente y el Presidente de la FEMP entregaron los Premios a las Buenas Prácticas Locales por el Clima Retrieved November 18, 2009, from http://www.femp.es/index.php/femp/noticias/gobierno_local/la_ministra_de_medio_ambiente_y_el_presidente_de_la_femp_entregaron_los_premios_a_las_buenas_pr_cticas_locales_por_el_clima

Federation of Municipalities and Provinces (FEMP). (2008). Contratación por concurso abierto: elaboración del informe "Las Entidades Locales Supramunicipales y el Cambio Climático". Retrieved November 18, 2009, from http://www.femp.es/index.php/femp/convocatorias_y_subvenciones/convocatorias/contrataci_n_por_concurso_abierto_elaboraci_n_del_informe_las_entidades_locales_supramunicipales_y_el_cambio_clim_tico

Fowler, B. (2001). Debating sub-State reform on Hungary's 'Road-to-Europe', *Working Paper 21/01*: Center for Russian and East European Studies, University of Birmingham.

Ghorra-Gobin, C. (n.d.). Climate change and public policies at the local scale: The complexity of the task, *Cities and climate change: Responding to an urgent agenda – Role of institutions, governance & urban planning: Improving management, coordination, and planning of cities to meet climate change challenges* (pp. 11). Paris: Centre National de la Recherche Scientifique.

Gobierno de Aragón. (2007). Cambio climático y administración aragonesa. Retrieved October 5, 2009, from http://portal.aragon.es/portal/page/portal/MEDIOAMBIENTE/cclimatico/Administracion

Gobierno de Aragón. (2008). *Climate Change and Clean Energy Strategy of Aragon (in Spanish).*

Gobierno Vasco. (2008). *Basque Plan to Combat Climate Change (in Spanish)*: Gobierno Vasco.
Government of South Australia. (2007). Tackling climate change 2007–2020 South Australia's Greenhouse Strategy (pp. 79): Department of the Premier and Cabinet.
Green and Blue Space - Adaptation for Urban Areas and Eco Towns (GRaBS). (2009). Retrieved May 2, 2010, from http://www.grabs-eu.org/
Halifax Regional Municipality. (2004). *Climate SMART*. Halifax NS: Halifax Regional Municipality.
Harrison, K. (1996). *Passing the buck: Federalism and Canadian environmental policy*. Vancouver: UBC Press.
Health Canada. (2008). *Human Health in a Changing Climate: A Canadian Assessment of Vulnerability and Adaptive Capacity*. Ottawa: Health Canada.
Hemerijck, A., & Visser, J. (2000). Change and immobility: three decades of policy adjustment in the Netherlands and Belgium. *West European Politics, 23*(2), 229–256.
Hulme, M., Neufeldt, H., Colyer, H., & Ritchie, A. (Eds.). (2009). *Adaptation and mitigation strategies: Supporting European climate policy. The final report from the ADAM project*. Norwich: Tyndall Centre for Climate Change Research, University of East Anglia.
ICLEI Oceania. (2008). Adaptive and Resilient Communities Program. Retrieved December 16, 2009, from http://www.iclei.org/index.php?id=adaptation-toolkit
Intergovernmental Climate Change Impacts and Adaptation Working Group. (2005). *National Climate Change Adaptation Framework*. Ottawa: Natural Resources Canada.
Interministerial Working Group on the Evaluation of Climate Change Impacts. (2009). *Evaluation du coût des impacts du changement climatique et de l'adaptation en France. Rapport de la deuxième phase*. Paris: ONERC.
Kecskemét Conference Statement. (2009). Az 'Aszály és a Szárazodás Magyarországon' Konferencia Állásfoglalása, October 7, 2009.
Kenniscentrum stedelijke vernieuwing (KEI). (2009). Urban regeneration in the Netherlands. Retrieved October 30, 2009, from www.urbanregeneration.nl
Kizos, T., Spilanis, I., & Mehmood, A. (2009). Climate change vulnerability: Planning challenges for small islands. In S. Davoudi, J. Crawford & A. Mehmood (Eds.), *Planning for climate change: Strategies for mitigation and adaptation for spatial planners* (pp. 94–104). London: Earthscan.
Knowledge for Climate (KFC). (2009). Hotspots. Retrieved November 12, 2009, from http://knowledgeforclimate.climateresearchnetherlands.nl/nl/25222799-Hotspots.html
KomPass. (2008). Kompetenzzentrum klimafolgen und anpassung. Retrieved October 3, 2009, from http://www.anpassung.net
KvVM. (2008). *National climate change strategy*. Budapest: Hungarian Ministry of Environment and
Waters.
Langeweg, F. (Ed.). (1989). *Concern for tomorrow: A national environmental survey 1985–2010*. Bilthoven: RIVM.
Lemmen, D., Warren, F., Bush, E., & Lacroix, J. (Eds.). (2008). *From Impacts to Adaptation: Canada in a Changing Climate 2007*. Ottawa: Government of Canada.
Lemmen, D. S., & Warren, F. J. (Eds.). (2004). *Climate change impacts and adaptation: A Canadian perspective*. Ottawa: Natural Resources Canada.
Massey, E. (2009). *Adaptation policy and procedures in Central and Eastern Europe*. Amsterdam: Institute for Environmental Studies.
Massey, E., & Bergsma, E. (2008). *Assessing adaptation in 29 countries*. Amsterdam: Institute for Environmental Studies.
Matczak, P., Flachner, Z., & Werners, S. E. (2008). Institutions for adapting to climate change in the Tisza River. *Klíma-21 Füzetek 55*, 87–1000.
Minister of Public Works and Government Services. (2006). *2006 Report of the Commissioner of the Environment and Sustainable Development to the House of Commons* Ottawa: Office of the Auditor General of Canada.

Ministry for Environment and Regional Policy. (1997). Hungary. Inventories, mitigation and scenarios of the greenhouse gas emissions and removals: Second national communication on the implementation of commitments under the United Nations Framework Convention on Climate Change. Budapest: Ministry for Environment and Regional Policy. Hungarian Commission on Sustainable Development.

Ministry for Environment and Regional Policy. (2002). Hungary. Third national communication for the UNFCCC. Budapest: Ministry for Environment and Regional Policy. Hungarian Commission on Sustainable Development.

Ministry for the Urban Planning Environment and Public Works. (2009). Special framework for the spatial planning and sustainable development of coastal areas and islands Retrieved November 18, 2009, from http://www.minenv.gr/download/2009/2009.08.28%20eidiko%20xorotaksiko%20plaisio%20paraktios%20nisia.doc

Ministry of Environment and Conservation Agriculture and Consumer Protection of North Rhine-Westphalia. (2009). Klimawandel in Nordrhein-Westfalen – notwendige anpassung als chance begreifen (in German). Retrieved September 28, 2009, from http://www.umwelt.nrw.de/umwelt/klimawandel/index.php

Ministry of Environment and Water. (2005). Hungary. The fourth national communication of the Republic of Hungary on climate change. Budapest: Ministry of Environment and Waters.

Ministry of Environment and Waters. (2005). Hungary. The fourth national communication of the Republic of Hungary on climate change. Budapest: Ministry of Environment and Waters.

Ministry of Housing Spatial Planning and the Environment (VROM). (2008). National Programme on Climate Adaptation and Spatial Planning *National Adaptation Strategy – Policy Memorandum*. Amsterdam: VROM.

Ministry of Rural Development and Food. (2007). Programme for the Rural Development of Greece: Section of Community Funds and Infrastructure, Ministry of Rural Development and Food, Athens.

Ministry of the Environment. (2007). St.meld. nr. 34. Norsk klimapolitikk. Oslo: Ministry of the Environment.

Ministry of the Environment. (2008). *Climate change adaptation in Norway: A government initiative on climate change adaptation (in Norwegian)*: Norwegian Ministry of the Environment.

Ministry of the Environment. (2009). About the Norwegian Climate Adaptation Programme. Retrieved October 5, 2009, from http://www.regjeringen.no/

Ministry of Transport Public Works and Water Management (MVW). (2009). Water and the future Delta Committee recommendations. Retrieved October 6, 2009, from http://www.verkeerenwaterstaat.nl/english/topics/water/water_and_the_future/delta_committee/

Ministry of Water Land and Air Protection. (2002). *Indicators of Climate Change for British Columbia 2002*. Victoria, BC: The Province of British Columbia.

Ministry of Water Land and Air Protection. (2004). *Weather, Climate and the Future: BC's Plan*. Victoria, BC: The Province of British Columbia.

Natural Resources Canada. (2009). Regional Adaptation Collaboratives. Retrieved October 11, 2009, from http://www.adaptation.nrcan.gc.ca/collab/index_e.php

Newman, P., & Thornley, A. (1996). *Urban planning In Europe. International competition, national systems and planning projects*. London and New York: Routledge.

Norwegian Pollution Control Authority. (2007). Klimatilpasninger – Veiledning om mulige tiltak i avløpsanlegg. SFT veileder 2317/2007. Oslo: STF.

Norwegian Research Council. (2006). *Nasjonal handlingsplan for klimaforskning, styrking av strategisk helhet og langsiktighet i klimaforskningen*. Oslo: Norwegian Research Council.

Nova Scotia Department of Environment. (2009). *Toward a greener future: Nova Scotia's climate change action plan*. Halifax: Nova Scotia Department of Environment.

Observatoire National sur les Effets du Rechauffement Climatique (ONERC). (2007). *Strategie Nationale d'Adaptation au Changement Climatique*. Paris: La Documentation Francaise.

Oficina Española de Cambio Climático, S. G. para la Prevención de la Contaminación y del Cambio Climático & Ministerio de Medio Ambiente. (2006). *National Plan for Climate Change Adaptation (in Spanish)*: Ministerio de Medio Ambiente.

Oficina Española de Cambio Climático, Secretaría de Estado para el Cambio Climático & Ministerio de Medio Ambiente y Medio Rural y Marino. (2009). *Plan Nacionál de Adaptacíon al Cambio Climático, Segundo Programa de Trabajo*

ONERC. (2009). Présentation. Retrieved November 20, 2009, from http://www.ecologie.gouv.fr/-Presentation,640-.html

Osterud, O. (2005). Introduction: The peculiarities of Norway. *West European Politics, 28*(4), 705–720.

Peters, B., & Pierre, J. (2005). Swings and roundabouts? Multilevel governance as a source of and constraint on policy capacity. In M. Painter & J. Pierre (Eds.), *Challenges to state policy capacity. Global trends and comparative perspectives* (pp. 38–51). Houndmills Basinstoke, UK: Palgrave Macmillan.

Planbureau voor de Leefomgeving. (2009). Milieubalans 2009. Bilthoven: Planbureau voor de Leefomgeving.

Queensland Government. (2007). ClimateSmart adaptation 2007-12. An action plan for managing the impacts of climate change (pp. 28): Department of Natural Resources and Water, The State of Queensland.

Red Española de Ciudades por el Clima. (2008). *Estrategia local de cambio climático*. Retrieved November 27, 2009. from http://www.redciudadesclima.es/index.php?action=detalle publicacion&modul = publicaciones&id_publicacion = 28&template = publicaciones&idioma List=.

Reiter, R. (2008). The 'European City' in the European Union. Comparing the implementation conditions for the EUs urban development agenda in Germany and France. In R. Atkinson & C. Rossignolo (Eds.), *The Re-creation of the European City. Governance, territory and polycentricity* (pp. 17–38). Amsterdam: Techne Press.

Rhônalpénergie-Environnement. (2008). Changement climatique – Comment s'adapter en Rhône-Alpes? Lyon: Rhônalpénergie-Environnement.

Ribeiro, M., Losenno, C., Dworak, T., Massey, E., Swart, R., Benzie, M., & Laaser, C. (2009). *Final report. Design of guidelines for the elaboration of regional climate change adaptation strategies*. Berlin: Ecologic Institute.

Schröter, D., Zebisch, M., & Grothmann, T. (2005). Climate change in Germany – Vulnerability and adaptation of climate-sensitive sectors: Klimastatusbericht

Skogstad, G. (1996). Intergovernmental relations and the politics nad environmental protection in Canada. In K. M. Holland, F. L. Morton & B. Galligan (Eds.), *Federalism and the environment: Environmental policymaking in Australia, Canada and the United States* (pp. 103–134). Westport: Greenwood Press.

Steen, R. (2003). Klimaendringer – Er beredskapen på plass? *PLAN, 5*, 18–23.

Swart, R., Biesbroek, R., Binnerup, S., Carter, T. R., Cowan, C., Henrichs, T., et al. (2009). *Europe adapts to climate change. Comparing national adaptation strategies*. Helsinki: Partnership for European Environmental Research.

The Council of the Federation. (2007). *Climate Change: Leading Practices by Provincial and Territorial Governments*. Manitoba: The Council of the Federation.

The Research Council of Norway. (2008). Climate change and its impacts in Norway (NORKLIMA). Retrieved October 5, 2009, from http://www.forskningsradet.no/

Toronto Environment Office. (2008). Ahead of the storm: Preparing Toronto for climate change. In Prepared in collaboration with the City of Toronto Climate Adaptation Steering Group & Clean Air Partnership (Eds.). Toronto: City of Toronto.

Wolf, J. (2006). *Climate change and citizenship: A case study of responses in Canadian coastal communities*. Unpublished PhD thesis, University of East Anglia, Norwich, UK.

Wolf, J., Brown, K., & Conway, D. (2009). Ecological citizenship and climate change: Perceptions and practice. *Environmental Politics, 18*(4), 503–521.

WWF. (n.d.). The New Vásárhelyi Plan: Catastrophic floods and the new governmental concept. Retrieved December 1, 2009, from www.wwf.hu

Chapter 8
Conclusion: The Development of Adaptive Capacity and Adaptation Measures in European Countries

E. Carina H. Keskitalo

Abstract This volume has related the development of adaptation policy and practice to a number of structural, context-based or procedural capacities; these include the extent of decentralisation, the type of planning systems, the institutionalization of environmental policy, and the occurrence of focusing events across multi-level governance systems. The case study material has consisted of nested case studies on national, regional and local levels in the UK, Finland, Sweden and Italy, supplemented by comparative cases viewed mainly within the context of the European Union. This chapter summarizes the results of the study with a focus on the parameters defined in the introduction and describes the significant variety in outcomes across the case study countries. Differences range from a comprehensive multi-level framework for adaptation in the centralised unitary UK state, to more limited approaches in Sweden and Finland, to the failure to thus far institutionalise adaptation policy in Italy. The case studies support identifying the parameters affecting adaptive capacity and the development of adaptation responses. However, no single factor in itself can readily explain the variety of responses to adaptation.

Keywords Adaptation · adaptive capacity · Europe · multi-level governance

8.1 Introduction

This volume has described four principal multi-level case studies of climate change adaptation policy development set against the context of relatively well-developed cases of multi-level adaptation policy in the EU and abroad. The studies show that a relatively comprehensive framework, including legislative and regulative requirements for adaptation on national, regional and local levels, has developed in the UK. This has taken place both through the development of e.g., regional stakeholder

E.C.H. Keskitalo (✉)
Political Science/Department of Social and Economic Geography, Umeå University, Umeå 90187, Sweden
e-mail: Carina.Keskitalo@geography.umu.se

E.C.H. Keskitalo (ed.), *Developing Adaptation Policy and Practice in Europe: Multi-level Governance of Climate Change*, DOI 10.1007/978-90-481-9325-7_8,

bodies and through the inclusion of adaptation in a performance assessment framework through which priorities and funding levels for local government are established. In Finland, a cross-sectoral national adaptation strategy has developed primarily through inter-ministerial cooperation and includes indicators for implementation in the country as a whole. Implementation is intended to be undertaken through the mainstreaming of these criteria into existing administrative bodies, but thus far without dedicated funding. In Sweden, responsibility for adaptation has been divided between different levels, with the state assuming responsibility for larger-scale measures while the municipalities are responsible for incorporating adaptation into existing risk response systems. In both Sweden and Finland, planning and building acts have been modified to require increased local responsibility for flood risk. Finally, a national conference during 2007 in Italy aimed to develop national adaptation policy, but such activities have not been carried forward by the subsequent government. The result has been the lack of a formal national adaptation policy, leaving regional and local levels to develop approaches as needed under existing frameworks.

The volume has demonstrated the role of multi-level governance system in adaption, including the role of political systems, decentralisation and institutionalised environmental policy at different levels. The national political system has been conceptualised as the political, administrative and planning system that makes up a specific country's policy style. Countries may also be further classified into such categories as planning family or the level of decentralisation. While different planning families or the level of decentralisation may play a role in adaptation (in this study specifically with regard to the steering capacities of centralised versus decentralised unitary states), such differentiations to some extent obscure the country-specific nature of governance systems. Network governance (Rhodes, 2000, for example), has been important also in highly centralised states. Environmental policy traditions are another field where general, national-level descriptions obscure a great variety of approaches with regard to the difference in prioritisation of environmental policy issues at both the national level and at regional and local levels. The volume illustrates cases where traditional followers in environmental policy within the EU context have developed relatively comprehensive adaptation policy frameworks. The volume also illustrates cases where a limited emphasis on environmental policy (together with unfavourable political and media contexts) may have limited progress on adaptation; and cases where those often considered leaders on environmental policy in the EU context have not prioritised adaptation as an issue.

This complex context of adaptation policy development suggests that what may be termed procedural or contextual factors, not only structural factors such as the level of decentralisation, are important for adaptation policy development. A number of such factors have been treated in the agenda-setting literature, among which the existence of political champions or leaders and focusing events such as floods and storms in particular may be especially crucial for drawing attention to climate change vulnerabilities. Beyond political leadership and focusing events, issue development may also be connected to other factors taken up in agenda-setting and related literature, such as the existence of supportive political, public and media climates (cf. Kingdon, 1995; Baumgartner & Jones, 1993). In addition, factors such as the sensitivity of ecosystems to climate change as it is presented in political processes may play a role in the perceived urgency of climate change adaptation (cf. Wurzel, 2002).

These considerations specify some of the factors and processes that may influence the development of adaptive capacity, defined here as the resources (political, financial, informational, and others) that determine the ability of a social unit (such as a nation or region) to adapt to climate change (cf. Smit & Wandel, 2006). While the complexity of adaptive capacity can only partly be captured through the application of the factors discussed in this volume, the framework highlights the multi-level and connected character of adaptation and provides a context for analysing adaptive capacity and the development of adaptation in less general terms than those of political, financial, institutional and other resources. This study has thus attempted to highlight capacities at levels from national to local that have supported the development of adaptation.

This concluding chapter outlines the ways in which each of the issue areas discussed has combined to shape adaptation at different scales, drawing upon the theoretical background described in the introduction (Chapter 1). The chapter is organised in such as way as to describe adaptation governance and the actors within the system, beginning with the national level and general political system features, followed by the regional and local levels, the role of the EU and finally, the role of actors beyond government and administration. Further, the chapter briefly discusses the different types of adaptation that may be used to distinguish the different cases. The chapter also discusses the potential for lessons learned with regard to adaptation and its diffusion across specific country-level and other contexts. The chapter ends with a note on the role of adaptation policy in general: while some see adaptation in the context of only smaller, technical changes, others perceive it as a phenomenon that will require substantial modifications to the way planning is undertaken. Given the differing levels of depth of the studies, this conclusion focuses primarily on the four main cases and less on the cases described in Chapter 7; however, these cases will be included in the discussion of main points where conclusive statements can be made with regard to each case.

8.2 National Level and General Political System Features

8.2.1 Structure of the Political and Planning System

This study was based on the assumption that regulatory styles differ across nations and can be differentiated partly based on such factors as the national political system or level of decentralisation (in both federal and unitary states). In turn, these factors may influence the scale at which planning is undertaken and the extent to which the state can steer and develop multi-level policy measures on adaptation. The study has aimed to investigate a variety of states and the way these may facilitate the development of adaptation on national and lower levels, given that considerable coordination (and potentially, dedicated funding) may be required and that a lack of institutionalisation of policies at the national level may otherwise impede possibilities for adaptation at lower levels. As some authors have suggested, decentralised or federal political systems may adjust more quickly to unexpected changes or be more accustomed to acting within a multi-level context (Glachant, 2001; Peters & Pierre, 2005).

In the four main case studies, the structure of the political system and the extent of (de)centralisation with regard to planning decisions in particular played a significant role. In the decentralised Swedish and Finnish systems, local governments possess considerable latitude to define their own adaptation agendas. However, this limits the ability of the state to enforce such an agenda and bring municipalities to a minimum common denominator. This was especially evident in the context of the local planning monopoly in Sweden. Conversely, central government in the centralised national context in the UK can both place and enforce requirements on local authorities; as a state committed to adaptation, the UK is therefore able to develop multi-level policy and foster practical development on adaptation. However, the feature of centralisation and the ability of the state to steer development would have a different impact in a state with a lesser focus on adaptation. In such a case, centralised steering could instead prevent or prohibit strong policy development on adaptation at the local level. In a state with less of a focus on adaptation, a decentralised system may thus be more beneficial to the development of adaptation in local cases where e.g., environmental sensitivities to climate change are identified, as a decentralised system may provide regions and/or local municipalities greater leeway to define local adaptation needs and enforce the measures to address them.

However, to draw distinctions between states with regard to centralisation requires some additional qualification. Among the four cases, the UK stands out not only as a centralised state, but as an example of a strong multi-level governance approach to adaptation with dedicated positions or bodies on adaptation at all levels. This is both in agreement with and contradictory to the UK's centralised state context. First, the UK has been able to introduce strong state measures that set the context for adaptation at all levels. Setting priorities for local government through a national framework is bringing local authorities to a lowest common denominator for adaptation. On the other hand, the UK has also exhibited strong development of a multi-level adaptation network. Though largely supported by the state level through the UKCIP and its regional approach, independent and bottom-up networks among local government such as the Nottingham Declaration partnership have played a significant role in developing acceptance for adaptation policy at the local level. The UK can thus be described not only as a centralised state – though this has been important in its dedicated development on adaptation – but also as a network polity (Rhodes, 2007; cf. Chapter 1) and a very specific regulatory system.

With regard to the issue of whether decentralised systems may respond more quickly to change, select local cases in such systems (e.g., Gothenburg and the Helsinki Metropolitan Area Council) have largely developed adaptation measures independently. This has taken place in response to locally identified sensitivities and exposure to events, and in relation to local adaptive capacities. However, in both of the decentralised (though unitary) systems of Sweden and Finland, the development of a lowest common denominator on adaptation and the establishment of adaptation as an issue to be taken into account at all levels may have been hindered, in Sweden for instance, by strong local planning rights that can render united and coherent measures across municipalities more difficult. In Sweden, the delegation of responsibility for adaptation measures beyond major infrastructural projects to the local level constitutes a considerable task for municipalities, which may have very different capabilities. While Gothenburg may be able to respond to such a demand, smaller municipalities may lack the financial, administrative, or staffing capacities. In comparison to Sweden, Finland has taken a more centralised approach in drawing up a national adaptation strategy intended for implementation through the existing horizontal and vertical political framework. As the Finnish national adaptation strategy does not conclude that immediate threats exist, national funding has not been allocated. As a result, implementation at the local level may be limited by the lack of designated financing and enforcement procedures.

The issue of decentralisation thus has different meanings and embodies vastly different approaches in the case studies than a state-level system description allows. Peters and Pierre (2005) note, for instance, that the unitary state of Sweden may even be described as nearing semi-federalism given the autonomy of the local level. On the other hand, the centralised UK state has been viewed as more of a network polity (Rhodes, 2000). In this study, these differences are apparent in how decentralisation impacts policy-making and in the strengths and weaknesses in the development of adaptive capacity and adaptation measures. In the UK, local level network policy development of adaptation including local leadership may have supported variation in the treatment of individual issues. On the other hand, the more centralised features of state-driven policy-making permit the setting of targets to bring municipalities to a minimum common denominator. In Sweden and Finland, relative decentralisation may contribute to the formation of leader municipalities over time. However, the focus on local planning may also cause smaller municipalities or those that do not perceive immediate vulnerabilities to lag behind in adaptation. Thus, despite being unitary systems, such states may be limited in the extent to which they are able to develop and implement coherent measures across local municipalities. In the federal cases described in this volume, federal and state processes have often allowed for states to design their own strategies under a broad common framework. However, in Canada, the absence of a national framework has resulted in the provinces acting on their own and thereby potentially in fragmented and uncoordinated response.

In all cases, the presence or absence of a national level framework on adaptation thus plays a significant role for lower levels and especially in establishing adaptation as an important policy area and promoting awareness. Examples of environmental policy-focused local governments that did not address adaptation until it became a national priority have been identified (in the case of Woking, UK), as well as those where adaptation is yet to be addressed at all given that it is not a national environmental policy priority (as in the case of Trollhättan, Sweden). In Italy, the development of local activities relevant to adaptation was considered difficult due to the lack of a national framework for adaptation and the lack of financial or political incentives. Similarly, in Greece, local development is seen as hampered by the lack of national frameworks for adaptation.

The Italian case specifically illustrates that the lack of a policy framework at the national level may impact the capacity of lower levels to respond to threats. The structure of Italian regional administrative federalism indicates that national priorities (or the lack thereof) may be limiting even in a system where considerable decision-making power is devolved to the regions, particularly due to strong national control over financial matters. Thus, while a region may act autonomously on an issue such as adaptation in lieu of a coordinated national strategy, funding or incentives for such action may not exist. Interviewees also noted that activities at the local level may be fragmented and piecemeal without coherent central direction. As a result, existing measures are undertaken under existing policy frameworks and without dedicated funding (unless provided by the region itself). While decentralisation makes it possible for local and regional bodies to act independently, actors noted that those authorities best able to cope with challenges outside established national policy may be those who can access funding either independently or from other levels, particularly that of the EU. However, this could also result in a significant differentiation between regions, prompting leaders to emerge based on existing internationalisation or Europeanisation and those without such connections to fall behind. The lack of integrated frameworks and funding may also lead to a situation in which organisationally well-developed municipalities or regions gain (somewhat in accordance with 'new regionalism' theory, cf. Veggeland, 2000), while others remain unable to develop such approaches in the absence of supporting networks or local capacities.

Other features of political systems also play a role in the development of adaptation policy. In both Italy and Greece, a fragmented political structure can be seen as a partial cause for the lack of adaptation policy development. The inability to develop an adaptation strategy in Italy after 2007, for example, was related to the unclear distribution of authority and inter-institutional competition at the national level, as well as to an unstable coalition government. In the absence of consistent coordination at the national level, regional and provincial bodies have drawn on existing national-level frameworks such as the National Action Plan to Combat Drought and Desertification developed under commitments to the UN Convention to Combat Desertification, Italy's National Plan for the Prevention of the Effects of Heat on Human Health, and EU directives. As a result, the EU and international level policy-making may play a larger role in such countries than in countries with more developed approaches to environmental policy or with stronger intervention from the national level. This suggests that EU policy development may impact laggards more than leaders who actively attempt to upload their existing policies and therefore experience less of an impact from changes at the EU level.

8.2.2 The Role of Environmental Policy Institutionalisation

This study has also inquired whether an established tradition in environmental policy may have supported the development of adaptation within the case study countries. An assumption here has been that adaptation, as a broad issue under the umbrella of sustainability, would fall within the remit of environmental policy. If identified as an environmental policy issue, the development of adaptation would therefore benefit from established institutional structures and issue champions in environmental policy and consequently suffer if such structures were less developed or prioritised. The level of institutionalisation of environmental policy would thus indicate some potential for political mobilisation and resource allocation for adaptation. If adaptation was framed within an environmental policy context, it would probably also mean that localities with high environmental policy aspirations would develop also this issue.

In most of the cases, adaptation has been framed as an environmental policy issue and was often linked to previous work on climate change, i.e. perceived as an extension of the mitigation policy field. However, adaptation was also in these cases generally understood as a broad issue that required integration into several policy fields and departments: while environment ministries were generally the leading bodies in developments, many countries exhibited cooperation bodies with multi-level and multi-stakeholder approaches. These included existing bodies that were given adaptation as an additional task as well as bodies that were created specifically dedicated to adaptation. Examples on national level include the Grenelle Environment in France, the interministerial adaptation group in Norway, and the National Climate Council in Spain. Adaptation was thus in a number of cases both recognized as an environmental and a cross-cutting issue. However, the difficulties of functional integration can be seen in the UK, for instance, where a number of interviewees at different levels noted difficulties garnering support in regional climate change partnerships from sectors other than those in the environmental policy arena. In particular, Finland constituted an exception to this pattern of viewing adaptation as based in the environmental issue area in that the lead ministry was the Ministry of Agriculture and Forestry.

The extent to which environmental policy development may have played a role thus differs largely between cases. This depends partly on the extent to which adaptation has been defined as relevant for national policy. For instance, as Sweden was perceived as less sensitive to the consequences of climate change at the national level, adaptation was often perceived as a limitation on environmental policy ('giving up' on mitigation; similar remarks were also given for the Netherlands, Hungary and Canada) or as an issue for developing countries. Consequently, despite a well-established environmental policy tradition in Sweden, adaptation has largely been seen as a marginal issue, gaining policy attention mainly following regional efforts to attract attention to climate vulnerabilities, focusing events, and changes in the international policy context. The way in which adaptation is perceived or framed in decision-making processes is thus significant in the Swedish national context. Given the limited institutionalisation of adaptation as an environmental issue, municipalities with environmental policy aspirations such as Trollhättan did not identify adaptation as a relevant issue. On the other hand, in municipalities such as Gothenburg adaptation has been independently identified as a requirement and has received support from an established environmental policy apparatus.

The UK has traditionally been seen as more of a follower in environmental policy on the EU scale. However, it has recently developed relatively ambitious targets both for mitigation of greenhouse gas emissions and for adaptation. Climate change may have thus become something of a focal issue for this otherwise reluctant EU member. However, this priority especially with regard to adaptation cannot be seen in isolation from a relatively high sensitivity to the impacts of climate change, or from the development of flooding as an issue in the UK. Thus, in the UK differentiation by issue and potentially also over time may be relevant considerations where environmental policy is concerned. The role of established environmental policy in a general sense may be especially notable in UK local authorities with an environmental focus such as Hampshire and Woking. The treatment of adaptation in these two cases differs. As mentioned previously, adaptation in Woking has come into focus to some extent as a result of the Council's general environmental ambitions. In Hampshire, it has also been strongly connected to focusing events and seen as a way of increasing the region's competitiveness (both with regard to adapting in the future and in relation to the EU policy arena).

In the Italian example, limitations on the development and institutionalisation of environmental policy, including those relevant to a bipartisan agreement on climate change, may have played a role in the limited development of adaptation policy. Such limitations may largely pertain to the reactive nature and late adoption of environmental policy in Italy, where policies are often employed to cope with and compensate for crises rather than prevent them. However, the institutionalisation of environmental policy (or lack thereof) cannot be treated in isolation from features such as the fragmentation in the political system at large or the comparatively limited media attention paid to climate change issues in Italy. More generally, this indicates the need to view adaptive capacity in specific cases as a result of multiple factors.

In the case of Finland, the development of an adaptation approach despite perceived limited sensitivity to the effects of climate change and few focusing events can potentially be linked to an institutionalised use of interministerial working groups for mainstreaming policy. Such an approach highlights the importance of viewing policy traditions and assumptions on the environment within the policy style of the country, which is only partly expressed through structural characteristics such as political system and level of decentralisation. Regulatory traditions (such as New Public Management approaches, cf. Pollitt & Bouckaert, 2000) and cultural differences may also be important. However, having an environmental policy agenda was considered a supporting structure for adaptation cooperative efforts in both the Espoo and Helsinki area. This may illustrate that adaptation, once it is developed as a priority, may be able to draw upon the institutions for issues it becomes linked to, but also that these issue linkages and the process of framing is crucial.

A number of examples at the local level thus in particular demonstrated the importance of institutionalised environmental policy. Cases such as the Helsinki Metropolitan Area Council, Gothenburg, Woking and Hampshire were all considered areas with long traditions in the environmental and air quality fields, highlighted by interviewees as an important support to the development of adaptation as an issue area. However, this was not in itself a sufficient motivator for action, for example, in the absence of development of adaptation as a priority on the national or local level, or in the absence of events linked to climate change.

8.2.3 The Role of Focusing Events and Other Contextual Features

The previous sections have highlighted that the broad structural and environmental policy factors treated here are relevant but not sufficient to explain the development of adaptation policy. In the political science literature, the emergence of new issues on the policy agenda has often been treated in the framework of agenda-setting (e.g., Kingdon, 1995; Baumgartner & Jones, 1993). This literature highlights that the emergence of an issue onto the political agenda is linked to: the confluence of participants or 'issue entrepreneurs' who push the issue; the existence of a positive political (and in some interpretations, media) environment; a beneficial policy context forwarding and linking the developing issue to other policy items; and focusing events that may make a condition into a problem for policy-making. These factors draw particular attention to the deeply contextual and process-based nature of how adaptation may develop in one country, region or locality but not another.

The role of focusing events (cf. Birkland, 1998) is demonstrated widely in the case studies: for adaptation through the 2001 floods and the Gudrun storm in Sweden; for risk response through drought and flooding incidences in Italy; for adaptation through floods on numerous occasions (marked by policy responses) in the UK; and through the 2003 European heat wave in a number of countries, with particularly devastating consequences for France. Data for Germany, Hungary, the Netherlands and Norway also indicate important focusing events, in particular with regard to flooding. However, as indicated by the case of Italy, the occurrence of flooding or drought events is not always sufficient to elicit an adaptation response. Especially in the Italian case, the potential focusing events were not framed or perceived as related to climate change. Instead, responses prompted a focus on crisis and risk management.

In Finland, on the other hand, adaptation came to feature on the political agenda – albeit without significant financial resources dedicated to its implementation – without major domestic focusing events. Finnish interviewees did not identify a strong national vulnerability to climate change, nor did they note large-scale focusing events of the same calibre as those that supported the emergence of adaptation on the agenda in Sweden and the UK. The speedy national mainstreaming of adaptation in Finland is thus somewhat puzzling in light of the role attributed to focusing events in many other cases.

This situation warrants some specific discussion of the Finnish case. Chapter 4 posits that the relatively early policy development of adaptation in Finland can be partly explained by the relative efficiency of the administration (potentially supported by the country's relatively small population), through which sub-national actors were made aware of national adaptation policy development early on. However, as focusing events seem to have had very limited impact on the Finnish policy process, a limited perceived urgency may have accounted for the absence of funding or modifications to national, regional or local roles. As mentioned above, Finland has largely treated adaptation policy as an additional issue to be mainstreamed into policy without any specific tools or specific funding for implementation. This suggests that it may have followed a 'business-as-usual' approach through which adaptation has followed the regular path of cooperation between representatives of Ministries (rather than a modification of decision-making systems to accommodate the specific nature of adaptation as an issue). In Finland, the main focus of the Uusimaa Regional Council, Helsinki Metropolitan Area Council, and KUUMA Municipal Cooperation climate strategies has, since their inception, been on mitigation. Adaptation was only developed later on.[1] The regular practice of mainstreaming policy issues through inter-ministerial cooperation may thus have been a factor in supporting adaptation policy development in Finland. While this may not detract from the political importance of focusing events (interviewees in Finland did note that if such events had occurred, they might have raised the priority on adaptation), this case illustrates again the importance of viewing adaptation policy development in the context of several factors that may together support adaptive capacity in governance systems.

In many ways, the development of an adaptation agenda at each of the levels can also be seen as linked to the development of political and media contexts favourable to the issue and to the existence of champions of the issue. This may in turn be the result of their historical establishment in the environmental policy area, lending a historical context seldom emphasised in agenda setting approaches. Issues emphasised in the agenda-setting literature as limitations to issue development such as the existence of a negative media environment, limited awareness among the public, and the perception of climate change as a partisan issue was among the main cases pronounced at the national level only in Italy where adaptation remains relatively undeveloped. These factors were also emphasised by interviewees in the Italian case as reasons for the country's limited development of adaptation policy. Among the supplementary cases, Greece, Canada and Hungary also noted instable governing coalitions or political cycles that have caused instability as well as a lack of sufficient national leadership on adaptation.

In many cases, leadership figures or champions at various scales were also highlighted in policy development, with notable reference to international policy and the international political environment – in particular Al Gore and his film *An Inconvenient Truth,* cited by interviewees in all four main case study countries. To some extent, the development of an adaptation agenda can thus also be seen as a result of international agenda-setting, particularly through the IPCC, the UNFCCC, and other processes that influence the national level. Despite relatively low EU involvement with the issue, adaptation has begun to emerge on the policy agenda

[1]The only Finnish area in the study that noted impacts from flooding events was Espoo, a city located close to Helsinki with considerable flood risks in highly developed areas, as well as a considerable environmental policy focus. As a result of these factors, Espoo developed a climate change preparedness strategy for the city within the Helsinki Metropolitan Area Council work and maintains an environmental security working group that has focused on climate change as a risk.

of a large number of EU countries (for a more detailed discussion, cf. Keskitalo, Westerhoff, & Juhola, in prep.).

Some cases also indicate, however, that not only international processes but agenda-setting at lower levels have influenced the central state in the development of policy, suggesting that the development of adaptation is a strongly multi-level issue. An example is the strong influence of regional and local lobbying from counties surrounding Lake Vänern on the development of state policy in Sweden. The UK provides another example where the state did not act in isolation from lower levels: actors in Hampshire County noted that they had lobbied and were able to influence both the UK and the EU on these issues. In Germany the Länder significantly contributed to voicing the need for a national and regional strategy for adaptation, and in Canada the provincial environment ministers have in many cases taken the lead on adaptation.

Finally, the way in which adaptation is organised with relevance for different levels and bodies is also significant. Differences exist between the countries in terms of how national level initiatives are developed: through a national conference in Italy, through inter-ministerial cooperation in Finland, and through national commissions or investigations in Sweden and to some extent in Norway and Spain. These differences in organisational development are not trivial, but impact the ways in which initiatives proceed and the specific actors are involved. In the UK, Germany, the Netherlands and Norway, a strong focus has also been placed on developing specific secretariats or bodies with some responsibility for developing adaptation at the regional and local levels (e.g., the German Competence Centre on Climate Impacts and Adaptation and the Climate Changes Spatial Planning programme in the Netherlands). Such bodies may contribute to institutionalising adaptation as a policy area. Many countries have also developed programmes that fund pilot projects with the aim of developing adaptation approaches, such as *Klimzug* in Germany, the *Climate Changes Spatial Planning* program in the Netherlands, and the Norwegian *Future Cities* programme. Such programmes, as well as best practice awards including adaptation goals in the UK and Spain, may help developing local government leadership on adaptation. In the future, such leadership may then serve as examples of best practice for the development of adaptation approaches.

While it may be early to produce conclusive statements on the importance of dedicated bodies on adaptation, the format of the UKCIP in the UK has been frequently cited as a model in other state contexts, and as especially important for communicating scientific conceptions of risk in a lay format for local and regional policy-making. The creation of specific bodies with dedicated funding may also serve to institutionalise adaptation as an issue – in effect, creating a polity with adaptation as its focus. Such institutionalised bodies may later lobby for a continued focus on adaptation in government – as the UKCIP did in the process towards the UK Climate Bill, thereby supporting continuation of part of its institutional focus as a policy priority. Similarly did the already established German Climate Protection Programme support an extension of the adaptation agenda in Germany.

Interviewees both within and outside the UK also cited the UKCIP as a crucial example of coordination between levels that serves to translate scientific information for decision-makers and officials through the promotion of specific tools for local vulnerability assessment. A particular example of this is the Local Climate Impacts Profile (LCLIP) tool intended to identify events and responses to support the development of adaptation measures. Elements of the UKCIP process and especially the LCLIP can also be seen in other countries. In Australia, a similar tool was developed by the NGO ICLEI Oceania. Its local government climate change adaptation toolkit is now used to support vulnerability assessment and the development of adaptation options.

Finally, most of the factors highlighted here also rest upon the existence of a strong scientific base of climate change knowledge. This has constituted the foundation for identifying potential impacts in several different areas. Chapter 7 notes that existing strategies on or towards adaptation at the national level are often founded on impacts and vulnerability assessments conducted and funded at national scale. The role of nationally funded climate science has thus been imperative at this stage in the development of adaptation, and the problems associated with the lack of such nationally funded and integrated climate science can be seen, for instance, in Italy and Greece. The studies indicate that access to climate change scenarios and impacts information among the four main cases was limited only in Italy, where adaptation was also the least developed. Whereas impacts and scenario information was coordinated nationally in the UK, Sweden and Finland, information was in Italy fragmented between several institutions, which made a coordinated approach to climate change difficult. Italian actors noted the need for greater resources dedicated to impacts and scenario research, as well as the need for the coordination and processing of such data within a national contact point – a function which the Italian Euro-Mediterranean Centre for Climate Change (CMCC) is currently taking up. By contrast, in the UK, Sweden and Finland, as well as many of the supplementary case study states, national research programmes and scenario development on national and occasionally sub-national scales exist, which makes the dissemination and application of research the significant issue in these cases.

8.3 Regional Level

While the factors discussed above impact regional and local levels and the processes there, it is important also to consider the special characteristics of these levels of administration. Drawing upon the framework described in the introduction (Chapter 1), the four main case study areas chosen at the regional level (for, among other things, their relatively well-developed approaches to adaptation) all turned out to be relatively highly populated and among the wealthiest regions in their respective countries. This lends some credibility to assumptions inherent in adaptive capacity and new regionalism literature that the ability to deal with issue areas such as adaptation may be differentiated between regions depending on, for instance, financial, administrative and information resources.

South East England (UK) has been considered one of the most vulnerable regions in England to climate change, as well as being one of the wealthiest and most pressured in terms of development. Within the region itself, the wealthy local authority of Hampshire has been a leader on adaptation. However, there are marked differences within the region as similarly (or potentially even more) vulnerable areas such as the city of Portsmouth have developed adaptation measures to a lesser extent. Similarly, in Sweden, the Västra Götaland County Administrative Board possesses considerable staffing and knowledge resources, given its relatively large administration as the implementing body of the state in one of the country's most populated regions. Large areas of the Västra Götaland County exhibit a high vulnerability to climate change, in part the result of development pressures in the area. The second largest city in Sweden, Gothenburg developed policy and practical measures for adaptation relatively early on. However, significant differences exist between Gothenburg and other Swedish case study municipalities with regard to their awareness and perceived vulnerability to climate change, as well as in their relative size and available resources for adaptation.

The Finnish study showed that relatively few Finnish regional councils had outlined specific measures for adaptation in their regional plans. Those that had were highly populated regions with high development pressures, such as the municipal cooperation in Espoo and Helsinki. However, as the Regional Councils in Finland are more of a coordinating body between the state and the municipalities, the regional level should not be expected to develop implementation measures to a significant extent. In both Finland and Sweden, regional reform processes are ongoing. Both the continued development of the Västra Götaland region, as well as broader reform and a stronger focus on the regional level for policy-making in Finland are likely to impact political power and issue development at the regional level in the future.

Finally, the Italian region of Emilia-Romagna is one of the country's regional leaders in environmental policy. It is relatively wealthy with supplemented funding from EU regional integration strategies and with an unusually high level of decentralisation to lower levels of administration. While this would make Emilia-Romagna difficult to compare directly with other Italian regions, the study highlights some of the problems that even a well-resourced region faces with regard to national political and planning structures. Adaptive capacity in Emilia-Romagna is supported by a relatively large and experienced regional environmental protection agency, a regional meteorological and climate service, and the region's participation in a number of EU projects. These in turn have supported the development of regional development plans (under the direction of the EU) that address adaptation in specific issue areas, including drought prevention and irrigation (for instance, by setting a consumption-based tariff).

Emilia-Romagna has thus been relatively committed to the achievement of environmental aims and has delivered timely regional plans to the EU well before several other regions, indicating that the impact of EU policies may differ substantially between regions also depending on the prioritisation of EU-related actions. However, regional adaptation in Emilia-Romagna has largely been developed in the context of sustainable development policy and has been handled in an ad hoc manner rather than in terms of potential future climate risks or within the framework of a comprehensive risk management approach. The extent to which and the capacities with which regions can act on adaptation thus vary significantly across the different administrative systems. Emilia-Romagna has large formal regional independence but still remains somewhat constrained by limited financial autonomy. By extension, this situation may limit adaptive capacity and the possibility of dealing with such a complex, cross-sectoral and multi-level issue as adaptation.

As the implementation of adaptation measures will impact future vulnerability and economic gains (e.g., in preventing flooding of infrastructure or safeguarding crops from drought), adaptation may also have more far-reaching political consequences. Speculatively, if such impacts led to an increased focus on adaptation within the political system, this could support regions that have been selected for coordinating functions developing along the lines of 'new regionalism' (cf. Veggeland, 2000). This may be particularly true for the UK, where select regional climate change partnerships have become significant bodies. Given the current discontinuation of England's Regional Assemblies, some focus may need to be placed on other bodies with broad regional representation, of which the Regional Climate Change Partnerships already account for a substantial number of regional stakeholders (cf. Sandford, 2005). The South East of England is often seen as an economic and demographic growth region, and the new management systems and national indicators that are negotiated regionally and locally may additionally provide the region with an increased informal role in spite of limited formal regional power. In Sweden and Finland, ongoing processes of regional reform may also impact the development of adaptation policy at this level.

8.4 Local Level

As described above, local level agency and potential decision-making paths at the local level are largely formed by the context of the political system, including the level of decentralisation for instance with regard to planning capacities and the level of autonomy afforded the local scale. Among the four main cases, Italy represents a country in which regions may act relatively independently from the state. Local authorities in Sweden and Finland illustrate locally highly decentralised planning systems. In the last two countries, the implementation of state requirements seems to vary across municipalities, to some extent as a result of extensive planning rights and a tradition of relatively large self-determination in local government (e.g., local comprehensive plans and vulnerability assessments are mandated by law in Sweden, but not enforced by the state). In both Sweden and Finland, the development of local policy and measures is thus to a large extent influenced by the local identification of sensitivity, exposure and adaptive capacity. Financial, political and institutional capacities such as direct funding, the size of local government, number of staff and the prioritisation of environmental issues and established environmental policy aims, as well as explicit leadership on adaptation, are important factors in determining adaptive capacity at local level. In the Finnish KUUMA municipal cooperation, important constraints on adaptive capacity, including time and human resources as well as the area's perceived limited environmental sensitivity, rendered adaptation less urgent (as KUUMA municipalities are not located along the coastline). An additional crucial aspect seems to be the extent to which policy-makers or councillors (not only civil servants) are aware of and address the issue, thus raising the degree of political prioritisation – an observation common to the case studies and expressed particularly in the UK and Italian cases.

As noted in the section on environmental policy traditions, the desire to be a policy leader in the environmental field can be seen as a motivating factor for some municipalities. Examples include the UK local authorities of Hampshire and Woking and to some extent Gothenburg in Sweden, where adaptation had been linked to environmental policy development. The Dresden Model Region in Germany could potentially also be seen as such an aspiring local leader. The cases also suggest that had adaptation been developed as a priority in environmental policy more generally in Sweden, municipalities such as Trollhättan with its environmental aspirations might have been more involved in adaptation. Less economically tangible or measurable benefits to local authorities such as recognised status may thus have an impact; such a system is explicitly institutionalised for instance in the UK through the Beacon Programme of the national Improvement and Development Agency (IDeA) which has attributed 'Beacon' status in climate change to applicant councils with particularly well-developed mitigation and adaptation approaches.

In the Italian case, the decision-making capacities of the local level were to a large extent formed by regional characteristics, although actors also noted that issues were often handled in a reactive manner and on an individual basis. This was in part seen as the result of the allocation of financing being tied to centrally- or regionally-mandated planning and management tasks (where adaptation is not a priority). The regional structure in Emilia-Romagna that decentralises activities to the provincial level was considered supportive of establishing contacts within the territory and among stakeholders. This structure was seen as based on historically strong vertical and horizontal relationships as well as a focus on subsidiarity. The Italian case study municipality also exhibited a relatively strong relationship to the EU, both through EU-funded projects that helped to develop adaptation and indirectly through regional plans that receive EU support. However, the lack of perceived impacts from climate change in the Municipality of Ferrara has limited action on adaptation, despite the municipality's developed focus on sustainability and energy issues.

Given the nature of local government, the study has considered cooperation networks and cooperation traditions across municipalities with similar problems important. To differing extents, municipalities or local governments in all case study countries have drawn upon such networks. In the UK, networking at the local level is mandated through Local Authority Agreements (LAA) that set up Local Strategic Partnerships (LSP) for cooperation between local government and other stakeholders. Networking is also supported through structures such as the Regional Climate Change Partnerships and the Beacon Programme mandated by the IDeA national agency. Within the context of environmental policy development at the local level, adaptation to climate change is also sometimes seen as a re-framing of sustainable development concerns. Adaptation policy and practice therefore become partly integrated into and draw upon existing networks in this area. In Sweden, the UK and Italy (in the latter, both at provincial and municipal levels) Local Agenda 21 (LA21) bodies have been involved with local adaptation-related actions. One actor in the UK even went so far as to suggest that the LAA approach had supplanted the LA21 approach (or the need for local stakeholder integration that LA21 targeted, thereby integrating stakeholder inclusion into current policy).

More local traditions of cooperation also seem to play a role at the local scale. The KUUMA cooperation in Finland noted that pooling resources made it possible for some of the smaller municipalities to consider adaptation. As smaller Swedish municipalities in the study noted limitations in terms of capacity to deal with adaptation, such cooperation could potentially also support Swedish municipalities. In countries such as the UK, Sweden, Finland and Spain, local government associations (i.e. LGA, SALAR, ALFRA and FEMP, respectively) were also relatively proactive and had begun to work with adaptation and (especially in the cases of the UK and Finland) develop adaptation policy or guidelines. In Sweden, SALAR was among the actors with the most nuanced understandings of adaptation and the potential problems with regard to integration of adaptation as an aim across multiple levels. However, in both Sweden and Finland, the main focus in these bodies was still placed on mitigation.

In general, independent or voluntary initiatives – as distinct from programmes imposed by the central government – have played a significant role at local level in the case study areas. In the Finnish case, all municipal cooperation initiatives were voluntary and were often funded through EU means, as no separate state funding exists. Actors noted, however, that it will be up to each individual municipality to determine the extent to which measures developed in these cooperative initiatives are taken forward at the municipal level. In Sweden, initiatives for the development of tools for adaptation at sub-national levels were developed in the research project Climatools, undertaken voluntarily by state agencies without a specific mandate to this effect. Even in the relatively strongly centralised context in the UK, interviewees noted that the national level could not have set up an indicator for adaptation with funding implications for local government had it not been preceded and prepared by the Nottingham Declaration partnership's voluntary commitment to mitigation and adaptation. Finally, given the absence of policy at the state level in Italy, regional, provincial and local initiatives have largely been based on established environmental policy and identified sensitivities to present exposure (e.g., flood and drought events). The provincial level in the Italian case study in particular noted that participation in voluntary networks (e.g., AMICA) had fostered understanding of impacts and potential adaptation strategies.

8.5 The Role of the EU

Given the requirement for member states to implement EU directives, it could be expected that the EU level would also be perceived as an influence on national, regional and local planning and natural resource use. The cases demonstrate that the

EU level has affected the development of adaptation at multiple levels in Member States, despite the lack of a defined EU policy on adaptation beyond the Green and White Papers. EU influences on adaptation at the national level have occurred principally in the form of directives relevant to land and water use as well as EU projects that have supported and funded adaptation-relevant development. In the UK, some interviewees noted that EU directives influenced the ability to engage in adaptation as a result of their being based on assumptions of a steady-state relationship to the environment whereby, for instance, the designation for protected areas would not need to change over time. Due to this assumption and the exclusion of climate change effects in many of the directives at present, it was thus considered that the EU limited action on adaptation. Some interviewees also noted that national interpretations of the directives may play a more or less important role in this respect, not only the directives themselves. Interviewees as well as recommendations in Chapter 2 noted that migration paths and cross-national linkages in the EU Natura 2000 network would have to become more developed in light of climate impacts.

Chapter 2 also asks whether a more centralised approach to adaptation policy at EU level should be developed, in particular to integrate sectors and limit the risks associated with differentiated approaches in individual countries depending on adaptive capacity (where some vulnerable southern European countries might for instance otherwise not act early on adaptation). The problem of integration is described as that many policy goals with relevance to adaptation existed previous to the current focus on adaptation and thus do not include this concern, that for instance water management is affected by many policies, and that many decisions on management are made not by water managers but by other stakeholders e.g. in the private sector. The chapter notes that strategies are often only partial or sectoral responses to what are potentially much larger, overarching questions. There may also exist goal conflicts, institutional divisions, and vested interests that result in policy fragmentation, for which the development of a separate institutional body on climate change such as a Climate Change Commission may support the integrity and integration of the issue (as suggested in Chapter 2).

Among the main case study countries, the EU framework played a significant role in particular in Italy, where the allocation of structural funds has incentivised the design and implementation of environmental measures, while EU projects have assisted in the development of climate data. In Sweden and Finland, the role of the EU was comparatively less pronounced, potentially due to the fact that the adaptation process has not yet progressed to a point where the integration of adaptation concerns has resulted in conflicting policy directions between the EU and national levels. Despite the limited existing policy development in the EU, however, some national processes refer to the Green and White Papers as important catalysts for national policy initiatives. In Austria and Italy, for example, the EU Green and White papers were seen as the basis for either preliminary or continuing work on adaptation.

Further, the role of EU projects or other forms of support was mentioned in all main cases and in several of the supplementary cases provided in Chapter 7. While EU projects may be easily identified as specific and measurable initiatives on adaptation, they still indicate a pronounced role for the EU level and EU support policies. In the UK (Hampshire and other partners in the ESPACE project), Sweden (Gothenburg), Finland (Uusimaa regional council, Helsinki Metropolitan Area Council) and Italy (province of Ferrara), EU projects were considered to have played some role in the development of adaptation: in the direct development of adaptation strategies in Uusimaa and Helsinki, and in placing the focus on climate change impacts such as rising sea level in Gothenburg. Municipalities in Finland further noted that given the absence of climate legislation and associated resources for adaptation, several municipal cooperation initiatives had drawn on EU project funding. In Hungary, where regional divisions in accordance with delineations for European regional development funding were seen as having played a large role, most future funding for adaptation was also expected to come out of EU structural, cohesion and rural development funding.

In Italy and Greece in particular, the EU may have had a large role in the absence of national priorities for adaptation and preventative environmental policy. In Italy, EU regional funding arrangements have supported mitigation efforts by regional and local governments. However, interviewees also noted that EU research projects focused on innovative methodologies rather than on developing baseline data, which constitutes a limitation for a country where such data has yet to be developed in a coherent fashion. Finally, regional development plans that refer to EU, national and regional goals and targets constitute important planning tools on the regional level. Thus, in the Italian case in particular, regions and local authorities may 'jump scale' and rely to a large extent on EU funding and policy rather than on the national government alone. The examples of the Italian regional and provincial context drawn upon in this study may to some extent be distinguished by their involvement with EU policies that allow access to funding beyond what would otherwise have been attainable through the national system. In this case may the EU system contribute some cohesion beyond what is possible outside the EU area, although the EU policy on adaptation per se has so far been relatively limited.

8.6 Actors Beyond Government and Administration

To develop adaptation as an issue area, it may be necessary to involve a large array of stakeholders beyond public administration, including industry, as these will impact the implementation and decision-making on adaptation-relevant aims. In the four principal cases treated in this volume, stakeholder involvement beyond government and administration differed greatly, although participatory mechanisms have often been important in the review of adaptation strategies. In the UK, the stakeholder approach is well integrated (as one might expect from a country with a developed New Public Management profile, cf. Pollitt & Bouckaert, 2000). For instance, the national adaptation indicator must be developed within stakeholder groups at the local level, while the regional climate change partnerships (RCCP) include a range of local as well as regional stakeholders. In Sweden and Finland, stakeholder integration beyond government and administration has been more limited. For instance, one interviewee in the national Swedish Commission on Climate and Vulnerability noted that it had focused on the public management system, although occasionally participants from the insurance industry and other stakeholders took part. With regard to water issues, an important stakeholder forum in the future may be the water council (drawing upon both Swedish traditional requirements for river-basin cooperation and the EU Water Framework Directive through which these are being modified). In Finland, stakeholders beyond government and administration were particularly included in the development of the Uusimaa regional council's climate change strategy. In the Italian case, stakeholder involvement was especially pronounced in the NGO-led AMICA project and in the regional Water Protection Plan, where it was considered an important factor in increasing the acceptance and implementation of the plan. In the supplementary cases, the role of stakeholders beyond government and administration was highlighted in France in particular, where it has been suggested that regional climate strategies be formulated to support both companies and communities with developing climate plans from 2011. A case of regional cooperation with the private sector was also highlighted in Canada, although it was considered a relatively unusual development. The Netherlands, Spain and Germany additionally expressed an ambition towards broader stakeholder involvement.

As stakeholder involvement differs across countries, interviewees reported rather divergent experiences. While industry is to be included in both LSP and RCCP in the UK, many interviewees noted that industry was a particularly difficult sector to involve. This is largely seen as a function of the way industry operates: on shorter timescales, under conditions of competition, and with relatively limited need for coordination between different sectors in comparison with the considerable coordination and cooperation needs of local authorities. In Sweden a focus on industry had so far not been an explicit aim (despite the noted value of the involvement of selected individuals from industry in the Commission on Climate and Vulnerability). The role of NGOs also varied between the cases, but was in general relatively limited in the material. Contrary to previous studies (e.g., Bulkeley & Betsill, 2005), the ICLEI and its Cities for Climate Protection Campaign (CCP) was only mentioned as a major factor in the KUUMA partnership in Finland. While the importance of the CCP was greater in the UK during a time when it received national funding, its importance is now less pronounced.

More significant roles for NGOs could be seen in the Australian case, where an NGO supported adaptation toolkit development, as well as in the Greek and Italian cases. In Italy, the impact of WWF Italia was significant in encouraging adaptation processes at the national level through its publication of a guideline document for national adaptation. These guidelines were also submitted to the 2007 National Climate Change Conference, an event designed to support development of a national adaptation strategy. In Greece, the WWF and the IUCN seem to have played a similar role in developing guidelines to support national level approaches. These findings could be seen to indicate that environmental NGOs could gain a prominent role particularly in the absence of strong national development of the issue and potentially in cases where NGOs occupy such a role in relation to environmental policy development in general. In the other main case studies, the role of independent development of strategies has instead been filled by municipalities (such as in Sweden) and local authorities (in the Local Government Association, the UK's Nottingham Declaration and to some extent the RCCP).

8.7 Types of Adaptation

The general analytical framework of Smit, Burton, Klein, and Wandel (2000) outlined in the introduction (Chapter 1) suggests the need for both identification of the actors and issues, as discussed above, and a description of the means for adaptation, or how adaptation is taking place. The introduction also details some of the relatively broad typologies existing for adaptation, suggesting that adaptation measures can be differentiated based on the existence of policy priorities on adaptation, binding measures such as legislation, and adaptive capacity-building features. Organisationally, adaptation may also be differentiated in terms of whether it is mainstreamed into existing organisational bodies, or whether issue-specific organisations have been introduced. Table 8.1 describes notable examples of these policy responses.

In most cases, policy (and therefore planned and deliberate measures) has been developed or is under development. Italy, Greece and Canada stand out in this study as countries still lacking national policy priorities on adaptation. The most extensive case of planned development of adaptation is found in the UK, where policy is now integrated through legislation and regulation (including planning frameworks) and performance assessment at multiple levels. The national framework ensures that adaptation is both mainstreamed and enforced through the centralised nature of the political system, by requiring planning and other bodies such as local government to take adaptation into account in decision-making. The integration or mainstreaming of considerations of future climate change vulnerability into decision-making, which at the same time institutionalises adaptation through committed bodies, may be the most significant contribution in the UK approach.

In general, while changes in legislation and regulation to integrate planned adaptation are less developed, such have been developed beyond in the UK to a more limited extent in Sweden, Finland and Norway. In those countries, modifications have been made, notably to Planning and Building Acts, requiring inclusion of either erosion, flood risk, or risk and vulnerability assessments. A review of legislation and regulation is also being undertaken with regard to minimum flow rates for rivers and streams with relevance for the Federal Water Act in Germany; likewise, review of legislation is underway in France, Spain and the Netherlands. Given the limited enforcement of adaptation measures, specific practical measures undertaken for implementing policy aims at the regional and local levels are relatively limited at this point in time. For instance, the most common measure adopted at the local level in Sweden and Finland was an increase in minimum building elevation. Other measures have included dredging and flood protection measures, often undertaken in response to flood events, minimising impacts of emergencies arising

Table 8.1 Examples of adaptation development in the case studies

	Level		
Type	National	Regional	Local
Policy priority	Policy priorities existing in all study countries except Italy, Greece and Canada	Mainstreaming of national priorities (UK, Sweden, Finland) Relatively independent regional approaches in German Länder and Spanish regions	Local Government Association strategies or guidance (Sweden, Finland, UK, Spain) Local authority climate change strategies (UK, to lesser extent Sweden, Finland), planned in France from 2011
Binding measures (legislation and regulation)	Comprehensive legislative framework including performance assessment system (UK) Planning and Building Act (Sweden, Finland, Norway) Reviews of relevant national legislation and regulation ongoing in Germany, the Netherlands, Spain and France	Mainstreaming of national priorities (UK, Sweden, Finland, Norway) Suggestion that French regional governments be obliged to develop adaptation strategies	NI 188 preparing to adapt indicator (UK) Raising of minimum building elevation (Sweden, Finland)[a]
Adaptive capacity-building measures	Risk assessments, cost benefit analyses, research, information measures (in all main case studies, in varying combinations) Research and monitoring emphasised in most supplementary case studies (although also other capacity-building measures may exist) National support for pilot projects or pilot municipalities (e.g. Norway, Germany, Netherlands, Australia) Best practice awards (UK, Spain)	Mainstreaming of national priorities, regional level awareness raising for cooperation (UK, Sweden, Finland, many supplementary case studies)	EU projects (all main case studies, some supplementary case studies) Municipal cooperation (different mixes in different countries) Some independent or policy-related development of practical measures such as included dredging and flood protection, development of more efficient water use

Table 8.1 (continued)

	Level		
Type	National	Regional	Local
Dedicated organisation (long-term scope)	National level committees, UKCIP (UK) Ministries or departments on climate change where adaptation is an integrated aim (e.g. Australia, Hungary) National dedicated organisations, e.g. in Germany, Norway, Austria, France and Australia, such as the Competence Centre on Climate Impacts and Adaptation KomPass (Germany), interministerial adaptation group with executive secretariat (Norway)	RCCP (UK) Regional centres for formalisation of adaptation plans (suggested for France) Climate Change service and commission (independently developed at regional level in particular case, Spain) Climate Change offices (Spain, Australia)	Commission of Enquiry, dedicated climate change political champion positions, community groups (UK)
Mainstreaming in existing organisations (rather than through a dedicated organisation)	Notable in Sweden and Finland (although existing as a policy aim in many countries)	Notable in Sweden and Finland	Cross-departmental or municipal cooperation (Sweden, Finland)

[a]Supplementary case studies (Chapter 7) provide too little detail to adequately assess potential developments in leading municipalities.

from climate change, and development of more efficient water use. In Sweden and Finland, these kinds of measures are decided at the local level as a result of the decentralised planning system.

In many countries, adaptation is thus at present in a phase of development which is marked more by the identification of approaches than by developing binding legislation and implementation. Most countries are at a similar level of what in the UK has been formalised as 'preparing to adapt': e.g. Austria is developing a database of adaptation-relevant activities and a vulnerability assessment, including a 2009 draft *Towards a national adaptation strategy* currently under review. In the Netherlands the focus is on beginning the 'institutionalisation of adaptation' through an adaptation agenda, and the mainstreaming of adaptation into planning aspects from 2015 onwards, and in Spain the national adaptation plan highlights the next steps needed to begin defining adaptation options (to then be implemented through work programmes). The relatively typical broad measures undertaken in existing good practice approaches can be exemplified by e.g. the Land of North Rhine-Westphalia, Germany, which has developed a regional adaptation strategy with the aim of raising public awareness, developing research, knowledge and adaptation measures, increasing overall adaptive capacity, and providing assistance to various sectors. Adaptive capacity-building measures are thus of particularly large significance in the present, and on the national level range from increased research and monitoring to the development of cost-benefit analyses and national support for pilot programmes.

Further has the institutionalisation of adaptation taken different paths in different countries. Dedicated organisations or bodies that institutionalise adaptation as a priority exist in several cases on several levels, while other countries have focused mainly on mainstreaming within existing organisations. While mainstreaming is a goal in most of the case study countries, Sweden and Finland are notable in the study as countries with developed policy priorities on adaptation that have chosen mainsteaming as their main approach without simultaneously focusing on the development of organisations dedicated to adaptation. A number of dedicated organisations such as the UKCIP in the UK and the National Climate Council in Spain explicitly (although with very different approaches) aim at multi-level and multi-actor integration. Future studies may here serve to indicate how well different choices for organisation on adaptation serve to support multi-level governance on the issue.

Finally, in some cases on different levels, adaptation has emerged largely out of existing practices under other frameworks and subsequently been re-defined as adaptation (e.g., in Woking and to some extent in the Province of Ferrara). In the Italian case in particular, adaptation has so far been seen as a response to current rather than future vulnerability, a reactive ‘planning for today’. This highlights both that the need to integrate adaptation with reference to long-term climate change may pose challenges to reactive approaches, but also that existing measures, although not constituting planned adaptations per se, may in some areas pose a basis for developing adaptation in the future.

8.8 Lessons for Current and Future Adaptation

As noted in the introduction (Chapter 1), the use of the terms ‘best practice’, ‘lesson drawing’ or policy transfer refers to the potential for utilisation and transfer of examples of adaptation that have developed in other countries or contexts. As ‘off the shelf’ solutions, they could be more quickly implemented than any domestic solutions that require development from scratch. However, given the cross-national differences described above, limitations on the transferability of approaches necessarily exist. Interviewees’ observations may thus mainly demonstrate the extent to which a relevant policy is seen as a model or as relevant to a given context, rather than indicate the direct transfer of solutions (Bandelow, 2007).

'Lesson drawing' or policy transfer is expressed differently within the different contexts of each of the cases. 'Lesson-drawing' is highly integrated within the domestic UK context, potentially as a result of the extent to which 'best practice' and NPM approaches are emphasised (cf. Wilson & Game, 2006). 'Best practice' transfer is considered a goal in national programmes such as the IDeA's Beacon Programme, and perceived as a tradition in local authority practices. Programmes for explicit lesson-drawing (or the sharing of experiences) also exist for instance through Germany's *Klimzug* programme and in the Netherlands' *Knowledge for Climate*, where municipalities are given the responsibility to undertake knowledge-sharing on adaptation according to the Dutch national adaptation strategy. Initiatives such as the *Future Cities* project for inter-municipal cooperation in Norway may also serve to support 'lesson drawing' within the national context. UK examples of development on adaptation are in particular referred to in other cases, with emphasis on the UKCIP approach to adaptation support for local and regional levels as a possible model for different contexts. In Sweden, the development of tools to support municipal adaptation in the Climatools project was inspired by the UKCIP's LCLIP, which thus potentially represents a clear and accessible 'off the shelf' solution that can be implemented in different contexts. In Finland, a focus on the development of tools for adaptation was largely lacking, with the exception of the Helsinki Metropolitan Area Council's suggestion that sectors should outline measures taken during unusual weather events.

More generally, existing networks for sustainability may also be utilised for 'lesson drawing'. Interviewees from both the Province and Municipality of Ferrara in Italy, for instance, emphasised participation in LA21 networks, the Sustainable Cities Campaign and EU projects as important for learning about 'good practice'. As noted above, several interviewees also highlighted the role of EU projects in the initiation of the development of adaptation or adaptation-relevant approaches (such as Gothenburg, Helsinki, and Hampshire), and thus in the provision of access to ideas from other contexts. Most supplementary case studies also describe networks at sub-national levels that have either prompted or extended regional and local adaptation.

These examples indicate that real-life situations may incorporate elements of both the specific, local context-dependency suggested in adaptive capacity literature, and certain elements of transferability identified by practitioners. In the context of governance, 'best practice' or 'good practice' may also relate to network means of governance, where novel municipal approaches undertaken within a given municipality may be inspired within networks or by other voluntary initiatives rather than imposed by the state. On the other hand, states may in a regulative context also use good practice examples as a means of establishing certain practices without having to develop more costly implementation and control measures. Bottom-up development of 'best practice' may for instance constitute the basis from which a government may establish binding measures once the voluntary application has become institutionalised. This was exemplified in particular by the development of the Nottingham Declaration local authority network in the UK, perceived by some interviewees as a strong precursor for the national NI 188 adaptation indicator later developed for local government.

Best practice examples can thus be differentiated first in terms of their domestic or international context, where best practice transfer within the domestic context may be supported by national programmes, whereas policy transfer between states may predominantly consist of inspiration from existing initiatives. To-date, given the early stage of issue development, indications of international policy transfer have been identified mainly within the LCLIP and to some extent the stakeholder-focused approach championed by UKCIP – itself inspired by a non-domestic case in early Canadian impact assessment. Second, best practice transfer can also be differentiated as either voluntary initiatives in the absence of state approaches, or as measures that supplement state initiatives. Such voluntary strategies could later be drawn upon by the state to set up similar but binding measures.

8.9 Conclusion: Does Adaptation Require a Change in Thinking?

On balance, this volume demonstrates very different adaptation development paths, where the development of adaptation policy as an issue area and its translation into practice depend upon a large number of factors. Beyond (and sometimes in the absence of) impacts perceived as relevant to climate change, these include a number of structural characteristics: the national political and planning system and the extent of decentralisation and decision-making power to different levels; institutionalised environmental policy and broader regulative traditions, to some extent manifested in national political systems; and a number of different resources such as financial and staffing resources (which may be related more broadly to environmental policy or to the capacities of the unit). Individual- and context-dependent measures that may impact the development of adaptation are pronounced. These include the presence of political champions that support policy-making on the issue, the public and media context that can help or hinder the development of adaptation as an issue, and events such as floods and droughts that may serve to focus attention on adaptation within a receptive policy environment. Differences across the development of these factors have resulted in the different treatment and integration of adaptation in each of the cases in this volume and have influenced perceptions of the importance of adaptation across a spectrum from it being seen as having a relatively limited impact to it being an issue with overarching impacts for planning systems.

Policy statements as well as interviewees in the UK and the Netherlands express a view of climate change as an issue that may fundamentally change the planning system, for instance, 'that the current approach to coastal defense may no longer be viable in the future' (VROM, 2008, p. 27, quoted in Chapter 7). In other areas, adaptation may be seen as an issue for less-demanding mainstreaming, designed without simultaneously allocating significant funding or dedicating institutional bodies for adaptation (or specific, well-funded adaptation roles in existing organisational bodies). In the Finnish case – one of the most explicit examples of adaptation where focusing events have been relatively unusual – one interviewee noted that it was still unclear whether adaptation would require only technical measures for implementation, or whether it would require changes in social organisation. However, several interviewees involved in sub-national strategies noted that their approach to adaptation differed from their approach to environmental issues in the past, mainly in the adaptation issue having a greater focus on cooperation between different levels and bodies.

As adaptation is an issue under development in many of the countries described in this volume, the largest contribution of this book may be the identification and discussion of the specific factors and policy contexts that influence the ways in which different multi-level systems may approach adaptation. Given the ongoing development of adaptation policies and measures, it is in some cases difficult to distinguish whether existing limitations on the development of adaptation refer to the capacity of the national level to steer adaptation, or whether these instead relate to the early stage of development itself. In the Swedish case, there are clear indications that the State is attributing significant responsibility for adaptation to the municipal level, given the municipal planning monopoly. In Finland, the extent to which the national level may fund local adaptation in the future remains undefined. However, as adaptation develops as an issue area, the limitations and particularities in each system may become increasingly apparent.

Some significant limitations on adaptive capacity can be noted in the studies, either historically or in the present. For instance, it seems that a state framework that does not define an adaptation policy (as in Italy or Greece), has not clearly distributed responsibilities (as in Sweden, up until the 2009 Bill) or does not require local authorities to address adaptation in a way that impacts measurable indicators or funding (as in the UK, up to the establishment of the NI188 adaptation indicator) may negatively impact the identification and implementation of adaptation measures at the local level. In general, the existence of qualitatively or quantitatively measurable and defined indicators or factors for the implementation and contextualisation of adaptation (preferably developed with some bottom-up component so that they match the needs of administrative or other units) may support the development of adaptation across scales. The need for incentives for local government to develop adaptation is also highlighted by the cases: for instance, the need to provide economic incentives for adaptation as within mitigation policy (noted by an interviewee in Finland) or to assure that adaptation is included within funding parameters.

While the need for adaptation policy and measures may become pronounced in the future given existing greenhouse gas emission levels, the many factors that impact the development of adaptation approaches in different contexts and on different scales may require differentiated development over time or depending on context. While certain countries have been able to progress on adaptation, differentiated national prerequisites will impact the continued institutionalisation of adaptation as a policy aim. Interviewees in the Italian case, for instance, noted that adaptation as a new policy issue could not be rushed as there had been a lack of preparation on the issue and the concept of adaptation had been poorly understood.

As such, it is possible that the basis for the development of a national adaptation strategy did not exist in 2007 when this development was attempted, given the media context and the partisan nature of the issue in Italy. In the UK, getting to a level where 'preparing to adapt' could be integrated as an aim at all levels of policy has taken over a decade since the inception of the UKCIP. Here, interviewees note that adaptation requires persistent conceptual and practical development in order to be more fully integrated into a context of organisational change. This is also the stage at which many nations, regions and localities find themselves today:

> The challenge for adaptation at this point is to understand what it means. We are still at an early stage of exploring what it means, what we mean by an adapted community or an adapted local government, what we mean by adaptive capacity or how you build capacity in an organisation to make it adaptation proof. (UK IDeA Agency/Nottingham Declaration, interview)

References

Bandelow, N. (2007). Health policy: Obstacles to policy convergence in Britain and Germany. *German Politics, 16*(1), 150–163.

Baumgartner, F.R., & Jones, B. (1993). *Agendas and instability in American politics.* Chicago: University of Chicago Press.

Birkland, T. A. (1998). Focusing events, mobilization and agenda setting. *Journal of Public Policy, 18*(1), 53–74.

Bulkeley, H., & Betsill, M. (2005). Rethinking sustainable cities: Multilevel governance and the 'urban' politics of climate change. *Environmental Politics, 14*(1), 42–63.

Glachant, M. (2001). Introduction: A policy perspective on the implementation of the Community environmental legislation'. In M. Glachant (Ed.), *Implementing European environmental policy. The impacts of directives in the Member State* (pp. 1–11). Cheltenham UK and Northampton, MA, USA: Edward Elgar.

Keskitalo, E. C. H., Westerhoff, L., & Juhola, S. (in preparation). Agenda-setting on the environment: The development of climate change adaptation as an issue in European States. 25p.

Kingdon, J. W. (1995). *Agendas, alternatives and public policies* (2nd ed.) New York: HarperCollins.

Peters, B. G., & Pierre, J. (2005). Swings and roundabouts? Multilevel governance as a source of and constraint on policy capacity. In M. Painter & J. Pierre (Eds.), *Challenges to state policy capacity. Global trends and comparative perspectives* (pp. 38–51). Houndmills Basinstoke: Palgrave Macmillan.

Pollitt, C., & Bouckaert, G. (2000). *Public management reform. A comparative analysis.* Oxford: Oxford University Press.

Rhodes, R. A. W. (2000). Governance and public administration. In J. Pierre (Ed.), *Debating governance. Authority, steering and democracy* (pp. 54–90). Oxford: Oxford University Press.

Rhodes, R. A. W. (2007). Understanding governance: ten years on. *Organisation Studies, 28*(8), 1243–1264.

Sandford, M. (2005). *The new governance of the English regions.* Houndmills Basingstoke: Palgrave Macmillan.

Smit, B., Burton, I., Klein, R., & Wandel, J. (2000). An anatomy of adaptation to climate change and variability. *Climatic Change, 45*, 223–251.

Smit, B., & Wandel, J. (2006). Adaptation, adaptive capacity and vulnerability. *Global Environmental Change, 16*, 282–292.

Veggeland, N. (2000) *Den Nye Regionalismen. Flernivåstyrning og europeisk integrasjon.* Fagbokforlaget, Bergen.

VROM (Ministry of Housing, Spatial Planning and the Environment) (2008). *National Programme on Climate Adaptation and Spatial Planning. National Adaptation Strategy – Policy Memorandum.* Amsterdam: VROM.

Wilson, D., & Game, C. (2006). *Local government in the United Kingdom* (4th ed.). Houndmills Basingstoke: Palgrave Macmillan.

Wurzel, R.K.W. (2002). *Environmental policy-making in Britain, Germany and the European Union. The Europeanisation of air and water pollution control.* Manchester: Manchester University Press.

Index

A
ACACIA project, 45
Accommodation, 26, 138
Adaptation
  definition, 182
  local level, 17–19, 23, 28, 30–31
  measures, 2–3, 5, 7–8, 25, 32, 52, 65, 135, 159–166, 168, 177–180, 189–228, 249–251, 271–332, 341–365
  research, 43, 47, 166–168, 174, 181, 258, 265, 306, 314–316, 319, 328
  web portal, 205
Adaptation Policy Framework (UK), 101, 104
Adapting to Climate Change Programme (ACC) (UK), 115–116
Adaptive capacity
  definition, 9, 179
  determinants of, 3, 8–10, 24, 142
  differentiation at local level, 33, 237, 340
Afforestation, 41, 71, 75, 81–86, 89, 91, 294
Agenda-setting, 6–7, 9, 15, 19–20, 226, 340–341, 347–349
Agriculture, 44, 46, 50, 52, 56–58, 60–64, 66, 71, 75–76, 79, 81, 85, 87, 114, 159, 163–165, 167, 171, 182, 191, 196, 204, 242–245, 250, 252, 254, 256, 259–261, 275, 277–279, 281, 283, 286, 288–289, 292, 294–295, 297–299, 303, 306, 308, 310, 313–316
Agricultural sector, 46, 63, 71, 124, 163, 169, 246, 254, 256, 259, 261–262, 268, 292, 315
Al Gore, 102, 211, 214, 263, 348
AMICA, 259–262, 266, 280, 290, 330, 354, 357
An Inconvenient Truth, 102, 211, 348
ARPA, 235, 237–238, 244–245, 252–254, 256–257, 264
Association of Finnish Local and Regional Authorities (ALFRA), 169–171, 354
Audit, 82, 100, 106, 108, 116, 119, 194, 320
Audit Commission, 106, 108
Australia
  Climate Change Adaptation Research Facility (NCCARF), 314
  Commonwealth Scientific and Research Organization (CSIRO), 314, 316
  Department of Climate Change, 314, 316–318
  Local Adaptation Pathways Program (LAPP), 317
  Local Government Climate Change Adaptation, 317, 350
  National Climate Change Adaptation Framework, 313–314, 316–317, 319
  National Climate Change Adaptation Programme, 313
Austria
  The Austrian Council on Climate Change (ACCC), 280
  AustroClim, 277
  BMLFUW, 277–280
  Bundesländer, 277–280, 323
  Kyoto Forum, 278–279

B
Baltic Sea, 171–172, 174, 220, 307
Beacon award, 129, 143
Benchmarking, 100
Berlusconi, 236, 246, 265
Best practice, 26–27, 105–106, 114, 116, 118, 122, 127, 129, 134–135, 140, 143, 170, 175, 210, 220, 261, 264, 293, 312, 349, 359, 361–363

E.C.H. Keskitalo (ed.), *Developing Adaptation Policy and Practice in Europe: Multi-level Governance of Climate Change*, DOI 10.1007/978-90-481-9325-7,

Biodiversity, 41, 44, 47–48, 54–56, 71, 73–86, 89, 122, 159, 167, 234, 242–243, 245, 279, 282–283, 292, 295, 308–309, 314–316
BirdLife International, 74, 82, 87
Birds Directive, 20, 74, 77, 131

C
Canada
  British Columbia, 273, 320–321
  Climate Change Impacts and Adaptation: a Canadian perspective, 319
  Climate Change Impacts and Adaptation Research Network (C-CIARN), 319, 321
  National Climate Change Adaptation Framework, 313–314, 316–317, 319
  Regional Adaptation Collaboratives (RAC), 319
  Toronto, 321
Carbon
  emissions, 163, 170, 173
  tax, 154, 236, 240
Case studies, 2–3, 16, 28–32, 130, 142, 172–173, 212–218, 266, 272, 342–343, 347, 353, 358–360, 362
Case studies, selection of, 272
Catchment Flood Management Plan (CFMP), 102, 110
Cities for Climate Protection (CPP), 26, 129, 133, 155, 171, 175, 211, 240, 252, 317, 357
Civil Protection Act, 199, 209
Climate Alliance, 211, 252, 259, 262–264, 266–267, 280, 290
Climate Change Act (UK), 2, 101, 103–105, 107, 113, 140
Climate change as an issue for developing countries, 192–193
Climate Change Commission of Inquiry (Sweden), 122–123, 125–126, 142
Climate Change Programme (UK), 40, 47, 100–101, 115–116, 267, 279, 291
Climate change research, 111, 157, 166–167, 241, 244, 246, 249, 275, 277, 280–281, 290, 303
Climate Commission, 209
Climate impacts, 2, 40–41, 43, 45, 49–52, 56, 60, 67, 76, 78, 90–91, 98, 103, 105, 107, 110–113, 124, 127, 166, 168, 179, 192, 195, 201, 205, 209–210, 216, 225–226, 241, 243–245, 247, 251, 253–255, 258, 263–264, 266, 275, 280–281, 286, 288–290, 295, 297–298, 303, 311, 313–316, 324, 330, 332, 349–350, 355
  See also Impacts of climate change
Climate information, 177, 258, 281, 314, 317, 322
Climate Investment Programme (Klimp) (Sweden), 210, 217
Climate Manifesto – a New Deal for Sustainable Adaptation and Environmental Security, 245
Climate Municipalities (Sweden), 211, 216, 223, 228
Climate South East, 98, 114–117, 122, 125, 134, 136, 143
Climatools, 195, 204–205, 211, 225, 228, 354, 362
CMCC, 244, 247–249, 265, 350
Coastal zone, 26, 234, 243, 245, 300, 308, 313
Coding, 32
Commission on Climate Change and Development, 193–194
Commission on Climate Change and Development interim report, 196, 199, 226
Commission on Climate and Vulnerability, 191–204, 208–209, 212, 219–222, 226, 228, 306, 357
Commission on Sustainable Development, 193–195
Committee on Climate Change (UK), 104, 134, 193
Common Agricultural Policy (CAP), 3, 33, 53, 65, 71, 81
Comparative studies, 27–28
Consensus-building, 150, 297, 323
Cooperation, 16, 21–22, 31–33, 44, 53, 72, 80, 85, 87, 110–111, 114–115, 120, 125, 130, 151, 155, 159, 161, 165, 167, 172–181, 193–195, 203, 205–207, 215–216, 218, 220, 222, 225, 228, 238, 257, 259, 261, 283, 285–288, 290, 293, 303, 305, 323, 326, 340, 345, 348–349, 351–354, 356–357, 359–360, 362, 364
Coordination, 4, 7, 41, 51, 54, 64–65, 85–91, 110, 115–116, 125, 128, 133–134, 136, 141–142, 156, 159, 165, 190, 194, 199, 206–209, 213–214, 216, 236–237, 244–245, 249, 255, 264–265, 284, 305, 309, 314, 317, 322–323, 326, 329, 331, 342, 344, 349–350, 357
Coping, definition, 8

Corporatism, 12, 291, 297
County Administrative Board, 190, 196–197, 199, 201–203, 205–209, 214, 218–225, 227, 229, 351
Creating a Better Place, 109
Crisis response, 6–7
  See also Emergency response

D
DAISIE project, 79
Decentralisation, 7, 12–14, 16–17, 19, 33, 87–88, 90, 190, 249, 261–262, 266, 271–272, 275, 293, 308, 323–327, 340–341, 343–344, 346, 351–352, 363
Department of Civil Protection (Italy), 250, 256
Department for Energy and Climate Change (DECC), 102
Department for Environment, Food and Rural Affairs (DEFRA), 98, 100–106, 108, 110–118, 120, 134–137, 141–143, 306
Desertification, 70, 84, 242–243, 245–246, 249, 275, 292, 344
Developing Policies & Adaptation Strategies to Climate Change in the Baltic Sea Region (ASTRA), 172
Domestic adaptation programme board (UK), 134
Dresden, 259, 290, 353
Drought, 2, 30, 40, 43, 54–61, 63–67, 69, 72, 91, 102, 124, 130, 150, 234, 243, 249, 255–257, 259–260, 267, 279, 286, 289, 291, 295, 308, 313, 344, 347, 351–352, 354, 363
Drought management, 66, 255
Drought Management Plans (DMP), 68, 255–256

E
The Economics of Ecosystems and Biodiversity (TEEB), 55, 84
Economic sector, 47, 157, 165, 171, 180, 295
Ecosystem services, 41, 55, 63, 73, 75, 82–85, 295
Emergency response, 51, 256
Emilia-Romagna, 31, 234, 252–258, 261, 266, 351–353
ENEA, 236, 238, 240, 244, 247, 249, 254
Energy
  efficiency, 40, 68, 154, 162–163, 171, 175, 179, 240, 259, 262–264
  national climate and strategy 2001, 154–155
  revised national climate and strategy 2005, 155–156
  revised national climate and strategy 2008, 156–157
  sector, 56–57, 61, 66, 89, 164, 192, 242, 304
England, 97–143, 226, 352, 352
Eni Enrico Mattei Foundation, 244, 248
Environment administration, 149, 153, 160–162, 165, 171, 179, 182, 212
Environment Agency, 98, 100–102, 104, 109–110, 122, 127, 130–131, 136, 139, 141
Environmental assessment, 155, 259, 297, 310
Environmental Impact Assessments (EIA), 78, 84–85, 159, 205, 236, 259, 310, 322
Environmental management, 159, 211, 260, 266, 268
Environmental policy
  institutionalisation, role of, 344–347
  state divisions according to, 13, 15
Environment sector, 161, 165, 175, 238, 254, 260, 309, 328–329
Equity, *see* Adaptive capacity, determinants of
Erosion, 70, 108, 123, 137, 191, 197–199, 202–203, 206, 212, 214, 243, 245, 259–260, 278, 358
ESPACE, 117, 122, 142, 274, 356
Espoo, 31, 173–174, 176–178, 180, 182, 346, 348, 351
2020 EU Climate Package, 168
EU Floods Directive, 55, 69, 288
EU Green Paper on adaptation, 4, 210, 220–221, 278
EU Habitats Directive, 21, 32, 77, 80, 131, 221, 228
EU Life +, 174
Euro-Mediterranean Centre for Climate Change (CMCC), 244, 247–249, 265, 350
European Climate Change Programme (ECCP), 40, 42–56, 81, 267
European Environment Agency (EEA), 2, 40, 46–47, 54–57, 62–64, 68–69, 74–77, 83, 86, 108, 240, 252, 274, 280
Europeanisation, 20, 27, 344

European Union (EU)
funded research and resources, 18–19, 29, 32–33, 40, 45–46, 50–51, 53–55, 71, 74, 82, 84, 106, 109–110, 115–116, 118–119, 121–122, 127–128, 130–133, 141–142, 151, 166–170, 172–174, 178, 181–183, 195, 197, 200–201, 204–206, 210, 214, 217–222, 227–228, 236, 239–240, 242, 244, 250, 254, 256–259, 261–262, 264, 266–267, 273, 277–282, 284–285, 287–294, 296–297, 300–307, 313–314, 317–319, 324, 330, 331, 340, 342–344, 348–357, 363–364
level impacts, 220–221
role in climate policy, 168–170
sectoral policies and warning systems, 51–56
stakeholders/NGOs, relationship, 329–330
EU Water Framework Directive, 90, 207, 227, 357
EU White Paper on adaptation, 44
Evaluation, 46, 74, 81, 108, 143, 162, 164, 166–168, 182–183, 193, 207, 224, 244, 247–248, 257, 274, 281, 283, 308, 311
Exposure, 7, 17, 142, 304, 342, 352, 354
Extreme Weather (Sweden), 40, 43, 49, 58, 60, 102, 125, 142, 177, 194, 197, 204, 213, 228, 243, 289, 295, 313

F

Families of states, 13
FAO, 169
FERN, 74, 82
Ferrara, Municipality of, 234, 252, 262–264, 267, 353, 362
Ferrara, Province of, 31, 252, 258–262, 356, 361
FINADAPT, 167
Finland
climate change adaptation, 149–183
environmental policy in, 152–153
research role in, 166–168
nested case studies, 30–31
First Carbon Initiative (Finland), 154
Flooding, 7, 10, 20–21, 24–25, 30, 40, 55, 69, 98, 101–103, 108, 113, 125, 127, 130, 136–137, 140, 142, 150, 161–162, 165, 176, 195–199, 202–204, 208, 212, 214–216, 218–219, 225, 227–228, 242–243, 256, 259–260, 267, 296, 299–301, 308, 321, 346–348, 352
Focusing events, 6, 19, 140, 142, 196, 226, 340, 345–348, 364
Forests, 44, 50, 52, 54, 69–71, 75–76, 79, 81–83, 163, 167, 282, 288, 308, 310
Fragmentation, 41, 77, 86, 88–89, 234, 238, 265, 346, 355
Framework directives (EU), 250, 292
France
French National Adaptation Strategy, 273, 282–284
French National Climate Plan, 284
French National Observatory of the Effects of Climate Warming (ONERC), 281–284, 328
GICC, 281
Grenelle, 283–284, 325, 329, 345
National Adaptation Plan, 272, 274, 276, 283–284, 287, 361
Funding, 44–46, 201, 224, 279, 290, 301, 317, 319, 324–325, 356

G

Germany
Dresden, 353
Federal Ministry for the Environment, Nature Conservation and Nuclear Safety, 286
German Climate Protection Programme, 286, 349
German Strategy for Adaptation to Climate Change (NAS), 286–289, 300, 323, 325, 328
Government-Länder Dialogue on Adaptation to Climate Change, 287
KomPass, 286
Länder, 285–290
REGKLAM, 290
GHG emissions, 40, 68, 90, 154, 239–241, 247, 252, 258, 262, 280, 309
Good practices, 27, 53, 99, 101, 133–135, 137, 171, 179, 217, 225, 261, 292, 361–362
Goro, 260
Göta River, 190–191, 196–198, 201–202, 207–208, 212, 214, 217
Gothenburg, 190–192, 201–202, 211–216, 221, 224, 227–228, 343, 345–346, 351, 353, 356, 362
Gothenburg Region Association of Local Authorities, 211

Governance
adaptive capacity connection, 22–28
adaptive capacity on different levels, 11–22
goal conflict and institutional confusion, 85–90
multi-level definition, 4–7
in Italian adaptation, 265–268
sub-national authorities, decentralisation/participation of, 325–327
Government Office (GO), 118–119
Government Office South East (GOSE), 105–106, 112, 114, 118–119, 137
Greece
ADAGIO, 292
Climate Change and Impact Research: the Mediterranean (CIRCE), 291, 330
Gudrun, 50, 139, 192–193, 196, 213, 347

H

Hampshire County Council, 98, 117, 121–126, 142, 205, 225
Health, 44, 51, 52, 108, 119, 135, 152–153, 157–158, 160, 167, 177, 190, 199, 204–206, 234–235, 237, 242–243, 245–246, 249–250, 252, 254, 264, 279–283, 286, 289, 295, 298, 301, 303, 308, 309–310, 313–315, 319, 344
Health sector, 51, 246
Heat/Health Watch/Warning System, 250
Heat wave, 2, 58–59, 72, 102, 108, 132, 234, 243, 249–250, 252, 256, 259, 264, 281, 347
Helsinki Metropolitan Area Council (HMAC), 173–182, 342, 346, 348, 356, 362
'Hollowed-out', 11, 17
Hungary
National Climate Change Strategy, 294
New Vásárhelyi Plan (VTT), 295–296
VAHAVA, 294
Hydro-Meteorological and Climate Service, 254

I

ICLEI-Local Governments for Sustainability, 240
Impacts of climate change, 2, 43, 51, 77, 111, 114, 158–161, 164–165, 168, 176–177, 181, 197, 203, 212, 241–242, 244, 257, 281, 287, 290, 294–295, 308, 310–311, 313, 315, 319–320, 346
Implementation
measures at municipal level, 155–156
of NAS, 160–162
NAS implementation, evaluation, 162–165
national environmental policy, relevance of, 327–329
Improvement and Development Agency for local government (IDeA), 112–113, 116–118, 129, 132–135, 138, 353, 362
Improvement and Efficiency South East (IESE), 114, 119
Incentive, 2, 21, 24, 68, 121, 136, 165, 200, 210, 224, 240, 246, 256, 258, 263, 267, 291–293, 326, 330–331, 344, 364
Indicator, 6, 54, 73–74, 98, 103, 105–106, 108, 110, 112–113, 117, 119–121, 125, 134, 138, 141–142, 160, 164, 169, 213, 215, 244, 282, 310, 312, 320, 340, 352, 354, 356, 359, 363–364
Infrastructure, *see* Adaptive capacity, determinants of
Institute for Environmental Protection and Research (ISPRA) (Italy), 236–237, 239, 244–245, 247–248, 253
Insurance, 47, 136, 158, 160, 195, 200, 219–220, 226, 228, 283, 304, 306, 308, 310, 357
Integration, 12, 17, 25, 51, 74, 80, 87, 99–100, 102–103, 107–110, 114, 121, 124, 128, 133–141, 150–151, 160, 165, 170–171, 178–179, 182–183, 192, 195, 204, 238, 247, 258, 260, 267, 287, 290, 294, 307, 309, 328, 345, 351, 354–356, 358, 361, 363
Intergovernmental Panel on Climate Change (IPCC), 1–2, 7–8, 42, 45–46, 102, 158, 180, 195, 213, 234, 242, 255, 321, 348
Interministerial Committee for Economic Planning (CIPE), 235–236, 239, 244, 249, 265
INTERREG, 53, 172, 181, 259, 267, 280, 284, 290, 293
Interview, 143–144, 183–184, 229, 268–269
methodology, 28
Invasive species, 40, 54, 78–80, 243
Invasive Species Act (IAS), 79
Irrigation, 60, 63–64, 69, 256, 279, 351
ISPRA, 236–237, 239, 244–245, 247–248, 253
ISTO, 167–168, 181

Italian National Agency for New Technologies, Energy and the Environment (ENEA), 236, 238, 240, 244, 247, 249, 254
Italy, 13, 14, 29, 31, 59, 233–268, 359

J
Järvenpää, 175
Jump scale', 3, 18, 28, 356

K
Kerava, 175–176, 180
KUUMA, 31, 173, 175–176, 179, 181, 348, 352–354, 357
Kyoto
  Club, 240, 264
  Protocol, 58, 88, 101, 152, 193, 239

L
Landslide, 70, 190–191, 196–200, 202–203, 208, 216–219, 222, 226–228, 243, 246, 252, 257, 303, 306
Law on Extraordinary Events, 209
Learning to Live With Rivers, 101
Lesson-drawing, 4, 26–27, 143, 225–226, 362
Local adaptive capacity (UK), 18
Local Agenda, 21, 130, 132, 211, 235, 237, 259, 261, 263, 354
Local Area Agreements (LAA), 106, 120–121, 123, 127, 132, 142, 353–354
Local Authority
  Hampshire, 121–124
  Portsmouth, 122, 125–127
  Surrey, 122, 127–128
  Winchester, 122, 125
  Woking, 122, 129–130
Local Climate Impacts Profile (LCLIP), 113, 126–128, 179, 205, 225, 228, 350, 362–363
Local Government Association (LGA), 102, 104–105, 107–108, 112, 120–121, 134, 138, 141, 354, 358–359
Local Investment Programme (LIP), 210, 217
Local planning, 18, 110, 120, 190, 223–224, 226–228, 237, 251, 302, 307, 342–343, 354
Local and Regional Adaptation Partnership Board (LRAP), 103, 105–106, 111–112, 115–116, 134, 140
Local Strategic Partnerships (LSP), 98, 106–108, 119, 125–127, 132, 136, 353
Lowest building level, 213, 215–216

M
Mainstreaming, 2, 25, 41, 44, 55, 91, 131, 133–134, 149–183, 222, 240, 267, 298–300, 322, 324, 331, 340, 346–348, 358–361, 363
Making Space for Water, 101, 108
Mäntsälä, 175
Media, 6–7, 9, 102, 113, 178, 195, 245, 248–249, 251, 282, 321, 340, 346–348, 363, 365
Mediterranean, 241, 291–293
Methodology, 4, 28, 99, 191, 204, 244, 260, 273, 321
Ministry of Environment, 153, 161, 182, 235–236, 239, 259, 261, 263–264, 274, 281, 289, 292, 294, 296, 304, 308–310, 312, 328
Ministry of Environment and Territory (MATTM) (Italy), 234, 236, 239–240, 242–246, 248–250, 255
Ministry of Health, 235, 249–250
Mitigation, 84, 109, 125, 173, 239–240, 247, 278, 298
  policies and programmes, 239–240
Mölndal, 31, 190, 212, 216–217, 228
Multi-level governance, definition, 4–7
Multi-National Corporations (MNC), 22
Municipality
  Espoo, 31, 173–174, 176–178, 180, 182, 346, 348, 351
  Ferrara, 262–264
  Gothenburg, 212–215
  Mölndal, 216–217
  Munkedal, 217–218
  Trollhättan, 217, 228
  See also Local Authority for local cases in general
Munkedal, 31, 191, 212, 216–218, 228

N
National Action Plan to Combat Drought and Desertification, 249, 344
National Action Plan (Italy), 109, 154, 239–240, 249, 292, 304, 344
National Adaptation Plan, 245, 272, 274, 276, 283–284, 287, 361
National Adaptation Programme (UK), 104, 312
National Adaptation Strategy, 2, 40, 105, 150, 159–160, 163, 183, 241, 245–248, 265, 265, 273–274, 276–279, 282, 284, 286–287, 289, 294–296, 298–300, 308, 322–323, 325, 327–328, 340, 343, 357, 360, 362, 365

National Adaptation Strategy (NAS), 2, 40, 105, 149–150, 159–160, 163, 183, 241, 245–248, 265, 268, 273–274, 276–279, 282, 284, 286–287, 289, 294–296, 298–301, 308, 322–323, 325, 327–328, 340, 343, 357, 360, 362, 365
National Association of Swedish Eco-municipalities (SEkom), 211
National Climate Change Conference (Italy), 234, 245–248, 266, 357
National Climate and Energy Strategy 2001 (Finland), 154–155
National Communication, 43, 45–46, 48, 53–54, 60–61, 234, 241–243, 249, 272, 275, 277–279, 285, 292, 294–295, 330
National Indicator (NI), 106, 110, 117, 120–121, 352
National Plan for the Prevention of the Effects of Heat on Human Health (Italy), 249
Natura 2000, 20, 44, 55, 71, 73–74, 78–84, 86, 130, 220, 225, 228, 330, 355
The "Natural Step", 211
Nested cases, 30–31, 172, 339
Netherlands
  adaptation agenda, 39–40, 45, 53, 55, 89, 162, 181, 276, 298, 300, 342, 348–349, 361
  Climate changes Spatial Planning (CcSP), 301, 349
  Delta Act, 299
  Delta Project, 299
  Knowledge for Climate, 298, 300–302, 362
  National Adaptation Strategy, 2, 40, 105, 149–150, 159–160, 163, 183, 241, 245–248, 265, 267, 273–274, 276–279, 282, 284, 286–287, 289, 294–296, 298–301, 308, 322–323, 325, 327–328, 340, 343, 357, 360, 362, 365
  National Programme on Climate Adaptation and Spatial Planning (ARK), 298
Network
  municipal climate change, 209–212
  voluntary and NGO, participation, 132–133
New Public Management (NPM), 12–13, 26, 119, 140, 143, 189, 346, 356, 362
NI, 106–108, 112–113, 120–121, 123, 125, 134, 142, 359, 363
Non-Governmental Organisation (NGO), 18, 21, 32, 66, 122, 126, 132–133, 157, 234, 240, 265, 272, 279, 293, 316, 330, 350, 357
Norway
  Bergen, 303, 306–307
  CICERO, 303, 306
  Climate Change: Impacts, Costs and Adaptation in the Baltic Sea Region (BaltCICA), 174, 179, 307, 330
  Directorate for Civil Protection and Emergency Planning (DSB), 305–307
  Future Cities, 306–307, 349, 362
  Klimatilpasning Norge, 305
  NorACIA, 303
  NORKLIMA, 303, 305
Nottingham Declaration, 112–113, 122, 129, 132–133, 135, 138, 140, 142, 342, 354, 358, 363
Nurmijärvi, 175

O

OECD, 169, 236–238

P

Partnership, 11, 21, 30, 33, 97–98, 100, 103, 105–108, 110–112, 114–123, 125–128, 134, 136, 140–143, 175–176, 181, 207, 257, 261–262, 307, 317–318, 342, 345, 352–354, 357
Partnership for Urban South Hampshire, 123, 126
Performance assessment framework for local government (UK), 134
Pitt Review, 102, 127, 140
Planned retreat, 26, 138
Planning and Building Act, 199, 222, 304, 306, 340, 358–359
  See also Planning and Building Law
Planning and Building Law, 199, 224
  See also Planning and Building Act
Planning families, 13, 277, 340
Planning monopoly, 18, 190, 206, 223–224, 226, 342, 364
Planning Policy Guidance (PPG), 103, 108, 114, 120, 129, 134
Planning Policy Statement (PPS), 102–103, 108–109, 114, 118, 134, 141
Planning system, 2–3, 13–15, 18, 28–30, 103, 108–110, 117–118, 134, 206, 223–224, 272, 275, 340–344, 352, 360, 363
  see Planning families

Policy
capacity, 10–11
context, 83, 138, 347, 349, 366
-makers, 6, 21–22, 111, 122, 226, 355
-making, 6–7, 20, 24, 26, 51, 87, 150, 157, 160, 166, 168, 178–179, 206, 226, 268, 343–344, 347, 349, 351, 363
styles, 13, 20, 27
transfer, 4, 26–28, 32, 133–140, 192, 225–226, 361–363
Political environment, 348
Political systems, 3, 5, 14, 33, 67, 190, 340–350, 352, 358, 363
Po River, 242, 245, 252–253, 256–258, 262
Pornainen, 31, 175–176
Portsmouth City Council, 99, 122, 125–127
'Preparing to adapt', 108, 120, 138, 141, 359–360, 365
See also NI
Private sector, 12, 32, 103, 136–137, 159, 175, 179, 218–220, 228, 240, 283, 287, 299, 305, 316, 321, 329, 355, 357
Prodi, Romano, 235, 239, 245–246, 266
Public research organisation, 166–167, 181
Public sector, 4, 44, 103–104, 137, 150, 159, 228, 305, 308

R
Ramsar Convention, 14, 43, 53, 70, 72–73, 75, 154, 239, 245, 249, 272, 275, 329, 344
Regional Assembly (UK), 114
Regional Climate Change Partnership (RCCP), 98, 103, 110–111, 114–117, 141, 345, 352–353, 356
Regional council, 151–152, 169, 171–174, 176, 178, 190, 348, 351, 356–357
Regional development, 32, 53, 151–152, 169, 171, 190, 205–207, 293–294, 307, 351, 356
Regional Development Agency (RAD), 116–118
Regional development plan, 151, 171, 293, 351, 356
Regional environment administration (RAS), 153, 248
Regional environmental protection agencies (ARPA), 235, 237–238, 244–245, 252–254, 256–257, 264, 266, 351
Regional federalism, 17, 250, 291
Regional Improvement and Efficiency Partnership (RIEP), 114, 118, 134, 141
Region, definition, 15–16
Regulation, 5, 11, 13, 15, 18, 21, 25, 27, 32, 54–55, 81, 99–100, 109, 113, 129, 131, 162, 169–170, 175–176, 189–190, 192, 194, 196–206, 208–209, 213–214, 219–221, 237, 285, 300–302, 304–305, 310, 326, 331, 358–359
Regulation letter, 198–204, 209
Regulatory state, 97, 119, 189
Reporting power, 134
Research, 45–46, 70, 161, 164, 166–168, 197–198, 210, 244–246, 281–283, 313–316, 359
Resilience, 8, 53, 73, 78, 81, 83, 85–86, 106–107, 113, 127, 132, 197, 298–299, 316, 321
Resources, 9, 20, 45–46, 68–71, 116, 159, 243, 246, 251, 308, 310, 315
See also Adaptive capacity, determinants of
Revised National Climate and Energy Strategy 2005 (Finland), 155
Revised National Climate and Energy Strategy 2008 (Finland), 156–157
Risk management, 23–24, 72, 108–110, 127, 177, 221, 236, 241, 250, 255, 258–260, 268, 298, 318, 321, 347, 352
Risk and Vulnerability Assessment, 206, 222, 227, 358
River Basin Authorities, 242, 249, 255, 257
River Basin Management Plans (RBMP), 44, 65–67, 84
Rural development, 46, 53–54, 71, 74, 80–81, 240, 254–256, 258–259, 291–292, 295–296, 356
Rural development plan, 240, 254, 256, 258–259

S
Scenario, 42, 66, 68, 105, 110, 123, 135, 150, 158–159, 163–168, 196, 204, 219, 226, 234, 247, 251, 258, 274, 284, 288, 294, 301, 303, 309–310, 319, 350
Sector, *see specific sectors*
Sensitivity, 7, 17, 30, 33, 52, 76, 91, 97, 126, 142, 216, 312, 341, 346, 352–353
Shoreline Management Plan, 110, 131
Societal tradition, 12
Soil, 55, 69, 70, 84, 123, 142, 158, 190, 208, 218, 236, 243–245, 248–249, 254–255, 257, 279, 288–289, 308–310

South East England, 31, 97–143, 350
South East England Development Agency
SEEDA, 104, 114–117, 119, 134, 136, 141
South East England Regional Assembly
SEERA, 110, 114, 117–118, 122, 129, 134
South East Plan, 117–118, 134
Spain
First Work Programme, 309–310
National Plan for Climate Change Adaptation (NCCAP), 276, 308–311, 327
Second Work Programme, 276, 309, 311
Spanish Climate Change Office (OECC), 308–309, 310, 312, 326–328
Spanish Network of Cities for Climate, 311–312, 330
Spatial planning, 2, 115, 117, 173, 251, 274, 276, 279, 286, 288, 292, 297–302, 307, 311, 328, 349
State traditions, 13
Steering, 11, 18, 24, 32, 105–106, 108, 119, 135, 152, 156, 162, 167, 181, 210, 212, 224, 272, 299, 323–325, 340, 342
Stern Report, 102, 123, 195
Storm surges, 150, 299, 321
Strategic Environmental Assessments (SEA), 78, 84–85, 259
Structural funds, 19, 151, 206, 221, 256, 258, 355
Sub-Committee on Adaptation, 113, 134
Surrey County Council, 99, 122, 127–130, 132, 142
Sustainability, 114, 117–119, 126, 129–130, 132–134, 152, 171, 210–211, 220, 237, 240, 256, 259, 261–265, 267, 317, 321, 345, 353, 362
Sustainable, 27, 52–53, 70, 75, 100, 109, 116–119, 123–124, 129–130, 133–134, 153, 157, 182, 192–195, 204, 210–211, 235–238, 240–241, 244–246, 249, 252–253, 258, 261–263, 267, 280–283, 287, 289, 292, 302, 315, 351, 353, 362
Sustainable Municipalities, 210
Sweden, 29, 31, 189–228, 361–362, 364
Swedish Association of Local Authorities and Regions (SALAR), 190, 205, 207, 210–212, 220–225, 354
Swedish Civil Contingencies Agency, 199–200, 205, 227
Swedish Emergency Management Agency, 200, 205
Swedish Environmental Quality Objectives, 199, 210
Swedish Geotechnical Institute (SGI), 198, 202, 226
Swedish Insurance Federation, 195, 219–220
Swedish Mapping, Castradal and Land Registration Authority, 198, 201
Swedish Rescue Services Agency, 198, 200, 203–205, 216, 218, 227
Switzerland, 14, 17, 59, 63–64, 251, 280

T

Thameswey, 130
Topographical database, 201–202, 219, 222
Tourism, 56, 66, 114–115, 126, 158, 160, 167, 197, 234, 242–243, 245–246, 251–252, 278–279, 281, 283, 286, 289, 296, 298, 305, 308, 313, 316
Transport, 10, 44, 46, 58, 86, 89, 100, 107, 119, 126–129, 137, 152–153, 157, 160, 163, 167, 169, 182, 192, 206, 237, 239–240, 254, 262, 264, 277, 279–280, 283, 286–287, 297, 299, 303–304, 307–308, 328
Transport sector, 163, 192, 239
Treasury Green Book, 103–104
Trollhättan, 31, 190, 209, 212, 216–217, 225, 228, 343, 345, 353
Tuusula, 31, 175–176
Types of adaptation, *see* Typology of adaptation
Typology of adaptation, 8

U

UK Climate Impacts Programme (UKCIP), 98, 103, 105, 110–113
UK Interregional Climate Change Group (UKIRCCG), 104, 111, 115–116, 136–137, 139, 141
United Kingdom (UK), 2, 12, 31, 97–143, 190
United National Convention to Combat Desertification, 275
United Nations Framework Convention on Climate Change (UNFCCC), 43, 45, 48, 52–53, 60, 154, 239, 241, 243, 272, 275, 277, 285–287, 292–295, 306, 328, 348
Uusimaa, 31, 150, 153, 162, 169, 171–174, 176, 178, 180, 182–183, 348, 356–357
Uusimaa Regional Council, 169, 172–174, 178, 180, 348, 356–357

V
Västra Götaland, 31, 189–228, 351
Västra Götalandsregionen, 205
Venice, 237, 251–252, 259
Voluntary initiative, 149–150, 176, 178, 354, 362–363
Vulnerability, 7–8, 195–198, 200, 219–222, 294, 303–304, 313–318, 350–352, 357–358

W
Water
authority, 176, 199, 206–207
court decision, 198, 203, 208, 221
court judgement, 197, 203, 208, 214
See also Water court decision
management, 32, 43, 62, 64–73, 161, 165, 169, 210, 227, 234, 236, 245–246, 249–250, 253–255, 263, 268, 278–280, 286, 288, 292, 295, 298–301, 303, 307–308, 313, 315, 328, 355
Water Framework Directive (WFD), 3, 20, 44, 55, 64–69, 71–72, 84, 90, 130–131, 207, 221, 227, 288, 357
Water Protection Plan (Italy), 257, 357
Winchester City Council, 99, 121–125, 142
Woking Borough Council, 98–99, 122, 127–130, 133
World Wildlife Fund (WWF), 152, 240, 245, 248–249, 293, 296, 330, 357

MIX
Papier aus verantwortungsvollen Quellen
Paper from responsible sources
FSC® C105338

If you have any concerns about our products,
you can contact us on
ProductSafety@springernature.com

In case Publisher is established outside the EU,
the EU authorized representative is:
Springer Nature Customer Service Center GmbH
Europaplatz 3, 69115 Heidelberg, Germany

Printed by Libri Plureos GmbH
in Hamburg, Germany